TREATISE ON ANALYTICAL CHEMISTRY

PART I
THERMAL METHODS
SECOND EDITION

TREATISE ON
ANALYTICAL CHEMISTRY

PART I

THERMAL METHODS

SECOND EDITION

VOLUME 13

Edited by JAMES D. WINEFORDNER

Department of Chemistry, University of Florida

Associate Editors:
DAVID DOLLIMORE

Department of Chemistry, University of Toledo

JEFFREY DUNN

School of Applied Chemistry, Curtin University of Technology, Perth, Western Australia

Editor Emeritus: I. M. KOLTHOFF

Department of Chemistry, University of Minnesota

AN INTERSCIENCE ® PUBLICATION.

JOHN WILEY & SONS, INC.
New York – Chichester – Brisbane – Toronto – Singapore

Library of Congress Cataloging in Publication Data:
(Revised for vol. 13)

Kolthoff, I. M. (Izaak Maurits), 1894–
 Treatise on analytical chemistry.

 Pt. 1, v. 8– : edited by Philip J. Elving;
associated editor, Edward J. Meehan.
 Pt. 1, v. 11– : edited by James D. Winefordner;
associate editor, Maurice M. Bursey.
 "An Interscience publication."
 Includes bibliographies and indexes.
 Contents: pt. 1. Theory and practice.
 1. Chemistry, Analytic. I. Elving, Philip Juliber,
1913–   . II. Meehan, Edward J. III. Title.
QD75.2.K64 1978   543   78-1707
ISBN 0-471-80647-1 (pt. 1, v. 13)

Printed in the United States of America

10 9 8 7 6 5 4 3 2 1

TREATISE ON ANALYTICAL CHEMISTRY

PART I
THERMAL METHODS

VOLUME 13

AUTHORS OF VOLUME 13

JAVED I. BHATTY   JOSEPH JORDAN
DAVID DOLLIMORE   MICHAEL READING
JEFFREY DUNN   JOHN SHARP
WILLIAM IRWIN   JOHN STAHL
MOSHE ISH-SHALOM

Authors of Volume 13

Javed I. Bhatty

Construction Technology Laboratories Inc.
Skokie, Illinois
Chapter 6

David Dollimore

Department of Chemistry
The University of Toledo
Toledo, Ohio
Chapter 1

Jeffrey Dunn

School of Applied Chemistry
Curtin University of Technology
Perth, Western Australia
Chapter 3

William Irwin

Drug Development Research Group
Pharmaceutical Science Institute
Aston University, Aston Triangle
Birmingham B4 7ET UK
Chapter 5

Moshe Ish-Shalom

Israel Ceramic and Silicate Institute
Technion City, Haifa, Israel
Chapter 4

Joseph Jordan

Professor Emeritus of Chemistry
The Pennsylvania State University
University Park, Pennsylvania
Chapter 2

Michael Reading

Research Section ICI Paints Division
Slough, Berkshire, UK
Chapter 1

John Sharp

Department of Engineering Materials
University of Sheffield, Sheffield, UK
Chapter 3

John Stahl

Associate Professor of Chemistry
Geneva College
Beaver Falls, Pennsylvania
Chapter 2

Preface to the Second Edition of the Treatise

In the mid-1950s, the plan ripened to edit a "Treatise on Analytical Chemistry" with the objective of presenting a comprehensive treatment of the theoretical fundamentals of analytical chemistry and their implementation (Part I) as well as of the practice of inorganic and organic analysis (Part II); an introduction to the utilization of analytical chemistry in industry (Part III) was also considered. Before starting this ambitious undertaking, the editors discussed it with many colleagues who were experts in the theory and/or practice of analytical chemistry. The uniform reaction was most skeptical; it was not thought possible to do justice to the many facets of analytical chemistry. Over several years, the editors spent days and weeks in discussion in order to define not only the aims and objectives of the Treatise but, more specifically, the order of presentation of the many topics in the form of a table of contents and the tentative scope of each chapter. In 1959, Volume 1 of Part I was published. The reviews of this volume and of the many other volumes of Part I as well as of those of Parts II and III have been uniformly favorable, and the first edition has become recognized as a contribution of classical value.

Even though analytical chemistry still has the same objectives as in the 1950s or even a century ago, the practice of analytical chemistry has been greatly expanded. Classically, qualitative and quantitative analysis have been practiced mainly as "solution chemistry." Since the 1950s, "solution analysis" has involved to an every increasing extent physicochemical and physical methods of analysis, and automated analysis is finding more and more application, for example, its extensive utilization in clinical analysis and production control. The accomplishments resulting from automation are recognized even by laymen, who marvel at the knowledge gained by automated instruments in the analysis of the surfaces of the moon and of Mars. The computer is playing an ever increasing role in analysis and particularly in analytical research. This revolutionary development of analytical methodology is catalyzed by the demands made on analytical chemists, not only industrially and academically but also by society. Analytical chemistry has always played an important role in the development of inorganic, organic, and physical chemistry and biochemistry, as well as in that of other areas of the natural sciences such as mineralogy and geochemistry. In recent years, analytical chemistry—often of a rather sophisticated nature—has become increasingly important in the medical and biological sciences, as well as in the solving of such social problems as environmental pollution, the tracing of toxins, and the dating of art and archaeological objects, to mention only a few. In the area of atmospheric science, ozone reactivity and persistence in the stratosphere is presently a topic of great priority; extensive analysis is required both for monitoring atmospheric constituents and for investigating model systems.

One example of the increasing demands being made on analytical chemists is the growing need for speciation in characterizing chemical species. For example, in reporting that lake water contains dissolved mercury, it is necessary to report in which oxidation state it is present, whether as an inorganic salt or complex, or in an organic form and in which form.

As a result of the more and less revolutionary developments in analytical chemistry, portions of the first edition of the Treatise are becoming—and, to some extent, have become—out-of-date, and a revised, more up-to-date edition must take its place. In recognition of the extensive development and because of the increased specialization of analytical chemists, the editors have fortunately secured for the new edition the cooperation of experts as coeditors for various specific fields.

In essence, it is the objective of the second edition of the Treatise, as it was of the first edition (whose preface follows this one), to do justice to the theory and practice of contemporary analytical chemistry. It is a revision of Part 1, which mirrors the development of analytical chemistry. Like the first edition, the second edition is not an extensive textbook; it attempts to present a thorough introduction to the methods of analytical chemistry and to provide the background for detailed evaluation of each topic.

I. M. KOLTHOFF
J. D. WINEFORDNER

Minneapolis, Minnesota
Gainesville, Florida

Preface to the First Edition of the Treatise

The aims and objective of this Treatise are to present a concise, critical, comprehensive, and systematic, but not exhaustive, treatment of all aspects of classical and modern analytical chemistry. The Treatise is designed to be a valuable source of information to all analytical chemists, to stimulate fundamental research in pure and applied analytical chemistry, and to illustrate the close relationship between academic and industrial analytical chemistry.

The general level sought in the Treatise is such that, while it may be profitably read by the chemist with the background equivalent to a bachelor's degree, it will at the same time be a guide to the advanced and experienced chemist—be he in industry or university—in the solution of his problems in analytical chemistry, whether of a routine or of a research character.

The progress and development of analytical chemistry during most of the first half of this century has generally been satisfactorily covered in modern textbooks and monographs. However, during the last fifteen or twenty years, there has been a tremendous expansion of analytical chemistry. Many new nuclear, subatomic, atomic, and molecular properties have been discovered, several of which have already found analytical application. In the development of techniques for measuring these and also the more classical properties, the revolutionary progress in the field of instrumentation has played a tremendous role.

It has been difficult, if not impossible, for anyone to digest this expansion of analytical chemistry. One of the objectives of the present Treatise is not only to describe these new properties, their measurement, and their analytical applicability, but also to classify them within the framework of the older classifications of analytical chemistry.

Theory and practice of analytical chemistry are closely interwoven. In solving an analytical chemical problem, a thorough understanding of the theory of analytical chemistry and of the fundamentals of its techniques, combined with a knowledge of and practical experience with chemical and physical methods, is essential. The Treatise as a whole is intended to be a unified, critical, and stimulating treatment of the theory of analytical chemistry, of our knowledge of analytically useful properties, of the theoretical and practical fundamentals of the techniques for their measurements, and of the ways in which they are applied to solving specific analytical problems. To achieve this purpose, the Treatise is divided into three parts: I, analytical chemistry and its methods; II, analytical chemistry of the elements; and III, the analytical chemistry of industrial materials.

Each chapter in Part I of the Treatise illustrates how analytical chemistry draws on the fundamentals of chemistry as well as on those of other sciences: it stresses for its particular topic the fundamental theoretical basis insofar as it affects the analytical approach, the methodology and practical fundamentals used both for the development of analytical methods and for their implementation for analytical service, and the critical factors in their application to both organic and inorganic materials. In general, the practical discussion is confined to fundamentals and to the analytical interpretation of the results obtained. Obviously then, the Treatise does not intend to take the place of the great number of existing and exhaustive monographs on specific subjects, but its intent is to serve as an introduction and guide to the efficient utilization of these specialized monographs. The emphasis is on the analytical significance of properties and of their measurement. In order to accomplish the above aims, the editors have invited authors who are not only recognized experts for the particular topics, but who are also personally acquainted with and vitally interested in the analytical applications. Only in this way can the Treatise attain the analytical flavor which is one of its principal objectives.

Part II is intended to be very specific and to review critically the analytical chemistry of the elements. Each chapter, written by experts in the field, contains in addition to a critical and concise treatment of its subject, critically selected procedures for the determination of the element in its various forms. The same critical treatment is contemplated for Part III. Enough information is presented to enable the analyst both to analyze and to evaluate a product.

The response in connection with the preparation of the Treatise from all colleagues has been most enthusiastic and gratifying to the editors. It is obvious that it would have been impossible to accomplish the aims and objectives cited in the Preface without the wholehearted cooperation of the large number of distinguished authors whose work appears in this and future volumes of the Treatise. To them and to our many friends who have encouraged us we express our sincere appreciation and gratitude. In particular, considering that the Treatise aims to cover all of the aspects of analytical chemistry, the editors have found it desirable to solicit the advice of some colleagues in the preparation of certain sections of the various parts of the Treatise. They would like at this time to acknowledge their indebtedness to Professor Ernest B. Sandell of the University of Minnesota for his interest and active cooperation in the organizing and detailed planning of the Treatise.

I. M. KOLTHOFF
P. J. ELVING

Minneapolis, Minnesota
Ann Arbor, Michigan

PART I. THERMAL METHODS

CONTENTS—VOLUME 13

2. Thermometric Titrations and Enthalpimetric Analysis

3. Thermogravimetry

6. Application of Thermal Analysis to Problems in Cement Chemistry

TREATISE ON ANALYTICAL CHEMISTRY

PART I
THERMAL METHODS
SECOND EDITION

Chapter 1

APPLICATION OF THERMAL ANALYSIS TO KINETIC EVALUATION OF THERMAL DECOMPOSITION

By David Dollimore, *Department of Chemistry,*
The University of Toledo, Toledo, Ohio,
and Michael Reading, *Research Section,*
ICI Paints Division, Slough, Berkshire, UK,

Contents

Treatise on Analytical Chemistry, Part 1, Volume 13,
Second Edition: Thermal Methods, Edited by James D. Winefordner.
ISBN 0-471-80647-1   © 1993 John Wiley & Sons, Inc.

I.   INTRODUCTION

A.   EXISTING LITERATURE

Several published reviews deal with the theory and practice of studying the kinetics of solid decomposition reactions of the type

$$A_{(s)} \longrightarrow B_{(s)} + C_{(g)}$$

(14, 35, 40, 86). The most recent of these (14) is particularly extensive and deals with most of the literature up to 1976. It is therefore intended in this review to provide a brief survey of the general information already well covered in the above-mentioned publications. Only more recent articles and those aspects of modern theory and practice are considered in detail. This review is divided into three sections. The first describes and classifies the experimental methods used to study the kinetics of solid state decomposition reactions. The second section discusses the current theories that seek to explain the kinetic behavior of these reactions. The third section explains how results obtained using the methods discussed in the first section can be interpreted within the framework of the theories discussed in the second to calculate values for kinetic parameters. It must be emphasized that other techniques should be used to gather information that is often required for the interpretation of reaction kinetics as rate data by themselves are rarely sufficient for a full understanding of reaction mechanism (13, 15, 24, 59).

B.   DEFINITIONS

The following quantities will be used throughout this work:

t = time

T = temperature (K)

R = universal gas constant = 8.314 34 J mol^{-1} K^{-1}

k = Boltzmann's constant = 1.380 62 $\times$ 10^{-23} J K^{-1}

h = Plank's constant = 6.626 20 $\times$ 10^{-34} J s

α = extent of reaction.

From a simple reaction that gives rise to only one product gas, α may be defined as a fraction of total volume of gas evolved or a fraction of total weight loss. Where two or more gases are evolved the extent of reaction is defined as the extent of reaction with respect to each individual gas, i.e., the fraction of total volume of, for example, CO_2 evolved is referred to as α_{CO_2}. α, therefore, always goes from zero at the start of a reaction to 1 at its end.

II. EXPERIMENTAL METHODS

Solid state decomposition reactions are usually studied using a family of experimental techniques known as thermoanalytical methods. Any given thermoanalytical method comprises three elements:

1. A method of temperature control
2. A method of measuring reaction rate
3. A method of controlling the reaction environment.

Generally any method of temperature control may be combined with any method of measuring reaction rate and any method of controlling the reaction environment, although in practice some combinations would be experimentally difficult to achieve. Each of these three elements may be considered separately and are dealt with under different headings in the first three parts of this section. Under the heading Methods of Measuring Reaction Rate the discussion is confined to mass loss measurements and infrared evolved gas analysis. Under the heading Methods of Temperature Control two types of temperature control are identified: program-determined temperature control and reaction-determined temperature control. The former is well established and the temperature programs that make this method possible are widely available. The latter, however, is much less well established and the ways in which this novel approach to thermal analysis has been achieved in practice are described in the final part of this section.

A. METHODS OF TEMPERATURE CONTROL

Two basic approaches to temperature control are used in thermoanalytical experiments:

1. Program-determined temperature control (PDTC)
2. Reaction-determined temperature control (RDTC)

In PDTC techniques the sample or sample environment is subjected to a predetermined temperature program that proceeds independently of any reaction undergone by the sample; thus, the controlled quantity is the temperature. In the latter case the temperature of the sample is altered in response to changes in some measured reaction parameter; thus, the controlled quantity, indirectly, is the reaction rate. If the sample does not react or the parameter being measured is insensitive to the progress of the reaction, then no true programming is achieved. Both of these types of programming may be used in a continuous mode, in which case any change in the controlled quantity can be described by a smooth continuous function, or in a jump mode, in which case the controlled quantity is changed in a sudden discontinuous manner. The division and subdivision of temperature programming methods can be described as follows.

1. Program-Determined Temperature Control

It is convenient to distinguish between nonisothermal and the special case of isothermal PDTC.

a. ISOTHERMAL PDTC

Methods in this category consist of heating the sample as near as possible instantaneously to a predetermined temperature and maintaining this temperature while in some way measuring the reaction rate.

(1) Continuous Isothermal PDTC

This method entails maintaining the sample temperature at a single unchanging value during the whole course of the experiment. Sometimes, if the time required to complete the reaction is unacceptably long, the temperature may be increased at the end of the experiment to bring the reaction to completion so that some final reaction parameter, such as final sample mass, may be obtained. Continuous isothermal PDTC is usually referred to as the isothermal method.

(2) Temperature Jump Isothermal PDTC

This method entails maintaining the sample temperature at a constant value over a certain period of time, then instantaneously jumping to another temperature and maintaining it at a second constant value over a further period of time. This procedure, jumping both up and down in temperature, may be repeated several times during one experiment. Temperature jump PDTC is usually referred to as the temperature jump method.

b. NONISOTHERMAL PDTC

Methods in this category subject the sample to a predetermined nonisothermal temperature program while the reaction rate is measured in some way.

(1)   Continuous Nonisothermal PDTC

This method entails subjecting the sample to a smooth continuous temperature program over the whole course of the reaction under study. The program may be described by any continuous function, e.g., linear, exponential, or hyperbolic. In practice, however, a linear heating program is almost always used. The continuous nonisothermal PDTC method is usually referred to as the rising temperature method.

(2)   Jump Nonisothermal PDTC

Methods in this category could conceivably be of various types, such as maintaining a linear temperature rise while introducing jumps into the program or jumping from a linear program to one of the alternative functions. However, methods of this kind have never been proposed as being of value and therefore shall not be further considered.

In the use of any form of PDTC one can adopt one of two strategies. Either the sample environment is subjected to a temperature program while the sample temperature is independently monitored using a thermocouple either near or, preferably, in the sample, or an attempt can be made to subject the sample itself to a temperature program and so detect and compensate for any self-cooling or heating effects due to the endothermic or exothermic nature of the reaction. Generally the latter is theoretically preferable.

2.   Rate-Determined Temperature Control

It is convenient to distinguish between those RDTC methods that control the rate of reaction in a nonconstant way and those that maintain the rate of reaction at a constant value.

a.   CONSTANT RATE RDTC

Methods in this category maintain some measured reaction parameter that is proportional to the reaction rate at a constant value, thereby keeping the reaction rate itself constant while measuring the sample temperature.

(1)   Continuous Constant Rate RDTC

This method entails maintaining the reaction rate at a single constant value throughout the whole experiment. Continuous constant rate RDTC is usually referred to as constant rate thermal analysis.

(2)   Jump Constant Rate RDTC

This method entails sudden changes of reaction rate, usually several up and down jumps during one experiment. Jump constant rate RDTC is usually referred to as cyclic constant rate thermal analysis or rate jump constant rate thermal analysis.

b. NON-CONSTANT RATE RDTC

Methods in this category alter the sample temperature in response to changes in some reaction parameter is such a way that, either due to the properties of the measuring device or of the chemical system under study, the rate of reaction is controlled in a nonconstant way while both reaction rate and sample temperature are measured.

(1)   Continuous Non-Constant Rate RDTC

This method entails controlling the reaction rate in such a way as to avoid any sudden discontinuous changes throughout the course of the whole experiment. Continuous nonconstant rate RDTC is usually referred to as controlled rate thermal analysis.

(2)   Jump Non-Constant Rate RDTC

This method entails sudden changes of nonconstant reaction rate, usually several up and down jumps during one experiment. Jump non-constant rate RDTC is usually referred to as cyclic controlled rate thermal analysis or rate jump controlled rate thermal analysis.

The nomenclature proposed above is a useful way of systematically classifying the different methods of temperature control. To conform to the usual practice adopted in the literature the generic names, which are also given above, will be used throughout.

B.   METHODS OF MEASURING REACTION RATE

1.   Thermogravimetry

The development and ready availability of reliable and accurate electronic microbalances (10, 56, 83) have led to their wide application in kinetic studies of the decomposition of solids. Such balances can give very high accuracy data, the main problems being buoyancy effects at higher pressures, thermomolecular flow effects, and accurate sample temperature calibration. Careful design of equipment and control of reaction conditions can minimize these difficulties.

The data obtained from mass loss curves are in an integral form. The reaction rate at any point may be calculated by drawing tangents or by electronic or mathematical derivation.

2.   Evolved Gas Analysis (EGA)

A typical unit for this purpose would be to obtain continuous nonaccumulatory infrared EGA under flowing nitrogen. This method can be applied only to those gases that absorb in the infrared region and care must be taken that the absorption bands of the different species evolved do not overlap. Ideally, the concentration of the gas being swept away from the sample at any instant is proportional to its rate of evolution; the experimental information is therefore in a differential form. If the swept volume of the gas cell is large, however, as is often the case for a multipass gas cell, a mathematical treatment of the results is required to correct the observed concentration to obtain the true value of the concentration of the gas entering the gas cell. This is not the place to provide such details. However, this correction is simpler to apply when the experimental curve is integrated, and for this reason EGA data are presented in an integral form. This point is further discussed during the description of the EGA apparatus. Typical results for isothermal, rising temperature, and constant rate thermal analysis are given in Fig. 1.

Nonisothermal results may be expressed in terms of a plot of extent of reaction, either mass loss or α, against temperature. This is a convenient way of presenting the relationship between these two quantities in a single diagram.

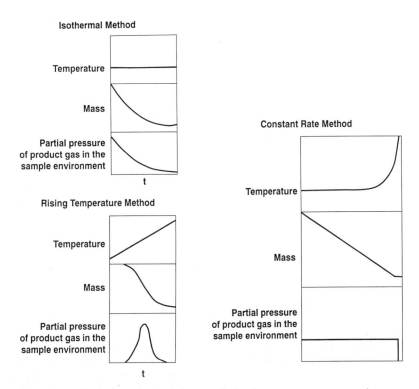

Fig. 1.   Typical results for various methods of temperature programming.

C. CONTROL OF REACTION ENVIRONMENT

There are two categories of approach to reaction environment control:

1. Vacuum experiments where the atmosphere around the sample consists only of gaseous reaction products
2. Experiments in which the sample is surrounded by an externally supplied atmosphere, when product gases are either removed or accumulated in the reaction vessel.

Product gases remaining within the sample environment may influence the reaction rate and give rise to complicating secondary reactions. For this reason the concentration of product gases above the reacting sample should be kept as low as possible. Methods that remove product gases are better suited to achieving this objective.

Typical results for isothermal, rising temperature, and constant rate thermal analysis methods using mass loss and EGA measurements are given in Fig. 1, together with graphs that show how product gas concentration near the sample changes during the experiment when continuous product gas removal is employed.

Temperature and pressure gradients within a sample are influenced by both instrumental and sample factors. Instrumental factors include the following:

1. Intrinsic temperature gradients across the reaction zone due to furnace design
2. Sample holder geometry
3. Method of control of reaction environment
4. Method of temperature control.

Factor 1 can be overcome by careful instrument design. Factor 2 can have an important influence on experimental results. Areas of the sample close to the sample holder walls may be more efficiently supplied with heat than the bulk of the sample. Alternatively, the proximity of the crucible wall may inhibit the decomposition reaction because it inhibits the removal of gaseous products. Barret (7) has shown that the advance of the reaction interface may conform to the geometry of the crucible rather than the geometry of the sample particles because of this effect. To overcome this problem Sestak (71) has recommended that powdered samples should be distributed as a thin film over a flat plate, and Oswald (60) has proposed the use of a flat crucible made from platinum gauze for crystal samples.

Factor 3 may also have a profound influence on reaction kinetics. Ideally, the pressure of product gases should be very low; however, at low pressures thermal transport is more difficult and may become rate limiting. The presence of an externally supplied atmosphere at high pressure, usually atmospheric, that does not influence the chemical nature of the reaction path is a possible solution to this problem. Both vacuum and atmospheric pressure methods are used and compared in this study.

Factor 4 may influence temperature gradients because changing the temperature of the sample necessarily implies creating a temperature gradient between those areas of the sample easily accessible to the heat source and those areas less accessible to the heat source. This correctly implies that an isothermal regime tends to minimize this problem; however, isothermal temperature control creates other difficulties. In practice, a sample cannot be heated to a predetermined temperature instantaneously; it may therefore start decomposing before the desired temperature is achieved throughout the sample. This problem may become particularly acute when examining the higher temperature behavior of a sample that decomposes in several poorly differentiated steps over a very wide temperature range.

Nonisothermal methods can overcome these problems but may give rise to greater temperature gradients within the sample over the course of the reaction as a whole. Each method of temperature control has its advantages and disadvantages and one of the aims of this review is to compare the possible methods of temperature control, although it may be said that generally low heating rates are preferable.

Sample factors include the following:

1. Thermal conductivity of reactant and product
2. Heat capacity of reactant and product
3. Sample packing (for powered samples)
4. Sample size
5. Decomposition rate.

Factors 1 and 2 are beyond the control of the experimenter, but their effects may be influenced by diluting the sample with an inert material if necessary. Factor 3 is connected with factor 2 of the instrumental set and the recommended sample distributions are given above. The influence of factor 4, sample size, on reaction kinetics has been established by several studies (1, 34, 79, 87). It is obviously more difficult to establish a zone of constant temperature within a large volume than a smaller one. Also, pressure gradients are more likely to occur within a larger sample than a smaller one where the product gases have less distance to travel from the center of the sample to the exterior. Therefore, sample sizes should be kept as small as possible within the sensitivity limits of the apparatus being used to measure reaction rate. Pressure and temperature gradients are a direct function of factor 5. Temperature changes resulting from the endothermic or exothermic nature of a reaction will be much reduced if the rate of reaction is slow. Similarly, pressure gradients between the interior and exterior of a sample are reduced by reducing the reaction rate. Consequently, experiments that keep the reaction rate as low as possible can be expected to yield better kinetic data.

In PDTC, product gases can be removed either by pumping or by using a flowing inert gas stream as described above. With PDTC, the rate of gas evolution changes during the course of the reaction. Therefore, if the rate of pumping or the rate of gas flow remains constant, then the pressure of the product gases in the sample environment changes as the reaction proceeds. Ideally, the pressure of product gases above the sample should remain low and constant. This may be achieved by controlled pumping, but Garn (36) has observed that between the source of gas and the pumping system a pressure gradient must exist, which will be different for different pumping rates. Another possible approach is to control the rate of flow of the gas stream, but this has never been achieved experimentally. In summary, it must be said that during any PDTC experiment that seeks to minimize the concentration of product gases in the sample environment by using continuous removal of product gases, it is very difficult to control the value of the pressure of the product gases above the sample in a precise manner. From what has been said about RDTC it is obvious that with a constant decomposition rate the pressure of product gases above the sample may be easily and accurately controlled.

D. RATE-DETERMINED TEMPERATURE CONTROL

RDTC has already been defined as a method of temperature control whereby the temperature of the sample is altered in response to changes in some measured reaction parameter. Ideally, this should be done in such a way as to maintain the reaction rate constant, i.e., maintain some measured reaction parameter that is proportional to the reaction rate at a constant value thereby maintaining the reaction rate itself constant. This approach to thermal analysis was first proposed in a general way in an article by Rouquerol (67) in which he said, "Instead of the usual control of the furnace heating to follow a temperature program, a quantity directly related to the decomposition rate is kept constant. This quantity may be, for instance, a gas flow, thermal flow or a signal of derivative thermogravimetry." In theory, there are many possible ways of implementing RDTC based on any of the numerous ways of measuring reaction rate, but to date only four types of RDTC apparatus have been constructed.

Paulik and Paulik (62) have modified an apparatus they developed called the derivitograph. The controlled reaction parameter in this case is the rate of mass loss that is obtained by electronically differentiating the mass signal. Thus, the rate of mass loss remains constant regardless of the nature of the gaseous species evolved. Paulik and Paulik call this method quasi isothermal quasi isobaric thermal analysis (QIQITA). It has the advantage that the sample may be surrounded by any atmosphere, although vacuum work is

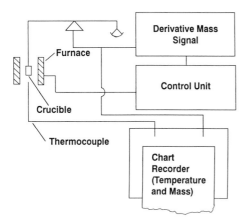

Fig. 2. Constant rate apparatus in which mass change is the controlled reaction parameter (62).

apparently not possible with the derivatographs. A schematic diagram of this apparatus is shown in Fig. 2.

Rouquerol chose as the controlled reaction parameter the pressure of product gas within a continuously evacuated reaction chamber. Any pumping system connected to a vessel in which the pressure is maintained constant should pump away gas at a constant rate. Thus, the pressure in the reaction chamber, and therefore above the reacting sample, is kept constant, which in turn results in a constant rate of gas evolution, and therefore reaction rate. Rouquerol (67) calls this method constant rate thermal analysis (CRTA). This system has been coupled to thermogravimetry (66, 67) so that the rate of mass loss can be simultaneously monitored. When only one gas is evolved, or more than one gas if the ratios of the rate of evolution of each gas remain constant, the rate of mass is constant. Thus, the results are essentially the same as for QIQITA, except that the experiment is carried out at low pressure under a self-generated atmosphere. The advantages of this approach are that it enables very low pressures to be accurately maintained above the sample, as low as 5×10^{-10} torr, and very low reaction rates can be obtained, as low as 1 μg/min (69). However, if more than one gas is evolved and the ratios of the rates of evolution of each gas do not remain constant, then the rate of mass loss changes during the course of the decomposition. This picture is further complicated because the Pirani and Penning pressure gauges employed by Rouquerol have different sensitivities to different gases. When the rate of decomposition is controlled in this nonconstant way the method is called nonconstant rate reaction determined temperature control, but Rouquerol uses the term controlled rate thermal analysis. A schematic diagram of this apparatus is given in Fig. 3.

It was to study more complex reactions that produce more than one product gas that the following two types of apparatus were developed. Stacey

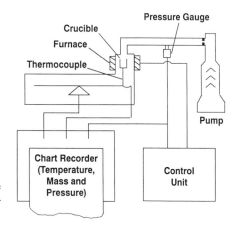

Fig. 3. Constant rate apparatus in which the gas pressure is the controlled reaction parameter (69).

used the partial pressure of water in a flowing gas stream as the controlled reaction parameter (81). He measured the partial pressure of water using an electronic dewpoint hygrometer, which was connected in series with a katharometer so that changes in the ratio between the concentrations of water vapor and any other gas evolved could be detected. This method has the advantage of controlling the partial pressure of water in the environment of the decomposing sample independently of the concentration of any other evolved species and of being able to use a variety of mediums as the vector gas. However, it is limited to controlling only the rate of evolution of water and cannot be used under vacuum. A schematic diagram of this apparatus is given in Fig. 4.

A similar method has been developed by the present authors except that two IR gas detectors connected in series are used. In theory, the rate of

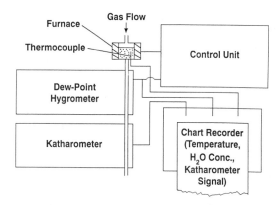

Fig. 4. Constant rate apparatus in which the partial pressure of water vapor in a flowing gas stream is the controlled parameter (81).

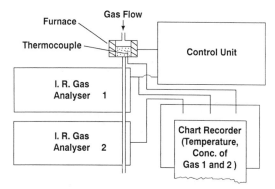

Fig. 5.   Constant rate apparatus using infrared detectors to monitor evolved gases.

evolution of any IR active gas may be controlled while simultaneously measuring the concentration of any other IR active product gas. The advantage of this method is that it enables the use of any gas as the vector gas, and even vacuum experiments are possible. A schematic diagram of this apparatus is given in Fig. 5.

The last three methods described above rely on control of the concentration of product gases. To use them in a jump mode, sudden increases in reaction rate may be induced by an increase in pumping speed for vacuum experiments or by increasing the rate of flow of the vector gas for experiments conducted under flowing atmospheres. It should be noted that the concentration of the product gas will remain the same before and after the rate jump. In contrast, a sudden increase in reaction rate brought about using QIQITA will result in an increase in the pressure of product gases above the sample unless the rate of flow of the vector gas is increased by the same factor as the rate of mass loss.

III.   THEORY

The relationship between rate of reaction and extent of reaction is generally expressed in the form

$$\frac{d\alpha}{dt} = f(\alpha)k \tag{1}$$

where α = extent of reaction

$\dfrac{d\alpha}{dt}$ = rate of reaction

$f(\alpha)$ = some function of α

k = a temperature-dependent quantity.

When the temperature is held constant, k is generally assumed to remain constant and is called the reaction rate constant.

Equation 1 is an incomplete description because reaction rate is often dependent upon the pressure of product gases above the reacting sample. Thus, a better rate model might be of the form

$$\frac{d\alpha}{dt} = f(\alpha)f(P)k \tag{2}$$

where $f(P)$ is some function of the pressure of product gases in the sample environment.

Reaction rate may also be influenced by the presence of some gases other than product gases. Thus, Eq. 2 may be further extended. To simplify the reaction system as much as possible, experimental conditions are often chosen to minimize the concentration of product gases in the reaction environment and the sample is surrounded by either a vacuum or an inert gas. The effects of $f(P)$ are then assumed to be negligible and Eq. 1 is assumed to be sufficient to describe the kinetic behavior of decomposing solids. Thus, two things must be determined:

1. $f(\alpha)$, i.e., the relationship between reaction rate and the extent of decomposition
2. How k changes with temperature, i.e., the relationship between reaction rate and the temperature.

These two points are discussed in the first two parts of this section. The final part deals with the possible effect on reaction rate of product gas pressure and other factors.

A. RELATIONSHIP BETWEEN REACTION RATE AND EXTENT OF REACTION

As mentioned above, when the temperature of a decomposing compound is maintained constant the influence of temperature on the reaction rate may be expressed in the form of a constant k, the specific reactions rate constant; thus, the shape of a plot of $d\alpha/dt$ against α is determined by $f(\alpha)$. Data for constant temperature (isothermal) experiments are often expressed in an integral form, which may be derived as follows:

$$\frac{d\alpha}{dt} = f(\alpha)k \tag{3}$$

$$\int_0^\alpha \frac{d\alpha}{f(\alpha)} = k\int_0^t dt = kt \tag{4}$$

When

$$\int_0^\alpha \frac{d\alpha}{f(\alpha)} = g(\alpha) \tag{5}$$

we may write

$$g(\alpha) = kt \tag{6}$$

Thus, the shape of a plot of α against t is determined by $g(\alpha)$. It should be noted that Eq. 3 applies equally to nonisothermal results. However, it is convenient to consider the relationship between reaction rate and extent of reaction in terms of the results obtained using the isothermal method, since Eq. 6 applies only under isothermal conditions and represents the simplest approach to discussing the nature of $f(\alpha)$ and thus $g(\alpha)$.

The decomposition of a solid usually starts with the formation of small localized areas of product called nuclei, which grow larger, forming an interface between product and reactant that proceeds into the bulk of the solid, gradually consuming the whole of the solid particle. Nuclei generally form on the surface of the reactant.

Fig. 6 shows a generalized α-t plot. The section A-B is variously attributed to surface desorption, surface decomposition, and nucleation. This may be followed by a period B-C during which the rate of reaction is scarcely measurable, called the induction period, which may be attributed to the slow formation of stable nuclei. The induction period is followed by C-D, a strongly acceleratory region up to the point of maximum reaction rate, point D, during which the nuclei are growing. The final part of the curve is deceleratory, during which the growing nuclei begin to overlap and thus the rate of reaction decreases until the reactant is totally consumed.

Nuclei form at energetically favored sites, such as points of crystallographic disorder on or near the surface of the reactant particle. It may be

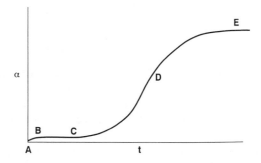

Fig. 6.   Generalized α-t plot for isothermal conditions.

argued (52) that for the general reaction

$$A_{(s)} \longrightarrow B_{(s)} + C_{(g)}$$

the first formed atoms of product B cannot be regarded as a distinct and separate phase, but, initially at least, tend to conform to the structure of, and retain their former positions with reference to, the reactant phase A. The tendency of accumulating molecules of B to assume the lattice configuration of the stable product, if it is different from that of the reactant, creates a strain energy.

As a consequence of this strain energy it can be shown that there is a critical size for product nuclei. Nuclei smaller than the critical size, called germ nuclei, are thermodynamically unstable and tend to revert to reactant, whereas larger nuclei, called growth nuclei, are stable and tend to grow.

The generation of growth nuclei establishes the existence of a reaction interface and the subsequent growth of the interface; thus, the form of $f(\alpha)$ and $g(\alpha)$, is governed by

1.   The type of nucleation
2.   The geometry of the reactant particle
3.   The influence of diffusion.

The type of nucleation depends upon the relative magnitudes of ΔG_n, the free energy for nucleation, and ΔG_g, the free energy for the growth of nuclei. When $\Delta G_g \ll \Delta G_n$, the growth of existing nuclei predominates over the formation of new ones, giving rise to discreet widely spaced individual nuclei which may be observed microscopically. In early studies the rate of growth of such nuclei was measured visually (38, 39, 50, 85). This type of nucleation gives rise to a sigma-shaped decomposition curve because as the nuclei grow the rate of decomposition increases until the growing nuclei begins to overlap, after which the rate of decomposition progressively decreases. One of the most important general expressions derived independently by Mampel (58), Avrami (3–5), and Erofe'ev (28), which deals with this type of behavior, is

$$\left[-\ln(1 - \alpha) \right]^{1/n} = kt$$

Expressions of this form are known as Avrami–Erofe'ev equations, where n usually has the values 2, 3, or 4; n may also take the value 1, and the equation is then known as a first-order equation. The first-order equation corresponds to the typical behavior of a unimolecular reaction in the gaseous or liquid phase. When found to apply to a solid state reaction it may be seen as an indication that the considerations of nucleation and reactant geometry do not apply and that individual molecules may decompose at random or that individual particles nucleate and rapidly decompose at random. This type of behavior may also arise because of the effects of particle size distribution.

If $\Delta G_g = \Delta G_n$, then a large number of diffuse nuclei form, none of which grow to a visible size. Thus, the acceleratory period is reduced or completely absent as the entire surface, or certain preferred crystal faces, of the reactant are rapidly covered with small nuclei. The interface then proceeds at a constant speed (under isothermal conditions) into the bulk of the solid. This behavior may be described by the equation derived by Mampel (58):

$$1 - (1 - \alpha)^{1/n} = kt$$

where n is the number of dimensions in which the interface advances and usually has the value of 2 or 3.

In the above equations the assumption has been implicitly made that the removal of product gases from the reaction interface is not a rate-limiting step. However, it is possible that the transport of gaseous reaction products from the reaction interface through the developing product layer is sufficiently difficult for it to become the rate-limiting process. A number of equations have been derived to model reactions governed by diffusion. For the simple case of one-dimensional diffusion the equation

$$\alpha^2 = kt$$

has been shown to apply (29). For two-dimensional diffusion out of a cylindrical particle the equation

$$(1 - \alpha)\ln(1 - \alpha) + \alpha = kt$$

(48) can be used. For three-dimensional diffusion out of a sphere the equation derived by Jander (53) may be used:

$$\left[1 - (1 - \alpha)^{1/3}\right]^2 = kt$$

or a modification derived by Ginstling and Brounshtein (42):

$$\left(1 - \tfrac{2}{3}\alpha\right) - (1 - \alpha)^{2/3} = kt$$

The $g(\alpha)$ values discussed so far are summarized in Table 1. It should be noted that the t referred to in these equations is measured from the start of the decomposition process, i.e., after the end of the induction period if one exists. Errors may arise because it is not always easy to accurately estimate the point from which t should be measured. Also, some $g(\alpha)$'s in Table 1 are multiplied by a constant. This is because the differential form of the equation, $f(\alpha)$, is assumed to give the correct value for k (as for heterogeneous

TABLE 1
Commonly Used Kinetic Equations

No.		$f(\alpha) = \dfrac{d\alpha/dt}{k}$	$g(\alpha) = kt$	Label
	Sigmoid Rate Equations			
1	Avrami–Erofe'ev	$(1 - \alpha)[-\ln(1 - \alpha)]^{1/2}$	$2[-\ln(1 - \alpha)]^{1/2}$	A2
2		$(1 - \alpha)[-\ln(1 - \alpha)]^{2/3}$	$3[-\ln(1 - \alpha)]^{1/3}$	A3
3		$(1 - \alpha)[-\ln(1 - \alpha)]^{3/4}$	$4[-\ln(1 - \alpha)]^{1/4}$	
	Deceleratory			
4	First order	$(1 - \alpha)$	$-\ln(1 - \alpha)$	F1
	Based on Geometric Models			
5	Contracting area	$(1 - \alpha)^{1/2}$	$2[1 - (1 - \alpha)^{1/2}]$	R2
6	Contracting volume	$(1 - \alpha)^{2/3}$	$3[1 - (1 - \alpha)^{1/3}]$	R3
	Based on Diffusion Mechanism			
7	One-dimensional diffusion	α^{-1}	$\frac{1}{2}\alpha^2$	D1
8	Two-dimensional diffusion	$[-\ln(1 - \alpha)]^{-1}$	$(1 - \alpha)\ln(1 - \alpha) + \alpha$	D2
9	Three-dimensional diffusion	$[1 - (1 - \alpha)^{1/3}]^{-1}(1 - \alpha)^{2/3}$	$^{3/2}[1 - (1 - \alpha)^{1/3}]^2$	D3
10	Ginstling–Brounshtein	$[(1 - \alpha)^{-1/3} - 1]^{-1}$	$^{3/2}[1 - \frac{2}{3}\alpha - (1 - \alpha)^{2/3}]$	D4

kinetics); thus, integrating $f(\alpha)$ according to Eq. 5 gives rise to a constant that must be included when analyzing data using the integral forms of the kinetic equations. The equations in Table 1 are classified according to the widely accepted convention as deceleratory or sigmoid and are labeled using the system proposed by Sharp et al. (78).

Hulbert (49) considered three types of nucleation: constant rate of nucleation, continuously decreasing rate of nucleation, and instantaneous nucleation, all of which give rise to an equation of the general form

$$-\ln(1 - \alpha) = kt^m$$

where the value of m is determined by the type of nucleation, the number of dimensions in which nuclei growth occurs, and whether the reaction is phase boundary or diffusion controlled. The results are summarized in Table 2. Inspection of Table 2 shows that different mechanisms give rise to the same value for m. Thus, for equations of this form, the reaction mechanism cannot be unequivocally identified using rate data alone and ancillary techniques such as microscopic examination should be used.

All of the expressions considered so far are based on the assumption that either one particle of a particular shape is decomposing or that a number of particles all of the same shape and size are all decomposing in a similar manner. A wide particle size distribution in the sample will affect the shape

TABLE 2
Summary of the Value of m for the General Equation $-\ln(1 - \alpha) = kt^m$ for Different Types
of Reaction Mechanisms

Model	Nucleation Rate	m Phase Boundary Control	m Diffusion Control
Three-dimensional growth	Constant	4	2.5
	Zero (instantaneous)	3	1.5
	Decreasing	3–4	1.5–2.5
Two-dimensional growth	Constant	3	2.0
	Zero	2	1.0
	Decreasing	2–3	1.0–2.0
One-dimensional growth	Constant	2	1.5
	Zero	1	0.5
	Decreasing	1–2	0.5–1.5

of experimental curves. Delmon (20) has shown that for a powdered sample
for which the individual particles each follow the contracting volume equation (Table 1) with a wide particle size distribution, the apparent $f(\alpha)$ will be
first order. This factor must be taken into consideration when trying to
determine the reaction mechanism.

Another assumption made in all of the above kinetic models is that the
reaction interface has no significant thickness. It may be argued that diffusion
of the product gas through the reactant lattice is possible. Thus, molecules
several molecular units distant from the reaction interface may decompose
and the product gas may diffuse through the intervening reactant layer and
escape beyond the interface. The interface may then be considered to have a
thickness that if significant compared to the diameter of the decomposing
particle, may have a significant affect on the apparent $g(\alpha)$ or $f(\alpha)$. Hill (46)
has considered the influence that diffusion may have at the reaction interface. He suggested that "a few ions or atoms from the product can move
through the dislocation network of the reactant and 'fertilize potential
nuclei'." He found that this process "can lead to exponential reaction-time
curves."

An attempt may be made to treat mathematically the more general type of
diffusion being proposed here. Consider an infinitely long cross section
through a plane of reactant of thickness a. Allow the y axis to bisect the
cross section. The situation may then be represented by Fig. 7. In the
reaction,

$$A_{(s)} \overset{k_1}{\underset{k_2}{\rightleftarrows}} B_{(s)} + C_{(g)}$$

assume that beyond $a/2$ and $-a/2$ the concentration of C is zero; this
corresponds to performing the experiment under a hard vacuum. Through

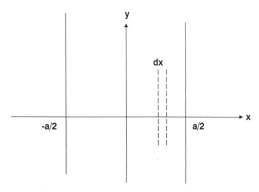

Fig. 7. Model for diffusion within the reactant.

the small interval dx along the x axis the rate of decomposition may be expressed as

$$\frac{\partial C_C}{\partial t} = D\frac{\partial^2 C_C}{\partial x^2} + k_1 C_A - k_2\left(C_{A_{t_0}} - C_A\right)C_C \tag{7}$$

where
$C_A = C_A(x, t) = $ concentration of A at x and t
$C_B = C_B(x, t) = $ concentration of B at x and t
$C_C = C_C(x, t) = $ concentration of C at x and t
$D \ \ = $ diffusion coefficient
$C_{A_{t_0}} = C_A(x, t_0) = $ constant $= $ concentration of A at $t = 0$
$k_2 = $ rate constant for back reaction
$k_1 = $ rate constant for forward reaction

$$C_{A_{t_0}} = C_A - \int_0^t \left[k_1 C_{A_{t_0}} - k_2\left(C_{A_{t_0}} - C_A\right)C_C\right] dt \tag{8}$$

This type of equation is analytically insolvable. It can be solved numerically using a computer program. However, it is possible to envisage the results of this type of diffusion in a qualitative way. If diffusion of C within the solid lattice of A is very easy, then nuclei may develop at energetically favored sites within the bulk of a small spherical particle of A. This may result in random decomposition of molecules of A and may correspond to first-order behavior. If, however, diffusion of C within A is impossible, then the reaction will have to begin on the surface of the spherical particle and proceed inward, corresponding to a contracting volume equation. If some intermediate position is imagined where diffusion is possible up to some depth within the particle, but not throughout the bulk of the particle, then some combination of contracting volume and first-order behavior will result.

For calcium carbonate, work by Haul and Stein (44), who have studied the diffusion of CO_2 within the lattice of $CaCO_3$, is of interest. Haul and Stein identified two types of diffusion: a rapid initial diffusion confined to a surface layer and a slow diffusion process within the lattice. The activation energy for diffusion within the lattice was found to be 242 kJ/mol, which is higher than that reported for bulk decomposition (14). It would thus seem that decomposition is energetically favored over diffusion within the bulk. However, Haul and Stein (44) did not evaluate the activation energy for the surface diffusion, although they observed that it was three to four times more rapid than diffusions within the lattice. Thus, it would seem that diffusion up to a certain depth within a calcium carbonate particle is comparatively easy; thus, the reaction interface may have a certain thickness. This thickness may become significant with respect to the size of the decomposing particles when finely divided powders are used.

In summary, equations have been derived to model a variety of idealized kinetic behaviors; a selection of the most commonly used expressions is given in Table 1. Deviations from such ideal behavior may result from temperature and pressure gradients within the sample, particle size distribution, and the interface having a thickness that is significant with respect to the size of the decomposing particle. The same kinetic behavior may also result from different reaction mechanisms. Thus, seeking the conformity of experimental rate data to various models is a useful process, but should not be taken as sufficient proof that a particular mechanism is operative in a particular case.

B. RELATIONSHIP BETWEEN TEMPERATURE AND REACTION RATE

It is generally true that the rate of any given chemical reaction will increase with increasing temperature. An empirical quantitative relationship between the reaction rate constant and temperature that is obeyed approximately by the majority of homogeneous reactions is the Arrhenius equation:

$$k = Ae^{-E/RT} \tag{9}$$

where k = the reaction rate constant
   E = a constant known as the activation energy
   A = a constant known as the pre-exponential factor.

Now,

$$\frac{dC_n}{dt} = f(C_1, C_2, \ldots, C_n)k \tag{10}$$

where C_n = the concentration of reactant n
   $f(C_1, C_2, \ldots, C_n)$ = a function of the concentration of the reactant that does not change with temperature.

This relationship has found a theoretical foundation that, stated simply, is based on the concept that a reaction must overcome an energy barrier, represented by E, to proceed. The fraction of the total number of molecules with sufficient energy to overcome this barrier can be described by the Boltzmann energy distribution function. The pre-exponential factor A is a measure of the effects of both the internal degrees of freedom of the reacting species and the necessary redistribution of that energy so that reaction ensues. This theoretical approach has been refined to produce the transition state theory. By analogy with homogeneous reaction kinetics, for any simple solid state decomposition reaction where the dissociation of the reactant molecules may be considered as a unimolecular process the following equation may be written:

$$\frac{d\alpha}{dt} = f(\alpha)k \tag{11}$$

that is,

$$\frac{d\alpha}{dt} = f(\alpha)Ae^{-E/RT} \tag{12}$$

Garn (37) maintains that in the solid state "the lack of a statistical distribution rules out the use of the Arrhenius equation." He assumes that "a substantial difference from the average energy is not achievable within the crystal." However, the Boltzmann distribution is at the basis of the successful statistical theory of heat capacity in the solid state (89). Also, the Arrhenius equation is used to model the kinetics of vaporization from the solid state (80). These would seem sufficient reasons for assuming that the Boltzmann energy distribution exists within the solid bulk and thus at the reaction interface, and consequently that the Arrhenius equation can be applied to the kinetics of solid state reactions. Deactivation of an excited molecule may be rapid, allowing little time for the internal energy of the excited species to be rearranged (so that sufficient energy is distributed into the reaction co-ordinate for reaction to ensure). However, excitation is also rapid; thus, in unit time the probability that a number of excited molecules will have their energy distributed in such a manner as to allow dissociation to occur is correspondingly high.

MacCallum and Tanner (57) have proposed that the degree of reaction α as a function of time and temperature should be considered as independent variables. Therefore, the total differential with respect to α is the sum of the effects of two partial differentials:

$$\alpha = F(T, t) \tag{13}$$

$$\therefore \quad d\alpha = \left(\frac{\partial \alpha}{\partial t}\right)_T dt + \left(\frac{\partial \alpha}{\partial T}\right)_t dT \tag{14}$$

$$\therefore \quad \frac{d\alpha}{dt} = \left(\frac{d\alpha}{dt}\right)_T + \left(\frac{d\alpha}{dT}\right)_t \frac{dT}{dt} \tag{15}$$

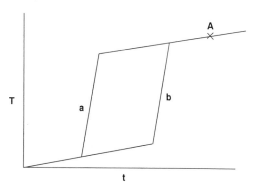

Fig. 8.   Two-path example for heating programs.

Thus, under nonisothermal conditions the instantaneous rate of decomposition at any value of α may be considered under this supposition not to be given by the differential form of the isothermal α function (Eq. 12) for the reaction at the instantaneous temperature T, but contains an additional quantity given by the second term on the right side of the above equation. Therefore, from this argument the simple Arrhenius expression can no longer be applied. This poses the question, What is the physical significance of the partial differential of α with respect to temperature at constant time? Eq. 14 suggests that an instantaneous temperature increase would result in a decrease in α over a zero time interval, which is an untenable position. Further, consider point A in Fig. 8.

At point A the temperature, time, and heating rate are the same for both heating programs a and b, yet heating program a would obviously result in a greater extent of decomposition, because more time is spent at a higher temperature than with heating program b. Thus, Eq. 13 cannot adequately describe the way in which α is influenced by both time and temperature. It is necessary to take account of the thermal history of the sample and how temperature changes with time. The temperature must therefore be considered as a function of time and Eq. 13 is then expressed as

$$\alpha = F[G(t), t] \tag{16}$$

where $T = G(t)$ $\tag{17}$

and where time and temperature cannot be considered as independent variables.

Fevre et al. (31) consider that the approach adopted by MacCallum and Tanner should be further extended in the manner

$$\alpha = F(T, t, b) \tag{18}$$

where b = the heating rate

$$\therefore d\alpha = \left(\frac{\partial \alpha}{\partial T}\right)_{tb} dT + \left(\frac{\partial \alpha}{\partial t}\right)_{Tb} dt + \left(\frac{\partial \alpha}{\partial b}\right)_{Tt} db \qquad (19)$$

The points made with regard to Eq. 13 apply to this extended approach and therefore cannot be considered a meaningful mathematical treatment. Indeed, the "two-path" example given above applies equally well to this supposedly extended form of the equation. Other authors have also criticized MacCallum and Tanners approach (43, 47).

Given that we can proceed with the normal homogeneous form of the Arrhenius equation, we may apply transition state theory in the usual way for a reaction of the type

$$A_{(s)} \longrightarrow B_{(s)} + C_{(g)}$$

In this instance we will take the molecule A to be BC, and assume that the reaction proceeds via some activated intermediate:

$$BC \longrightarrow [B ---- C]^{\ddagger} \longrightarrow B + C$$

Postulating a pseudo equilibrium between the reactant and the activated intermediate we obtain

$$BC \underset{K^{\ddagger}}{\rightleftharpoons} [B ---- C]^{\ddagger} \longrightarrow B + C \qquad (20)$$

where $K^{\ddagger}$, the pseudo equilibrium constant, is given by

$$K^{\ddagger} = \frac{[B ---- C]^{\ddagger}}{[BC]} \qquad (21)$$

From transition state theory we may write

$$\text{Rate} = w\nu[B ---- C]^{\ddagger} \qquad (22)$$

where w = some transition coefficient between 0.5 and 1
ν = classical vibration in the reaction coordinate

From Eq. 21 we may write

$$[B ---- C]^\ddagger = K^\ddagger[BC] \qquad (23)$$

$$\therefore \quad \text{Rate} = w\nu K^\ddagger[BC] \qquad (24)$$

Continuing with pseudo thermodynamic arguments we may write

$$\Delta G^{\circ\ddagger} = \Delta H^{\circ\ddagger} - T\Delta S^{\circ\ddagger} \qquad (25)$$

where $\Delta G^{\circ\ddagger}$ = standard Gibbs free energy for the formation of the activated complex (from reactant)
$\Delta H^{\circ\ddagger}$ = standard enthalpy of formation of activated complex
$\Delta S^{\circ\ddagger}$ = standard entropy of formation of activated complex

and

$$\Delta G^{\circ\ddagger} = -RT \ln K^\ddagger \qquad (26)$$

$$\therefore \quad K^\ddagger = e^{(-\Delta G^{\circ\ddagger}/RT)}$$

$$\therefore \quad \text{Rate} = w\nu \, e^{(-\Delta G^{\circ\ddagger}/RT)}[BC] \qquad (27)$$

Let the reaction rate constant = k; then

$$k = w\nu \, e^{(-\Delta G^{\circ\ddagger}/RT)} \qquad (28)$$

$$\therefore \quad k = w\nu \, e^{(\Delta S^{\circ\ddagger}/R)} e^{(-\Delta H^{\circ\ddagger}/RT)} \qquad (29)$$

where $\Delta H^{\circ\ddagger} = E$ \qquad (30)

From statistical mechanics we may write

$$e^{(\Delta S^{\circ\ddagger}/R)} = \frac{\bar{k}T}{h\nu} \frac{Q^\ddagger}{Q} \qquad (31)$$

where $Q^\ddagger$ = the partition function for the activated complex, excluding
that for the reaction coordinate
$\dfrac{\bar{k}T}{h\nu}$ = the partition function for the classical oscillator assumed to
be the reaction coordinate
Q = complete partition function for the reactant

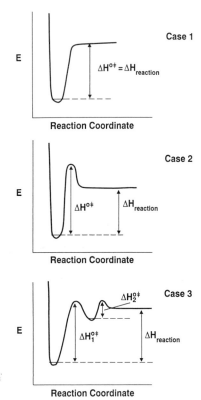

Fig. 9. Possible reaction pathways for solid state decomposition reactions.

Thus, we obtain the well-known expression

$$k = w\frac{\overline{k}T}{h}\frac{Q^{\ddagger}}{Q}e^{(-\Delta H^{\circ\ddagger}/RT)} \tag{32}$$

Three types of reaction pathways may be postulated, cases 1, 2, and 3 given in Fig. 9. In the first case $\Delta H^{\circ\ddagger}$ may be identified with ΔH° for the decomposition reaction. This may apply to certain simple processes that do not involve a chemical step such as the sublimation of solids.

Beruto and Searcy (9) have proposed that the Hertz–Knudsen–Langmuir (HKL) equation, which was originally developed for the study of the evaporation of solids, should be applied to solid state decomposition reaction:

$$J = \alpha\frac{P_{eq}}{(2\pi MRT)^{1/2}} = \alpha\frac{e^{(-\Delta G_{d}^{\circ}/RT)}}{(2\pi MRT)^{1/2}} \tag{33}$$

where J = steady state flux
            α = vaporization coefficient
            P_{eq} = equilibrium carbon dioxide pressure
            M = molecular weight
            ΔG_d° = standard Gibbs free energy of the physical reaction at
                   temperature T.

The vaporization coefficient is the ratio between the observed flux and the maximum possible flux, which is equal to the value of the equation when α equals one. This equation implicity identifies the activation energy with ΔH° for the decomposition reaction. Even for the simple case of the vaporization of a solid Somarjai and Lester (80) have pointed out that the HKL equation is generally inadequate to describe the kinetics of vaporization and it is unlikely that it could find a wide application to the more complex case of decomposition reactions.

The second case shown in Fig. 9 is more likely to apply to any reactions involving a bond-breaking step; thus, the kinetic parameter $\Delta H^{\ddagger \circ}$ cannot be identified with a measurable thermodynamic quantity. The third case shown in Fig. 9 postulates the existence of some intermediate, possibly adsorbed, state between dissociation from the solid lattice and escape from the solid surface.

Early attempts to evaluate the pre-exponential constant A, assuming the second case from Fig. 9 applies, resulted in the Polanyi–Wigner expression (65)

$$\frac{dx}{dt} = \nu \frac{2E}{RT} x e^{-E/RT} \tag{34}$$

where $\dfrac{dx}{dt}$ = rate of interface advance
         x = the incremental advance for unit reaction
         ν = an appropriate vibration frequency.

The term $2E/RT$ arises because the critical energy required can be achieved through three degrees of vibrational freedom. Shannon (77) examined the relevant data for 29 reactions and found that only one-third gave order of magnitude agreement. He concluded that the weakness of the Polanyi–Wigner equation lay in its failure to take into account rotational and other degrees of freedom. Shannon used Eq. 32 for a more complete analysis of the pre-exponential term of calcium carbonate and magnesium carbonate and postulated two possible cases for the reactant: the CO_3^{2-} ions can undergo either rotation or torsional oscillation:

$$Q_1 = f_L^3 f_A^3 f_{vib}^6 \tag{35}$$

or

$$Q_2 = f_L^3 f_{rot} f_{vib}^6 \tag{36}$$

where Q_1 = the complete partition function for a carbonate ion undergoing torsional oscillation

Q_2 = the complete partition function for a rotating carbonate ion

f_{vib} = partition function for internal vibrations

f_L = partition function for lattice vibrations

f_λ = partition function for torsional vibration

f_{rot} = rotational partition function for a linear molecule with one
degree of freedom.

Vibrational partition functions were calculated using

$$f_{vib} = [1 - e^{-(h\nu/RT)}]^{-1} \tag{37}$$

where ν = vibrational frequencies.

Rotational partition functions are calculated for a one-dimensional rotor
using

$$f_{rot} = \left(\frac{8\pi^2 IkT}{\sigma^2 \lambda}\right)^{1/2} h \tag{38}$$

where I = moment of inertia about the axis of rotation

σ = the symmetry number or the number of indistinguishable
orientations of the molecule.

The vibrational frequencies used for the calculation of these partition
functions were given by Herzberg (45) and Schroeder (72) and are listed in
Table 3. Shannon (77) postulated that the activated complex consisted of a
CO_2 molecule parallel to the plane of the solid surface rotating freely about
an axis at right angles to the O=C=O axis. The molecule has three lattice
vibrations modes, one normal to the solid surface and two parallel to the
surface, and its normal internal vibrational modes. The complete partition
function for the activated complex, excluding that of the reaction coordinate,
is

$$Q^{\ddagger} = f_L^2 f_{L'}^2 f_\lambda f_{rot} f_{vib}^4 \tag{39}$$

where f_L = the partition function for the lattice modes arising from the
Ca–O bond

$f_{L'}$ = the partition function for the lattice modes arising from the
O–CO_2 bond.

TABLE 3
Vibrational Frequencies and Partition Functions for $CaCO_3$ at 1100 K

	Frequency (cm^{-1})	Partition Function (Shannon)	Recalculated Partition Function
CO^{2-}			
Lattice vibration	$\nu_1 = 367$	2.64	2.62
	$\nu_2 = 330$	2.85	2.85
	$\nu_3 = 106$	7.80	7.73
Rotation libration		6.73×10^4	6.73×10^4
	$\nu_1 = 36$	29.40	21.78
	$\nu_2\nu_3 = 100$	8.40	8.16
Internal vibrations	$\nu_1\nu_2 = 712$	1.65	1.65
	$\nu_3\nu_4 = 1460$	1.16	1.77
	$\nu_5 = 881$	1.46	1.46
	$\nu_6 = 1070$	1.33	1.33
$O^{2-}-CO_2$			
Lattice vibrations	$\bar{\nu}_{CaO}(normal) = 690$	1.68	1.68
	$\bar{\nu}_{CaO}^{x,y}(tan) = 364$	2.65	2.64
	$\nu_{OCO}^{x} = 350$	2.72	2.72
Librations	$\nu_a = 60$	13.00	13.26
	$\nu_b = 80$	10.00	10.07
Rotation		31.5	31.5
Internal vibrations	$\nu_1 = 1345$	1.21	1.21
	$\nu_2\nu_3 = 667$	1.71	1.72
	$\nu_4 = 2349$	1.04	1.05
Q_1 (torsional oscillation)		86×10^3	6.10×10^5
Q_2 (free rotation)		28.2×10^6	2.84×10^7
$Q_a^{\ddagger}$		13.1×10^4	1.35×10^5
$Q_b^{\ddagger}$		10.1×10^4	
		1.03×10^5	
$C = Q^{\ddagger}/Q$			
C_1			
a		1.54	2.22×10^{-1}
b			1.68×10^{-1}
C_2			
a		0.46×10^{-2}	4.78×10^{-3}
b		0.36×10^{-2}	3.63×10^{-3}

The reaction coordinate was assumed to be the component of the mode of vibration perpendicular to the surface and represents a bond between the potential O_2^{2-} atom in the solid and the CO_2 molecule just before it is broken. The partition function of the activated complex does not include the partition function corresponding to this degree of freedom. The lattice vibration for CaO are given by Born and Huang (11). Torsional frequencies of the CO_2 molecule were taken from the spectra determined on thick films of solid CO_2 deposited an AgCl plates. Calculations were made assuming a lower limit of 60 cm^{-1} (a) and an upper limit of 80 cm^{-1} (b). The two lattice modes arising from the $O-CO_2$ activated complex were estimated assuming a bond order of $\frac{1}{2}$, which led to an estimate of a bond length of 1.57 Å from a chart showing bond length against bond order (constructed from C–O bond lengths of CO, acetone, and methanol). From this bond length a bond frequency of 300–400 cm^{-1} was derived by extrapolating from a table correlating C–O vibration frequencies with bond lengths made by Fassel et al. (30). The lattice frequency used in this calculation was 350 cm^{-1}.

In Table 3 the vibrational frequencies are given, together with values for the partition functions calculated by Shannon (77) and by the present authors and coworkers. It can be seen that Shannon (77) made errors resulting in an order of magnitude discrepancy for the case of torsional oscillation. For this case, only one value for C_1^* using the lower vibration frequency of 60 cm^{-1} was quoted. The rotational partition functions were assumed to be correct. Approximate agreement was claimed between the calculated value of C^* assuming free rotation and an experimental value obtained from the results of Britton et al. (12), where $C_{experimental}^*$ is given by Shannon as 6×10^{-2} $(C^* = Q^{\ddagger}/Q$, see Table 3). However, the data from Britton et al. (12) gives values of $k = 7.45 \times 10^5$ s^{-1}, $T = 1041$ K, and $E = 148$ kJ/mol. For a series of experiments using lumps of calcite we obtain $C_{experimental}^* = 0.917$. Shannon (77) performed similar calculations for MgCO$_3$, but used a temperature of 600 K instead of the 600°C used by the experimenters whose results he quotes (12).

Many criticisms of Shannon's work are possible apart from the mathematical errors. Perhaps the most serious in terms of the magnitude of A is that the value of C^* was calculated assuming that decomposition occurs freely throughout the sample, whereas the reaction almost certainly occurs through the advance of a reaction interface (12), which, as observed by Cordes (19), affects the value of the pre-exponential factor by a number of orders of magnitude.

When considering the carbonate ion in a solid lattice a judgment must be made as to what extent external degrees of freedom, including lattice motions, must be included in the calculations of the partition function; i.e., to what extent the energies associated with these external degrees of freedom may be considered to be available, at least in part, to the reaction process. Shannon (77) implicitly assumes the involvement of lattice and external modes (rotation or libration). At the other extreme, it is possible to make corresponding calculations for the hypothetical gas phase carbonate ion decomposition where

$$Q^{\ddagger} = f_{vib}^5 \qquad (40)$$

and

$$Q = f_{vib}^6 \qquad (41)$$

The one-dimensional degree of freedom about the O_2^{2-}–C axis must be replaced by further vibration. This will be a loose vibration and may be assigned an approximate frequency of 250 cm^{-1}, from which $C = 1.84$. The value of the real C factor may lie at some intermediate position between these two extremes.

If reaction occurs only at a reaction interface then the pre-exponential factor must be related to the area of the interface. Cordes (18) found that for spherical particles with a ratio of surface to volume of 10^{-6}, the pre-exponential factor is decreased by 10^4. If the interface is considered to have a certain thickness greater than one molecule, then the value of A will be further altered, and must then be considered in terms of the area of the reaction interface multiplied by the apparent thickness. Thus, even if the exact nature of the activated complex may be assumed, several considerations would still make the calculation of the pre-exponential factor uncertain. However, the models proposed by Shannon represent the most comprehensive treatment to date, and can be used to examine how the pre-exponential factor might change with temperature and thus affect the calculation of E. Accordingly, the value for C for a series of temperatures from 1000 to 1100 K were calculated together with values for the specific rate constants, assuming an activation energy of 210.0 kJ/mol. The results are given in Table 4. These values were then used to make an Arrhenius plot from Table 5 to give Fig. 10. The results of these Arrhenius plots, assuming the usual form of the Arrhenius equation ($k = A e^{-E/RT}$), are given in Table 6.

As can be seen from Table 6 the temperature dependence of the pre-exponential factor can give rise to errors of about 5%. Inspection of Table 4 shows that for reaction model 1 the quantity C increases as the temperature decreases. These two effects tend to cancel each other out and so the pre-exponential factor remains approximately constant and the calculated activation energy in Table 6 is very close to the actual value of 210.0 kJ/mol. However, for reaction model 2, C remains approximately constant; in fact, it decreases slightly with decreasing temperature. Thus, for this case the

TABLE 4
Change of Specific Reaction Rate Constant (R) with Temperature Following Shannon's
Reaction Models for $CaCO_3$

Temperature (K)	C_1		$C_2 \times 10^2$		k_1		k_2	
	a	b	a	b	a	b	a	b
1100	0.222	0.168	0.478	0.362	542.6	410.7	11.68	8.849
1090	0.225	0.171	0.479	0.364	441.5	335.5	9.399	7.142
1080	0.228	0.173	0.487	0.370	357.7	271.4	7.640	5.804
1070	0.231	0.175	0.478	0.363	228.5	218.6	5.971	4.534
1060	0.232	0.177	0.476	0.362	229.8	175.3	4.714	3.585
1050	0.236	0.179	0.477	0.362	184.5	140.0	3.729	2.830
1040	0.238	0.181	0.474	0.361	146.2	111.2	2.913	2.218
1030	0.240	0.183	0.474	0.361	115.4	87.98	2.279	1.735
1020	0.243	0.185	0.473	0.361	90.87	69.25	1.771	1.351
1010	0.247	0.188	0.474	0.360	71.65	54.53	1.375	1.044
1000	0.249	0.190	0.473	0.360	55.69	42.50	1.058	0.805

TABLE 5
Value for $1/T$ and $\ln k$ from Table 4 from Which Fig. 10 is Plotted

$1/T \times 10^3$ (K)	$\ln k_1$		$\ln k_2$	
	a	b	a	b
0.9091	6.2965	6.0178	2.4583	2.1803
0.9174	6.0901	5.8157	2.2406	1.9660
0.9259	5.8796	5.6036	2.0333	1.7586
0.9346	5.6648	5.3872	1.7868	1.5116
0.9439	5.4371	5.1665	1.5506	1.2768
0.9524	5.2177	4.9413	1.3162	1.0404
0.9615	4.9853	4.7115	1.0691	0.7967
0.9709	4.7482	4.4771	0.8236	0.5513
0.9804	4.5105	4.2378	0.5713	0.3011
0.9901	4.2718	3.9988	0.3184	0.0433
1.0000	4.0198	3.7494	0.0563	−0.217

Arrhenius equation would more accurately take the general form

$$k = T^n A e^{-E/RT} \tag{42}$$

where $n = 1$. Failure to take this into account results in overestimating the activation energy by some 5%. The linear regression coefficients show that in all cases good straight lines are obtained. Thus, it is not possible, given that there will always be a certain amount of experimental error, to detect this

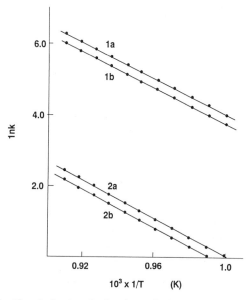

Fig. 10. Arrhenius plot for theoretical data from Table 5.

TABLE 6
Arrhenius Parameters from the Data Given in Table 5, (assuming the usual form of
the Arrhenius equation)

$Q^{\ddagger}/Q$	E (kJ/mol)	A (s^{-1})	r
1 a	208.5	4.331×10^{12}	0.999 99
b	208.0	2.090×10^{12}	0.999 98
2 a	220.5	3.475×10^{11}	0.999 97
b	220.1	2.539×10^{11}	0.999 97

apparent temperature dependency of the pre-exponential factor and thus obtain a value for n.

In summary, it would appear reasonable to assume that the Arrhenius equation would give, to a first approximation, a quantitative relationship between reaction rate and temperature for solid state decomposition reactions for both isothermal and nonisothermal experiments. It would appear that a more appropriate modification of the Arrhenius equation would be of the form given by Eq. 42, but it is not possible at present either experimentally or theoretically to assign a value of n with confidence. It should be borne in mind that if a reaction occurs at an interface, then the pre-exponential factor will be reduced by several orders of magnitude. For this reason and because the interface may be considered to have a certain thickness, attempts to relate experimental values to theoretical calculations are very difficult. It would seem reasonable to assume that the type of reaction pathway followed by solid state decomposition reactions would be of either type two or three given in Fig. 8. It is difficult to differentiate between these two pathways on the basis of the value of the pre-exponential factor for the reasons given above and because work done by Cordes (19) shows that if a type three pathway is assumed, corresponding to complexes that have free translation on the surface, calculations do not give rise to an easily differentiable group of values.

C. INFLUENCE OF REACTION ENVIRONMENT

When the atmosphere surrounding a sample contains an appreciable amount of the gaseous product of a reversible reaction, e.g., CO_2 from carbonate decompositions, then the kinetic behavior may show a dependence on partial pressure. Various theories attempt to explain the way in which decomposition rate is influenced by pressure of the product gas (8, 51, 63, 73, 88). The approach proposed here, which is based on transition state theory, is given at the end of this section. Generally, there is a decrease in reaction rate with increasing product gas pressure. To study the simplest and most fundamental process occurring in a decomposition reaction, experimental conditions are often chosen that minimize the concentration of product gases in the reaction environment to avoid the influence of disturbing phenomena such as diffusion or secondary reactions. To achieve this product gases are continuously removed either by pumping, i.e., carrying out the experiment under vacuum, or by using a flowing atmosphere. However, any method of removing product gases will establish a pressure gradient between the reaction interface and the external environment and this pressure gradient may change significantly with the rate of reaction and have a corresponding perturbing influence on experimental results. Constant rate thermal analysis attempts to overcome this problem by keeping the rate of reaction low and constant; indeed, this is the main advantage of using this type of experimental approach. When more convential methods are used it is normally assumed that the use of vacuum or flowing atmospheres is sufficient to make the influence of product gas pressures negligible. Such an assumption could be justified by comparing results using conventional methods with results obtained using constant rate thermal analysis.

Another possible effect of product gas partial pressure is the effect it may have on the porosity and surface area of the solid product. The profound influence that quite small changes of pressure may have has been shown by Rouquerol et al. (68, 70) for the case of Gibbsite where a change in pressure of one torr may change the specific surface area accessible to nitrogen from 56 to 320 m^2/g. It may be reasonably assumed that there is an associated influence upon the form of $f(\alpha)$. Again, whether meaningful kinetic measurements are possible without precise control over the partial product gas pressure can be determined only by further experimental studies.

Problems of thermal transport become particularly acute when experiments are carried out under vacuum. Garn (36) has stated that "decomposition kinetics as learned in hard vacuum may have little meaning." The use of an externally supplied atmosphere may overcome these problems, but may alter apparent kinetic characteristics by hindering the removal of gaseous products. Using an externally supplied atmosphere that reacts with either the reactant or solid or gaseous products, may obviously affect the kinetic behavior of the reaction.

It is not yet clear what experimental conditions, whether vacuum or under a flowing externally supplied atmosphere, yield the most reliable results. Nor is it clear whether constant rate thermal analysis or conventional methods give better results.

To understand the effect that significant concentrations of product gases in sample environment may have, the transition state theory may be extended to include the effects of the reverse reaction in the following manner. Consider Eq. 27:

$$\text{Forward rate of reaction} = w_f \nu_f \, e^{-\Delta G_f^{\circ\ddagger}/RT}[\text{BC}] \qquad (27)$$

We may postulate that the reverse reaction

$$\text{B} + \text{C} \longrightarrow [\text{BC}]$$

proceeds via the same activated intermediate as the forward reaction, thus,

$$B + C \rightleftharpoons [B ---- C]^{\ddagger} \longrightarrow BC \qquad (43)$$

and in the same manner as that given in the preceeding section arrive at Eq. 27:

$$\text{Backward rate of reaction} = w_b \nu_b \, e^{-\Delta G_b^{\circ \ddagger}/RT}[C][B] \qquad (44)$$

Thus,

$$\text{Net rate of reaction} = w_f \nu_f \, e^{-\Delta G_f^{\circ \ddagger}/RT}[BC] - w_b \nu_b \, e^{-\Delta G_b^{\circ \ddagger}/RT}[C][B] \quad (45)$$

It is convenient to consider the reaction rate in terms of the specific reaction rate constant for the forward reaction:

$$\text{Rate of reaction} = k[BC]$$

$$\therefore \quad k_f = w_f \nu_f \, e^{-\Delta G_f^{\circ \ddagger}/RT} - w_b \nu_b \, e^{-\Delta G_b^{\circ \ddagger}/RT} \frac{[C][B]}{[BC]} \qquad (46)$$

Now

$$\Delta G^{\circ} = \Delta G_f^{\circ \ddagger} - \Delta G_b^{\circ \ddagger} \qquad (47)$$

where ΔG° = standard Gibbs free energy for the reaction

$\Delta G_f^{\circ \ddagger}$ = standard Gibbs free energy for the formation of activated intermediate from the reactant

$\Delta G_b^{\circ \ddagger}$ = standard Gibbs free energy for the formation of the activated intermediate from the products

$$-\Delta G_b^{\circ \ddagger} = \Delta G^{\circ} - \Delta G_f^{\circ \ddagger} \qquad (48)$$

$$\therefore \quad k_f = w_f \nu_f \, e^{-\Delta G_f^{\circ \ddagger}/RT} - w_b \nu_b \, e^{-\Delta G_f^{\circ \ddagger}/RT} \, e^{+\Delta G^{\circ}/RT} \frac{[C][B]}{[BC]} \qquad (49)$$

Assuming that $w_f \nu_f = w_b \nu_b$, then

$$k_f = w_f \nu_f \, e^{-\Delta G_f^{\circ \ddagger}/RT} \left[1 - e^{\Delta G^{\circ}/RT} \frac{[C][B]}{[BC]} \right] \qquad (50)$$

Now

$$\Delta G = \Delta G^{\circ} + RT \ln \frac{[C][B]}{[BC]} \qquad (51)$$

where ΔG = change in Gibbs free energy

$$\therefore \quad \ln\frac{[C][B]}{[BC]} = \frac{\Delta G - \Delta G^\circ}{RT} \tag{52}$$

$$\therefore \quad \frac{[C][B]}{[BC]} = e^{-(\Delta G - \Delta G^\circ)/RT} \tag{53}$$

$$\therefore \quad e^{-\Delta G^\circ/RT}\frac{[C][B]}{[BC]} = e^{\Delta G^\circ/RT}\, e^{(-\Delta G^\circ + \Delta G)/RT} = e^{\Delta G/RT} \tag{54}$$

From Eqs. 54 and 50

$$k_f = w_f \nu_f\, e^{\Delta G_f^{\circ\ddagger}/RT}(1 - e^{\Delta G/RT}) \tag{55}$$

For the case of $AB_{(s)} \to A_{(s)} + B_{(g)}$

$$\Delta G = \Delta G^\circ + RT \ln P_B \tag{56}$$

P_B = pressure of gas B
Now

$$\Delta G^\circ = -RT \ln P^* \tag{57}$$

where P^* = equilibrium pressure of the gas B at temperature T

From Eqs. 56 and 57 we may write

$$\frac{\Delta G}{RT} = \ln\frac{P_B}{P_B^*} \tag{58}$$

For Eqs. 55 and 58 we may write

$$k_f = w_f \nu_f\, e^{\Delta G_f^{\circ\ddagger}/RT}\big[1 - e^{\ln(P_B/P_B^*)}\big] \tag{59}$$

$$= w_f \nu_f\, e^{\Delta G_f^{\circ\ddagger}/RT}\left(1 - \frac{P_B}{P_B^*}\right) \tag{60}$$

Following the same procedure as that adopted in Section III.B to obtain Eq. 29 we may obtain from Eq. 60

$$k_f = w_f \nu_f\, e^{\Delta S_f^{\circ\ddagger}/R}\, e^{-\Delta H_f^{\circ\ddagger}/RT}\left(1 - \frac{P_B}{P_B^*}\right) \tag{61}$$

Now

$$\left(1 - \frac{P_B}{P_B^*}\right) = e^{\ln(1 - P_B/P_B^*)} \tag{62}$$

$$\therefore \quad k_f = w_f \nu_f\, e^{\Delta S_f^{\circ\ddagger}/R}\, e^{-[\Delta H_f^{\circ\ddagger} - RT\ln(1 - P_B/P_B^*)]/RT} \tag{63}$$

Thus, the apparent activation is given by

$$E_{\text{apparent}} = \Delta H_f^{\circ\ddagger} - RT \ln\left(1 - \frac{P_B}{P_B^*}\right) \tag{64}$$

Equation 64 may be used to examine whether any experimental measurements were made too close to the equilibrium gas pressure.

Rearrangement of Eq. 61 gives

$$\frac{k}{1 - P_B/P_B^*} = w_f \nu_f \, e^{\Delta S_f^{\circ\ddagger}/R} \, e^{-\Delta H_f^{\circ\ddagger}/RT} \tag{65}$$

Thus,

$$\ln\left(\frac{k}{1 - P_B/P_B^*}\right) = \ln\left(w_f \nu_f \, e^{\Delta S_f^{\circ\ddagger}/R}\right) - \frac{\Delta H_f^{\circ\ddagger}}{RT} \tag{66}$$

A plot of

$$\ln\left(\frac{k}{1 - P_B/P_B^*}\right) \text{ against } \frac{1}{T}$$

leads directly to a gradient of $\Delta H^{\circ\ddagger}/R$, where $\Delta H_f^{\circ\ddagger} = E$, the activation energy.

IV. CALCULATION OF KINETIC PARAMETERS

In Section III it was implied that the best starting point for modeling the kinetic behavior of decomposing solids is the Arrhenius type equation:

$$\frac{d\alpha}{dt} = f(\alpha) A T^n \, e^{-E/RT} \tag{42}$$

Four unknowns need to be determined: $f(\alpha)$ or $g(\alpha)$, E, A, and n.

It was stated earlier that experimental rate data are rarely sufficiently accurate to enable the determination of the value of n and consequently a value of n has to be assumed. Given that there is no well-established theory from which n may be reliably predicted in many experimental studies reported in the literature, the simplifying assumption is made that $n = 0$. In this section, however, the more general form of the above equation is used so that experimental results can be analyzed if n is assigned some value other than 0.

In principle, $g(\alpha)$ can be found without knowing E or A, and E can be found without knowing $g(\alpha)$ or A. However, both $g(\alpha)$ and E must be known before A can be calculated. The first part of this section deals with methods of measuring the activation energy without making any hypothesis about $g(\alpha)$ or knowing A. The second part describes how $g(\alpha)$ may be determined. The third part indicates how $g(\alpha)$, E, and A may all be determined from a single nonisothermal experiment. The final part discusses ways of estimating the value of A.

A. MEASUREMENT OF ACTIVATION ENERGY

Typical experimental results are illustrated in this section using mass loss measurements as the means of reaction rate measurement.

1. Jump Methods

a. TEMPERATURE JUMP METHOD

Typical results from this method are given in Fig. 11. Just before time t when the temperature jump occurs we have the controlled parameter T_1 with the measured rate of decomposition $(d\alpha/dt)_1$. Just after time t we have the controlled parameter T_2 with the measured rate of decomposition $(d\alpha/dt)_2$. Assuming no significant change in α between these two points, we may write

$$\left(\frac{d\alpha}{dt}\right)_1 = f(\alpha_t)\,AT_1^n\lambda\,e^{-E/RT} \tag{67}$$

$$\left(\frac{d\alpha}{dt}\right)_2 = f(\alpha_t)\,AT_2^n\,e^{-E/RT} \tag{68}$$

$$\therefore\quad \frac{(d\alpha/dt)_1}{(d\alpha/dt)_2} = \left(\frac{T_1}{T_2}\right)^n e^{[-E/R]\left(\frac{1}{T_1} - \frac{1}{T_2}\right)} \tag{69}$$

$$\therefore\quad \ln\left[\frac{(d\alpha/dt)_1}{(d\alpha/dt)_2}\right] = n\ln\left(\frac{T_1}{T_2}\right) - \frac{E}{R}\left(\frac{T_2 - T_1}{T_1 T_2}\right) \tag{70}$$

$$-E = \frac{R\ln[(d\alpha/dt)_1/(d\alpha/dt)_2]/(T_1/T_2)^n}{(T_2 - T_1)/T_1 T_2} \tag{71}$$

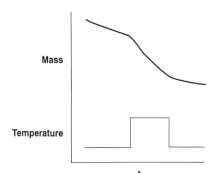

Mass

Temperature

Fig. 11.   Typical temperature jump results.

t

If n is assumed to be 0 this simplifies to

$$-E = \frac{R \ln[(d\alpha/dt)_1/(d\alpha/dt)_2]}{(T_2 - T_1)/T_1 T_2} \tag{72}$$

Thus, E may be easily calculated.

b. RATE JUMP CONSTANT RATE THERMAL ANALYSIS

Typical results for this method are given in Fig. 12. Just before time t we have the controlled parameter $(d\alpha/dt)_1$ with the measured parameter T_1. Just after time t we have the controlled parameter $(d\alpha/dt)_2$ with the measured parameter T_2. Again assuming no significant change in α between these two points, we may use the same derivation as above to obtain Eq. 71 and thus Eq. 72.

The advantage of jump constant rate thermal analysis is that, as discussed in Sections II.B, II.C, and III.C, the pressure above the sample may easily be controlled to ensure that it remains the same both before and after the rate jump. Also, because this type of experiment is conducted on a single sample, the surface conditions on the sample should remain unchanged during the small time interval required to effect the rate jump. Thus, the only changed quantities are the rate and temperature. This method was first proposed by Rouquerol. The temperature jump method shares many of these advantages, but it is much more difficult to keep the pressure of product gas above the sample low and constant (see Section II.C). This method has been developed in greater detail by Flynn and Dickens (21, 22, 23, 33).

For the temperature jump method several temperatures should be used. Using only two temperatures does not confirm the applicability of the Arrhenius equation; it merely assumes it. If different activation energies are obtained for different temperature intervals, then this suggests the relationship between reaction rate and temperature is not given by the Arrhenius equation. Similarly, for the rate jump method several rates and rate jump ratios should be used.

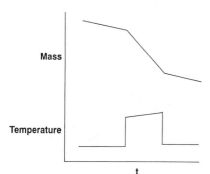

Fig. 12. Typical rate jump constant rate thermal analysis results.

2. Nonjump Methods

a. Isothermal Method

Typical results for a series of isothermal experiments at different temperatures are given in Fig. 13. From Eq. 42 we may write

$$\ln\left(\frac{d\alpha/dt}{T^n}\right) = \ln[f(\alpha)A] - \frac{E}{RT} \tag{73}$$

If we take the same extent of decomposition in each experiment α_i (see Fig. 13), then the first term on the right side of Eq. 73 becomes constant. Thus, a plot of left side of this equation against $1/T$ should give a straight line, from which E can be determined. Assuming n to be zero simplifies the equation to

$$\ln\left(\frac{d\alpha}{dt}\right) = \ln[f(\alpha)A] - \frac{E}{RT} \tag{74}$$

Alternatively, take the integral form of Eq. 3 for isothermal experiments, i.e., Eq. 6,

$$g(\alpha) = kt \tag{6}$$

$$\text{where} \quad k = AT^n e^{-E/RT} \tag{42}$$

$$\therefore \quad g(\alpha) = tAT^n e^{-E/RT} \tag{75}$$

When T is constant,

$$\therefore \quad -\ln(tT^n) = \ln\left[\frac{A}{g(a)}\right] - \frac{E}{RT} \tag{76}$$

Again, take the same extent of decomposition in each experiment α_i (see Fig. 13). The first term in the right side of Eq. 76 becomes a constant; thus a plot of the left side of this equation against $1/T$ should give a straight line from which E can be determined. Assuming n to be zero simplifies the

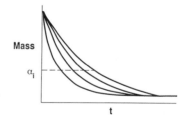

Fig. 13.   Typical results for a series of isothermal experiments.

equation to

$$-\ln(t) = \ln\left[\frac{A}{g(\alpha)}\right] - \frac{E}{RT} \tag{77}$$

The above forms of the equation can be applied to isothermal experiments when the experimental data is in an integral form, as is often the case for mass loss measurements, and accurate differential data is unavailable.

b. CONSTANT RATE THERMAL ANALYSIS

Typical results for a series of constant rate thermal analysis experiments carried out at different decomposition rates are given in Fig. 14. In each experiment we have a series of controlled parameters, $(d\alpha/dt)$ values, and T as the measured parameters. Again Eqs. 73 and 74 apply and a plot of the left sides of these equations against $1/T$ should give a straight line from which E can be determined.

An integral approach is also possible:

$$T = T_0 + bt \tag{78}$$

where T_0 is the ambient temperature and b is the linear heating rate dT/dt. Thus,

$$\frac{d\alpha}{dt} = \frac{d\alpha}{dT}\frac{dT}{dt} = \frac{d\alpha}{dT}b \tag{79}$$

Substitution into Eq. 42 gives

$$\frac{d\alpha}{dT} = \frac{A}{b}f(\alpha)T^n e^{-E/RT} \tag{80}$$

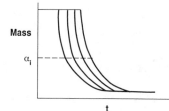

Fig. 14. Typical results for a series of constant rate experiments.

Now consider the integration of this expression from $\alpha = 0$ and $T = T_0$ to some α and T:

$$\int_0^\alpha \frac{d\alpha}{f(\alpha)} = \frac{A}{b} \int_{T_0}^T T^n e^{-E/RT} dT \tag{81}$$

From Section III,

$$\int_0^\alpha \frac{d\alpha}{f(\alpha)} = g(\alpha) \tag{5}$$

Suitable rearrangement gives

$$g(\alpha) = \frac{A}{b} \int_{T_0}^T \frac{e^{-E/RT}}{(R/E)^n (E/RT)^n} d\left(\frac{E}{RT}\right)\left(\frac{-RT^2}{E}\right) \tag{82}$$

$$= \frac{A}{b} \int_{T_0}^T \frac{e^{-E/RT}}{(R/E)^n (E/RT)^n} \frac{(-1) d(E/RT)}{(R/E)(E/RT)^2} \tag{83}$$

$$\therefore \quad g(\alpha) = \frac{A}{b} \int_{E/RT}^{E/RT_0} \frac{e^{-E/RT} d(E/RT)}{(R/E)^{n+1}(E/RT)^{n+2}} \tag{84}$$

$$= \frac{AE}{bR}\left[\frac{E}{R}\right]^n \int_{E/RT}^{E/RT_0} \frac{e^{-E/RT} d(E/RT)}{(E/RT)^{n+2}} \tag{85}$$

Putting $x = E/RT$ and $x_0 = E/RT_0$ we obtain

$$g(\alpha) = \left(\frac{AE}{bR}\right)\left(\frac{E}{R}\right)^n \int_x^{x_0} \frac{e^{-E/RT}}{x^{n+2}} dx \tag{86}$$

Putting

$$P_n(x, x_0) = \int_x^{x_0} \frac{e^{-E/RT}}{x^{n+2}} dx \tag{87}$$

If $x_0 = E/RT_0 \gg 1$, this can be approximated to

$$P_n(x, x_0) \approx P_n(x, \infty) = P_n(x) \tag{88}$$

and therefore we can write

$$g(\alpha) = \left(\frac{AE}{bR}\right)\left(\frac{E}{R}\right)^n P_n(x, x_0) \tag{89}$$

$$\approx \left(\frac{AE}{bR}\right)\left(\frac{E}{R}\right)^n P_n(x) \tag{90}$$

Integration by parts eventually yields

$$P_n(x) = \frac{e^{-x}}{x^{n+2}}[\ln(E, T)]$$

$$\text{where} \quad \ln(E, T) = 1 - \frac{n+2}{E/RT} + \frac{(n+2)(n+3)}{(E^2/RT)} - \cdots$$

Several alternative series may be used to calculate the exponential integral (6, 32, 54, 64, 75, 82) and some given better approximations using fewer terms over certain ranges of values for E/T. However, this advantage is no longer important since with modern programmable calculators and computers as many terms as desired may easily be used. The advantage of the series given above is that nonzero noninteger values for n may be used, which is not the case for the alternative series. Continuing with the derivation we may write

$$\therefore \quad g(\alpha) = \left(\frac{AE}{bR}\right)\left(\frac{E}{R}\right)^n \frac{e^{-x}}{x^{n+2}}I(E, T)$$

$$= \left(\frac{AE}{bR}\right)\left(\frac{E}{R}\right)^n \frac{E^{-E/RT}}{(E/RT)^{n+2}}I(E, T)$$

$$= \frac{AR}{bE}T^{n+2} e^{-E/RT}I(E, T) \tag{91}$$

Rearranging gives

$$\ln\left[\frac{b}{T^{n+2}I(E, T)}\right] = \ln\left[\frac{AR}{Eg(\alpha)}\right] - \frac{E}{RT} \tag{92}$$

Again taking the same extent of reaction α_i in a series of experiments carried out at different heating rates (see Fig. 15), the first term on the right side of the equation becomes a constant and we are left with a series of values for b with corresponding values for T. Thus, a plot of the left side of the equation against $1/T$ should give a straight line from which E can be

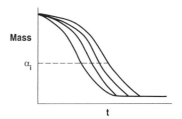

Fig. 15.   Typical results for a series of rising temperature experiments.

determined. The problem arises as to how to evaluate $I(E, T)$, which requires foreknowledge of E. Equation 92 was first derived by Whitehead et al. (84), assuming n to be zero. They proposed that the function $I(E, T)$ performs a correcting role only; that a suitable starting point for a series of iterative plots is $(E, T) = 1$, which from a plot of $\ln(b/T^2)$ against $1/T$ gives the first approximation to the activation energy. Substituting this first value of E into $I(E, T)$ and plotting $\ln b/T^2 I(E, T)$ against $1/T$ gives the second approximation, and so on until the difference between successive values of E is insignificantly small. As implied above, assuming n to be zero simplifies Eq. 92 to

$$\ln\left[\frac{b}{T^2 I(E, T)}\right] = \ln\left[\frac{AR}{E g(\alpha)}\right] - \frac{E}{RT} \qquad (93)$$

and $I(E, T)$ becomes

$$I(E, T) = 1 - \frac{2!}{E/RT} + \frac{3!}{E/RT^2} - \cdots \qquad (94)$$

The computer program that carried out this iterative procedure has two iterative loops. One loop resubstitutes values of E into $I(E, T)$ until a sufficiently accurate value of E is obtained, and a second loop truncates the series used to calculate $I(E, T)$ when successive terms change the value of the function by an insignificantly small amount. Rapid convergence toward E is the rule, and usually no more than four loops are required. This method is an improvement on a method proposed by Ozawa (61) who used an approximation to the $p(x)$ function

$$p(x) = P_n(x) \qquad \text{where} \quad n = 0$$

and that made by Doyle (25), viz.,

$$\ln p(x) = -5.330 - 1.0516 E/RT \qquad (95)$$

$$\text{for values of } \frac{E}{RT} > 20$$

This approximation is based on the observed approximate linearity of plots $p(x)$ against $1/T$ (25, 26, 27).

From Eq. 90, assuming $n = 0$, we may write

$$g(\alpha) = \frac{AE}{bR}p(x) \tag{96}$$

Rearrangement gives

$$b = \frac{AE}{g(\alpha)R}p(x) \tag{97}$$

and substituting yields

$$\ln b = \ln \frac{AE}{g(\alpha)R} - 5.330 - 1.0516 \frac{R}{RT} \tag{98}$$

Taking the same extent of decomposition α_i in a series of experiments carried out at different heating rates (see Fig. 15) the first and second terms on the right side of the above equation are constants and we are left with a series of values for b with corresponding values for T. Thus, a plot of the left side of the above equation against $1/T$ should give a straight line from which E may be determined. This method has the advantage of simplicity, but the Doyle approximation may not always be accurate and the iterative method using Eq. 93 may be more easily adapted if a nonzero value for n is desired.

The classical method of finding E from a series of isothermal experiments is described in the opening part of Section IV.B.

B. DETERMINATION OF THE FUNCTION OF α

Consider Eq. 6:

$$g(\alpha) = kt \tag{6}$$

for isothermal experiments. As discussed in Section III.B the influence of temperature embodied in k becomes constant and it becomes possible to study the form of $g(\alpha)$ without knowing E or A.

One method is to test the linearity of plots of $g(\alpha)$ against time. The function that gives a good straight line fit may be assumed to be the correct function. The slope of the line is then k and the reaction rate is constant. A series of experiments carried out at different temperatures should all fit the same $g(\alpha)$, but yield different values for k. Thus, a series of values of k with corresponding values for T are obtained. From Eq. 42, we obtain

$$k = AT^n e^{-E/RT} \tag{42}$$

$$\therefore \quad \ln\left(\frac{k}{T^n}\right) = \ln A - \frac{E}{RT} \tag{99}$$

A plot of the left side of this equation against $1/T$ should give a straight line from which E and A may be determined. Assuming n to be zero simplifies the equation to

$$\ln k = \ln A - \frac{E}{RT} \tag{100}$$

This is the classical method evaluating A, E, and $g(\alpha)$.

A reduced time method may also be used. Taking the time elapsed up to $\alpha = 0.9$ to be $t_{0.9}$, we may write

$$\frac{g(\alpha)}{g(0.9)} = \frac{t}{t_{0.9}} \tag{101}$$

A series of master plots of α against $g(\alpha)/g(0.9)$ may be drawn for the different $g(\alpha)$'s. Example of these plots for the equations in Table 1 are given in Fig. 16. Experimental reduced time plots of α against $t_i/t_{0.9}$ may then be superimposed upon these master plots, and the equation that gives the closest fit can be selected. This method was proposed by Sharp et al. (78); it has the advantage that data from experiments carried out at several different temperatures may be collected onto a single plot. Thus, any change in $g(\alpha)$ with temperature can be discerned. A disadvantage inherent in this method, as discussed by Sharp et al. (78), Geiss (41), and others is that it involves the comparison of curves. Jones et al. (55) have proposed a plot of

$$\frac{g(\alpha)}{g(0.9)} \quad \text{against} \quad \frac{t}{t_{0.9}}$$

When experimental data are in good agreement with the theoretical expression they should produce a straight line with a slope of unity that passes through the origin. The most significant problem in the use of any of these methods is the accurate evaluation of t when allowance must be made for an induction period. Similar methods based on the differential form of the

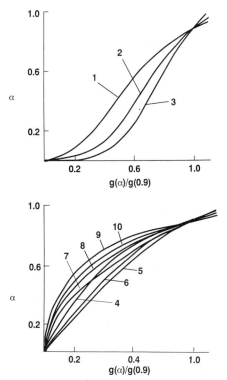

Fig. 16. Reduced time plots. Numbers are from Table 1 and refer to specific kinetic equations.

isothermal rate equation,

$$\frac{d\alpha}{dt} = f(\alpha)k \tag{1}$$

are possible by using the quantity reduced rate instead of reduced time. Taking the rate of decomposition at $\alpha = 0.9$ to be $(d\alpha/dt)_{0.9}$, we may write

$$\frac{d\alpha/dt}{(d\alpha/dt)_{0.09}} = \frac{f(\alpha)}{f(0.9)} \tag{102}$$

This approach has the advantage that it overcomes possible errors due to the presence of an induction period. However, very high quality differential data must be available before it can be employed.

An alternative and widely used method of preliminary identification of the rate law, which provides the most satisfactory fit to a set of data, is through a plot of the form (56),

$$\ln[-\ln(1 - \alpha)] = n \ln t + \text{constant} \tag{103}$$

Magnitudes of n have been empirically established for those kinetic expressions that have found most extensive applications. For example, values of n for diffusion-limited equations are usually between 0.53 and 0.58 for the contracting area and volume relations are 1.08 and 1.04, respectively; for the Avrami-Erofe'ev equation, values are 2.00, 3.00, etc. Once the range of possible mechanisms has been narrowed down, the final selection may be made using the linear plot method described above. However, as a means of preliminary identification this method has few advantages over the reduced time plot.

This is especially true because a reduced time plot of all the experimental data for different temperatures is an advisable first step to verify that the results are all isokinetic, i.e., follow the same $g(\alpha)$. The procedure recommended is as follows

1. A reduced time plot of all the data for experiments carried out at different temperatures should be drawn on the same graph to investigate whether the experimental results were isokinetic.
2. A comparison of reduced time plots with master curves should be made to select the most likely $g(\alpha)$'s.
3. This is followed by calculation of mean values for reduced time coordinates using the formula

$$\frac{\sum_0^n t_i/t_{0.9}}{n} = \frac{t_i}{t_{0.9}}$$

(where n = number of experiments) for several values of α_i.
4. Finally, plot

$$\frac{g(\alpha_i)}{g(0.9)} \quad \text{against} \quad \frac{t_i}{t_{0.9}}$$

using the already selected most probable forms of $g(\alpha)$. As stated above these plots will give a straight line with a slope of unity that passes through the origin. The equation that gives the best straight line fit is chosen as the most probable $g(\alpha)$.

Simple inspection of the experimental curve (see Fig. 16) is sufficient to differentiate between sigmoid rate equation and deceleratory rate equations. Thus, deciding whether nucleation plays a rate-limiting role in the decomposition, as represented by a sigmoid shaped curve, or does not may easily be established without numerical analysis of the results.

Constant rate thermal analysis results may be treated as follows:

$$\frac{d\alpha}{dt} = f(\alpha) A E^{-E/RT} = C \tag{104}$$

assuming $n = 0$, when $C =$ some constant. Then

$$\ln C = \ln f(\alpha) + \ln A - E/RT \tag{105}$$

$$\therefore \quad \frac{1}{T} = \frac{R}{E}\left(\ln f(\alpha) + \ln \frac{A}{C}\right) \tag{106}$$

Taking $T_{0.9}$ to be the temperature at $\alpha = 0.9$ and $T_{0.3}$ to be the temperature at $\alpha = 0.3$ we may write

$$\frac{1/T - 1/T_{0.9}}{1/T_{0.3} - 1/T_{0.9}} = \frac{\ln f(\alpha) - \ln f(0.9)}{\ln f(0.3) - \ln f(0.9)} \tag{107}$$

$$\therefore \quad \frac{a - T_i}{aT_i/b} = \frac{\ln f(\alpha_i) - q}{d} \tag{108}$$

where $a = T_{0.9}$

$b = \dfrac{T_{0.9} - T_{0.3}}{T_{0.9}T_{0.3}}$

$q = \ln f(0.9)$

$d = \ln f(0.3) - \ln f(0.9)$

and a, b, q, and d are constants.

Thus, in a similar manner to reduced time plots, a series of "reduced temperature" master plots of $\ln f(\alpha) - q/d$ against α can be drawn. Examples of these plots using the $f(\alpha)$ functions in Table 1 are given in Fig. 17. Plots drawn from experimental data of

$$\frac{a - T_i}{aT_i/b} \quad \text{against} \quad \alpha_i$$

can be superimposed upon master plots and the most suitable equation can be selected. It should be noted that it is not possible to distinguish between two functions $f(\alpha)$ and $f'(\alpha)$ using this method if

$$\ln f(\alpha) = a \ln f'(\alpha)$$

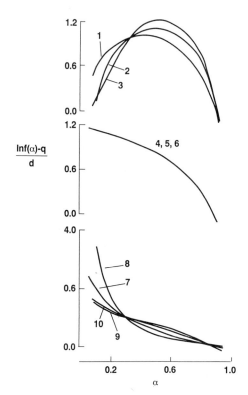

Fig. 17. Reduced temperature plots. Numbers are from Table 1 and refer to specific kinetic equations.

Thus, forms of $f(\alpha)$ of the type $(1 - \alpha)^a$ all give the same reduced temperature plot and the value of a cannot be determined. Having selected the most suitable values for $f(\alpha)$ by comparison of the experimental reduced temperature plot with the master plots, a plot of

$$\frac{a - T}{aT/b} \quad \text{against} \quad \frac{\ln f(\alpha) - q}{d}$$

should give a straight line that passes through the origin with a slope of unity. The recommended procedure is as follows:

1. The experimental reduced temperature plot should be compared with the master plot and the most probable $f(\alpha)$ selected.

2. A plot of

$$\frac{a - T}{aT/b} \quad \text{against} \quad \frac{\ln f(\alpha) - q}{d}$$

should be constructed for the most probable $f(\alpha)$'s, and the equation that gives the best straight line fit selected as the correct $f(\alpha)$.

If $f(\alpha) = (1 - \alpha)^a$ is the selected function, the procedure outlined in the next section for constant rate thermal analysis experiments that conform to this type of behavior could be carried out to evaluate a.

The master plots (Fig. 17) reveal that the functions of α fall into three easily distinguishable groups: Avrami Erofe'ev expressions, geometric expressions, and diffusion controlled expressions. Thus, inspection of the reduced temperature plot is sufficient to determine whether nucleation or diffusion plays a rate-limiting role in the reaction mechanism or whether the reaction is simply governed by geometric considerations.

C. ANALYSIS OF RESULTS FROM SINGLE NONISOTHERMAL EXPERIMENTS

Two different types of methods are used for the analysis of results obtained from a single nonisothermal experiment: differential and integral.

1. Differential Methods

Methods of this type are based on the differential form of the rate equation:

$$\frac{d\alpha}{dt} = f(\alpha) A T^n e^{-E/RT} \tag{109}$$

Rearrangement gives

$$\frac{d\alpha/dt}{T^n f(\alpha)} = A e^{-E/RT} \tag{110}$$

$$\ln\left[\frac{d\alpha/dt}{T^n f(\alpha)}\right] = \ln A - \frac{E}{RT} \tag{111}$$

A series of values of $d\alpha/dt$ with corresponding values for T may be collected from a single nonisothermal experiment. Thus, a plot of the left side of this equation against $1/T$ should give a straight line from which E and A can be determined from the slope and intercept, respectively. However, the problem arises that $f(\alpha)$ must be known. One possible solution is to draw plots based on the above equation for a number of probable functions of α and select the equation that gives the best straight line fit as the true $f(\alpha)$. E, A and $f(\alpha)$ may thus all be determined from a single nonisothermal experiment. Assuming n to be zero simplifies Eq. 111 to

$$\ln\left[\frac{d\alpha/dt}{f(\alpha)}\right] = \ln A - \frac{E}{RT} \tag{112}$$

This method can be used only when accurate differential data can be obtained.

Assuming n to be zero and applying Eq. 104 to results obtained using constant rate thermal analysis yields

$$-\ln f(\alpha) = \ln\frac{A}{C} - \frac{E}{RT} \tag{113}$$

where C = a constant

Consider the two functions $f(\alpha)$ and $f'(\alpha)$ where

$$\ln f(\alpha) = a \ln f'(\alpha) \tag{114}$$

Using Eqs. 113 and 114 we may write

$$-\ln f'(\alpha) = \frac{\ln[A/C]}{a} - \frac{E}{aRT} \tag{115}$$

Thus, if $f(\alpha)$ is the correct function of α, then both $f(\alpha)$ and $f'(\alpha)$ when plotted against $1/T$ will give equally good straight lines. It is not possible, therefore, to distinguish between functions that in their differential form are of the type

$$f(\alpha) = (1 - \alpha)^a \tag{116}$$

A plot $\ln(1 - \alpha)$ against $1/T$ will yield the quantity E/a. It is then necessary to establish E using one of the methods described in Section IV.B. Once E is known, a may be calculated; then the true $f(\alpha)$ is given by Eq. 116.

Another possible approach for constant rate thermal analysis results assuming $f(\alpha) = (1 - \alpha)^a$ may be derived as follows:

$$\frac{d\alpha}{dt} = Ae^{-E/RT}(1 - \alpha)^a = C \tag{104}$$

where C = a constant

$$(1 - \alpha)^a = \frac{C}{A} e^{-E/RT} \tag{117}$$

$$(1 - \alpha) = \left(\frac{C}{A}\right)^{1/a} e^{-E/RTa} \tag{118}$$

$$\alpha = -\left(\frac{C}{A}\right)^{1/a} e^{-E/RTa} + 1 \tag{119}$$

$$\frac{d\alpha}{dT} = \frac{E}{T^2 aR} \left(\frac{C}{A}\right)^{1/a} e^{-E/aRT} \tag{120}$$

$$\ln\left(\frac{d\alpha}{dT} T^2\right) = \ln\left[\frac{E}{aR}\left(\frac{C}{A}\right)^{1/a}\right] + \frac{E}{aRT} \tag{121}$$

The interest of this result lies in the fact that it is not necessary to know the absolute value of α. The quantity $d\alpha/dT$ can be inaccurate by some constant and a plot of the left side of the above equation against $1/T$ will still give a gradient of E/a. This approach may be useful for studying decompositions that occur in a number of poorly defined steps where the $f(\alpha)$ may be approximated to using $(1 - \alpha)^a$, but the exact value for α at each step may be difficult to assess. E must be determined using one of the methods in Section IV.A; then a can be found using the above equation.

2. Integral Methods

Integral methods are based on the integral form of the rate equation derived in Section IV.A:

$$g(\alpha) = \left(\frac{AE}{bR}\right)\left(\frac{E}{R}\right)^n P_n(x) \tag{90}$$

It should be noted that this equation applies only to linear rising temperature experiments. In Section IV.A it was shown that from the above equation we may derive

$$g(\alpha) = \left(\frac{AR}{bE}\right) T^{n+2} e^{-E/RT} I(E, T) \tag{91}$$

and therefore

$$\ln\left(\frac{g(\alpha)}{T^{n+2}I(E,T)}\right) = \ln\left(\frac{AR}{bE}\right) - \frac{E}{RT} \tag{122}$$

For a linear heating rate the first term on the right side of this equation becomes a constant. Thus, a plot of the left side against $1/T$ should give a straight line from which E and A can be determined. Again, a number of probable $g(\alpha)$'s can be tried and the same iterative procedure as described in Section IV.A can be used to evaluate $I(E,T)$ and therefore E. The $g(\alpha)$ that gives the best straight line fit may be taken to be true $g(\alpha)$. If n is assumed to be zero, Eq. 122 simplifies to

$$\ln\left[\frac{g(\alpha)}{T^2I(E,T)}\right] = \ln\left(\frac{AR}{bE}\right) - \frac{E}{RT} \tag{123}$$

An alternative approach using the Doyle approximation (Eq. 95) and Eq. 90, assuming n to be zero, gives

$$\ln g(\alpha) = \ln\left(\frac{AE}{bR}\right) - 5.330 - 1.0516\frac{E}{RT} \tag{124}$$

Thus, a plot of $\ln g(\alpha)$ against $1/T$ should give a straight line from which E and A may be found from the slope and intercept, respectively. This method suffers from the disadvantage that the Doyle approximation is inaccurate for certain values of E over certain temperature ranges, which affects the values of E and A and the linearity of the plot.

A popular method proposed by Coats and Redfern (17, 18) consists of limiting the asymptotic series given for $I(E,T)$ to only two terms. Again assuming n to be zero and using Eqs. 91 and 94, we may write

$$\therefore \qquad g(\alpha) = \frac{ART^2}{bE}e^{-E/RT}\left(1 - \frac{2}{E/RT}\right) \tag{125}$$

$$\therefore \quad \ln\left[\frac{g(\alpha)}{T^2}\right] = \ln\left[\frac{AR}{bE}\left(1 - \frac{2RT}{E}\right)\right] - \frac{E}{RT} \tag{126}$$

Assuming the first term on the right side of the equation to be constant over a short temperature interval, a plot of the left side against $1/T$ should yield a straight line from which E and A can be determined. This method suffers from the disadvantage that using only two terms from the $I(E,T)$ series (see Section IV.A) does not always provide a sufficiently accurate value for this function and the values of E and A and the linearity of the plot may be affected.

Zsako (90) takes logarithms of Eq. (90), assuming n to be zero:

$$\ln g(\alpha) = \ln\left(\frac{AE}{bR}\right) - \ln p(x)$$

As stated in Section IV.A, plots of $\ln p(x)$ against $1/T$ are approximately linear. Thus, as $\ln p(x)$ and $\ln g(\alpha)$ differ by a constant, a plot of $\ln g(\alpha)$ against $1/T$ should also be linear and parallel, the two lines being separated by a distance equal to $\ln(AE/bR)$. Therefore, $\ln g(\alpha)$ is plotted against $1/T$ for a variety of probable $g(\alpha)$'s and the function giving the best straight line fit is assumed to be the correct $g(\alpha)$. Then a plot of $\ln p(x)$ against $1/T$, where $\ln p(x)$ is calculated using the correct value of E, should be parallel to the plot of the selected $g(\alpha)$. Zasuko et al. (87) proposed a trial and error method to find E. Satava and Skvara (71) used an iterative procedure carried out by a computer program. Once E is known, A may be calculated from the distance between the two parallel lines. This method has the advantage that any series taken to as many terms as desired may be used to calculate the value of $p(x)$ and it may easily be adapted to nonzero values for n. However, the method suffers from the disadvantage that the assumption that a plot of $\ln p(x)$ against $1/T$ is linear may not always be true. This fact is pointed out by Zsako (90), who recommended that high heating rates should be used to shift the temperature range over which a decomposition occurs upward to higher values, because plots of $\ln p(x)$ against $1/T$ tend to be more linear at higher temperatures. Unfortunately, high heating rates yield poor experimental results (see Section III.C).

An important point to note about methods that use poor approximations to the $p(x)$ function or incorrectly assume plots of $p(x)$ against $1/T$ to be linear is that they introduce errors that affect the linearity of the plots from which the correct $g(\alpha)$ is selected when the criterion used to identify the correct $g(\alpha)$ is the linearity of its plot. Thus, even given perfect experimental data, there is the possibility that the correct $g(\alpha)$ may not give the best straight line fit. Even using the iterative method based on Eq. 123 or a differential method to overcome these difficulties, studies have shown (2, 13) that selection of $g(\alpha)$ or $f(\alpha)$ based on a statistical measurement of the degree of linearity of some kinetic plot is an uncertain procedure because of the inevitable effect of experimental errors. In response to this Clarke and Thomas (16) have suggested the use of a single isothermal experiment to provide additional information with which $g(\alpha)$ or $f(\alpha)$ may be more confidently identified. Nevertheless, it should be remembered that the proposed functions of α (see Table 1) are idealized models that will never be exactly conformed to in reality and may be influence by the experimental conditions. The true function of α may not be among those tested. In addition, an incorrect choice for the value of n will influence the apparent function of α.

An alternative procedure is to base the treatment upon the integral method, as follows:

1. Initially plots should be made of

$$\ln\left[\frac{g(\alpha)}{T^2}\right] \quad \text{against} \quad \frac{1}{T}$$

    (effectively a Coats and Redfern plot) for the initial selection of the more probable $g(\alpha)$ and the rejection of obviously nonlinear plots.
2. The iterative procedure should then be applied to the selected near linear plots from step 1. This allows selection of $g(\alpha)$ and therefore E and A, giving due weight to both the quality of the straight line fit given by the data and the results of isothermal experiments carried out under similar experimental conditions.

In addition, Eq. 114 may be applied to the results of some constant rate thermal analysis experiments.

D. THE PRE-EXPONENTIAL FACTOR

As discussed in Section III.B, the value of the pre-exponential factor is almost certainly a function of temperature, but to a first approximation it may be considered to be a constant. Also, as discussed in Section III.A, the physical significance of the value of this parameter is not well understood. However, it is useful to ascribe some formal value to this quantity. Methods based on Eqs. 111, 112, 122, and 123 automatically determine E, $f(\alpha)$ (or $g(\alpha)$), and A when the correct $f(\alpha)$ (or $g(\alpha)$) is selected on the basis of the criteria outlined in Section IV.C. Methods using Eqs. 71 and 73 can only be used to determine the value of E. The use of Eqs. 76 and 77 give as the intercept the quantity $\ln[A/g(\alpha)]$. When $g(\alpha)$ has been determined using the methods described in Section IV.B, a plot of $g(\alpha)$ against $g(\alpha)/A$ should give a straight line with a slope of $1/A$; thus, A may be determined. Methods based on Eqs. 73 and 74 give as the intercept the quantity $\ln[f(\alpha)A]$. Once $f(\alpha)$ has been determined using one of the methods described in Section IV.B, a plot of $f(\alpha)$ against $f(\alpha)A$ should give a straight line with a slope of A. Similarly, methods based on Eqs. 92 and 93 give as their intercept the quantity $\ln[AR/Eg(\alpha)]$, at the same time giving a value for E from the gradient of the plot. The intercept gives the value of $A/g(\alpha)$. Once $g(\alpha)$ is determined using one of the methods described in Section IV.B, a plot of $g(\alpha)/A$ against $g(\alpha)$ should give a straight line with a slope $1/A$; thus, A may be determined.

TABLE 7
How $f(\alpha)$ and $g(\alpha)$ Change with α

α	0.1	0.5	0.9
Eq. 1 $f(\alpha)$	0.2921	0.4163	0.1517
$g(\alpha)$	0.6492	1.6651	3.0349
Eq. 2 $f(\alpha)$	2.2789	3.6498	0.1744
$g(\alpha)$	1.4169	2.6550	3.9615
Eq. 3 $f(\alpha)$	0.1664	0.3798	0.1869
$g(\alpha)$	2.2789	3.6498	4.9274
Eq. 4 $f(\alpha)$	0.9000	0.5000	0.1000
$g(\alpha)$	0.1054	0.6931	2.3026
Eq. 5 $f(\alpha)$	0.9487	0.7071	0.3162
$g(\alpha)$	0.1026	0.5858	1.3675
Eq. 6 $f(\alpha)$	0.9322	0.6300	0.2154
$g(\alpha)$	0.1035	0.6189	1.6075
Eq. 7 $f(\alpha)$	10.0000	2.0000	1.1111
$g(\alpha)$	0.0050	0.1250	0.4050
Eq. 8 $f(\alpha)$	9.4912	1.4427	0.4343
$g(\alpha)$	0.0052	0.1534	0.6697
Eq. 9 $f(\alpha)$	27.0111	3.0537	0.4021
$g(\alpha)$	0.0018	0.0638	0.4310
Eq. 10 $f(\alpha)$	27.9766	3.8473	0.8662
$g(\alpha)$	0.0017	0.0551	0.2768

Note. Equation numbers from Table 1.

It is clear from what has been said that it is not possible to find the value of A without knowing $f(\alpha)$ or $g(\alpha)$. Examination of experimental data shows that in many cases it is not always possible to unambiguously determine $f(\alpha)$. It may alternatively seem preferable to determine the limits within which A lies. Inspection of Table 7 shows how the values for $f(\alpha)$ and $g(\alpha)$ change with α. At $\alpha = 0.9$ the disparity between the various possible values of the function of α are at a minimum for both $f(\alpha)$ and $g(\alpha)$, lying between the values 1.1111 and 0.1000 for $f(\alpha)$ and 4.9274 and 0.2768 for $g(\alpha)$. Thus, if the value for the intercept for differential methods based on Eqs. 73 and 74 at $\alpha = 0.9$ is found, we may then assume $f(\alpha)$ to be equal to some value between 1.1111 and 0.1000. Taking some intermediate value, 0.5 (which is smaller than half the maximum and greater than twice the minimum), we may write

$$\text{Antilog of intercept at } \alpha = 0.9 = f(0.9)A = 0.5A \qquad (127)$$

to within better than two orders of magnitude. Thus,

$$2 \times (\text{Antilog of intercept at } \alpha = 0.9) = A \times 10^{\pm 1} \qquad (128)$$

For integral methods based on Eqs. 92 and 93 the value of $g(\alpha)$ at $\alpha = 0.9$ may be assumed to be equal to some value between 4.9274 and 0.2768. Taking some intermediate value, 1 (which is smaller than five times the minimum and greater than one-fifth of maximum), we may write

$$\text{Antilog of intercept at } \alpha = 0.9 = \frac{AR}{E\,g(0.9)} = \frac{AR}{E \times 1} \qquad (129)$$

within better than two orders of magnitude; thus,

$$\frac{E}{R} \times (\text{Antilog of intercept at } \alpha = 0.9) = A \times 10^{\pm 1} \qquad (130)$$

where R is known and E may be found from the slope of the plots arising from Eqs. 92 and 93 at $\alpha = 0.9$. This method is simpler when applied to methods based on Eqs. 76 and 77, where the intercept gives the value $A/g(\alpha)$. Assuming that $g(\alpha) = 1$ at $\alpha = 0.9$ gives

$$\text{Antilog of intercept at } \alpha = 0.9 = A \times 10^{\pm 1}$$

While the accuracy with which the values of A are determined is theoretically better than plus or minus an order of magnitude, taking into account experimental errors and the fact we have considered only a selection of the possible $f(\alpha)$'s and $g(\alpha)$'s, it is inadvisable to claim any better accuracy. Indeed, the true value may lie outside these limits. However, the range of possible values determined in this way can be useful in understanding the possible mechanism of the reaction (see Section III.A).

It should be remembered that when n is assumed to have a value other than zero and this additional function of temperature is included in the calculation of the values plotted on the g axis, then the value obtained for the pre-exponential factor is correspondingly altered.

REFERENCES

1. Achar, B. N. N., G. W. Brindley, and J. H. Sharp, *Proc. Int. Clay Conf.*, Jerusalem, **1**, 67 (1967).

2. Arnold, M., G. E. Veress, J. Paulik, and F. Paulik, *J. Therm. Anal.* **17**, 507 (1979).

3. Avrami, M., *J. Chem. Phys.* **7**, 1103 (1939).

4. Avrami, M., *J. Chem. Phys.* **8**, 212 (1940).

5. Avrami, M., *J. Chem. Phys.* **9**, 177 (1941).

6. Balarin, M., *J. Therm. Anal.* **12**, 169 (1977).

7. Barret, P., *Proc. 4th Int. Symp. React. Solids*, Amsterdam, 178 (1960).

8. Barret, P., *C. R. Acad. Sci., Ser. C* **266**, 856 (1968).

9. Beruto, D., and A. W. Searcy, *J. Phys. Chem.* **70**, 2145 (1974).

10. Bevan, S. C., S. J. Gregg, and N. D. Parkyns, *Prog. Vac. Microbalance Technol.*, 2 (1973).

11. Born, M., and K. Huang, *Dynamical Theory of Crystal Lattices*, Oxford, UK: Clarendon, 1954, pp. 80, 85.

12. Britton, H. T. S., S. J. Gregg, and G. W. Winsor, *Trans. Faraday Soc.* **48**, 63 (1952).

13. Brown, M. E., and A. K. Galway, *Thermochin. Acta* **29** 129 (1979).

14. Brown, M. E., D. Dollimore, and A. K. Galway, *Comprehensive Chemical Kinetics*, C. H. Bamford and C. F. Tipper (Eds.), Amsterdam: Elsevier, 1980.

15. Brunauer, S., P. H. Emmett, and E. Teller, *J. Am. Chem. Soc.* **48**, 690 (1926).

16. Clarke, T. A., and J. M. Thomas, *Nature* **219**, 1149 (1968).

17. Coats, A. W., and J. P. Redfern, *Analyst* **88**, 906 (1963).

18. Coats, A. W., and J. P. Redfern, *Nature* **201**, 68 (1964).

19. Cordes, H. F., *J. Phys. Chem.* **72**, 2185 (1968).

20. Delmon, B., *Introduction a La Cinetique Heterogene*, Paris: Technip, 1969.

21. Dickens, B., *Thermochim. Acta* **29**, 41 (1979).

22. Dickens, B., *Thermochim. Acta* **29**, 57 (1979).

23. Dickens, B., *Thermochim. Acta* **29**, 87 (1979).

24. Dollimore, D., P. Spooner, and A. Turner, *Surf. Technol.* **4**, 121 (1976).

25. Doyle, C. D., *J. Appl. Polym. Sci.* **6**, 639 (1962).

26. Doyle, C. D., *Nature* **207**, 290 (1965).

27. Doyle, C. D., *Techniques and Methods of Polymer Evaluation*, Vol. 1, P. E. Slade and L. T. Jenkins (Eds.), London: Arnold, 1966, pp. 113.

28. Erofe'ev, B. V., *C. R. Dokl. Akad. Sci. USSR* **52** 511 (1946).

29. Evans, U. R., *An Introduction to Metallic Corrosion*, London: Arnold, 1948.

30. Fassel, V. A., E. M. Layton, and R. D. Kross, *J. Chem. Phys.* **25**, 135 (1956).

31. Fevre, A., M. Murat, and C. Comel, *J. Therm. Anal.* **12**, 429 (1977).

32. Figgis, B. N., and J. Lewis, *Techniques of Inorganic Chemistry*, Vol. 4, H. B. Jonassen and A. Weissberger (Eds.), New York: Interscience, 1965, p. 137.

33. Flynn, J. H., and B. Dickens, *Thermochim. Acta* **15**, 1 (1976).

34. Gallagher, P. K., and D. W. Johnson, *Thermochim. Acta* **6**, 67 (1973).

35. Galway, A. K., *Chemistry of Solids*, London: Chapman & Hall, 1967.

36. Garn, P. D., *Crit. Rev. Anal. Chem.* **3**, 65 (1972).

37. Garn, P. D., *J. Therm. Anal.* **13**, 581 (1978).

38. Garner, W. E., *Proc. R. Soc. Ser. A* **189**, 508 (1947).

39. Garner, W. E., and W. R. Southern, *J. Chem. Soc.* 1705 (1935).

40. Garner, W. E., *Chemistry of the Solid State*, New York: Academic, 1955, pp. 417.

41. Geiss, E. A., *J. Am. Ceram. Soc.* **46**, 374 (1963).

42. Ginstling, A. M., and B. I. Brounshtein, *Zh. Prikl. Khim.* **23**, 1327 (1950).

43. Gyula, G., and J. Greenhow, *Thermochim. Acta* **5**, 481 (1973).

44. Haul, R. A. W., and L. H. Stein, *Trans. Faraday Soc.* **51**, 1280 (1955).

45. Herzberg, G., *Molecular Spectra and Molecular Structure*, New York: Van Nostrand, 1945; *Z. Phys.* **109**, 586 (1938); **110**, 760 (1938).

46. Hill, R. A. W., *Trans. Faraday Soc.* **54**, 685 (1958).

47. Hill, R. A., *Nature* **227**, 703 (1970).

48. Holt, J. B., J. B. Cutler, and M. E. Wadsworth, *J. Am. Ceram. Soc.* **45**, 133 (1962).

49. Hulbert, S. F., *J. Br. Ceram. Soc.* **6**, 11 (1969).

50. Hume, J., and J. Colvin, *Proc. R. London Soc. Ser. A* **125**, 635 (1929).

51. Hyatt, E. P., I. B. Cutler, and M. E. Wadsworth, *J. Am. Ceram. Soc.* **41**, 70 (1958).

52. Jacobs, D. W. M., and F. C. Tomkins, *Chemistry of the Solid State*, W. E. Garner (Ed.), London: Butterworth, 1955, Chap. 7; Jacobs, D. W. M., *Mater. Sci. Res.* **4**, 37 (1969).

53. Jander, W., *Z. Anorg. Allg. Chem.* **163**, 1 (1927); *Angew, Chem.* **41**, 79 (1928).

54. Janzen, E. G., *J. Phys. Chem.* **76**, 157 (1972).

55. Jones, L. F., D. Dollimore, and T. Nicklin, *Thermochim. Acta* **13**, 240 (1975).

56. Keattch, C. J., and D. Dollimore, *An Introduction to Thermogravimetry*, London: Heyden, 1975.

57. MacCallum, J. R., and J. Tanner, *Nature* **225**, 1127 (1970).

58. Mampel, K. L., *Z. Phys. Chem. Abt. A* **187**, 43, 235 (1940).

59. Moore, W. J., *Physical Chemistry*, London: Longman, 1974.

60. Oswald, H. R., *Proc. of the 2nd European Symp. on Thermal Anal.*, Vol. 1, D. Dollimore (Ed.), Aberdeen: Heyden, 1981.

61. Ozawa, T., *Bull. Chem. Soc. Jpn.* **38**, 1881 (1965); *J. Therm. Anal.* **2**, 301 (1970).

62. Paulik, J., and F. Paulik, *Anal. Chim. Acta* **56**, 328 (1971).

63. Pavlyuchenko, M. M., and E. A. Prodan, *5th International Symp. on Reactivity of Solids*, G. M. Schwab (Ed.), Amsterdam: Elsevier, 1965, p. 407.

64. Pavlyuchenko, M. M., M. P. Gilevich, and A. K. Potapovich, *5th Int. Symp. on Reactivity of Solids*, G. M. Schwab (Ed.), Amsterdam: Elsevier, 1965, p. 488.

65. Polanyi, M., and E. Wigner, *Z. Phys. Chem. Abt. A* **139**, 1347 (1928).

66. Rouquerol, J., *Thermal Analysis*, F. W. Schwenker and P. D. Garn (Eds.), New York: Academic, 1969, p. 281; *J. Therm. Anal.* **2**(2), 123 (1970).

67. Rouquerol, J., *J. Therm. Anal.* **5**, 205 (1973).

68. Rouquerol, J., F. Rouquerol, and M. Gauteaume, *J. Catal.* **36**, 99 (1975).

69. Rouquerol, J., and M. Gauteaume, *J. Therm. Anal.* **11**, 201 (1977).

70. Rouquerol, J., F. Rouquerol, and M. Gauteaume, *J. Catal.* **57**, 223 (1979).

71. Satava, V., and F. Skvara, *J. Am. Ceram. Soc.* **52**, 591 (1966).

72. Schroeder, R. A., C. E. Weir, and E. R. Lippencott, *J. Res. Nat. Bur. Stand.* **66A**, 407 (1962).

73. Searcy, A. W., and D. Beruto, *J. Phys. Chem.* **82**, 163 (1978).

74. Sestak, J., *Talanta* **13**, 567 (1966).

75. Sestak, J., *Thermochim. Acta* **3**, 150 (1971).

76. Sestak, J., S. Satava, and W. W. Wendlandt, *Thermochim. Acta* **7**, 333 (1973).

77. Shannon, R. D., *Trans. Faraday Soc.*, **60**, 1902 (1964).

78. Sharp, J. H., G. W. Brindley, and B. N. N. Achar, *J. Am. Ceram. Soc.* **49**, 379 (1966).

79. Simon, J., *J. Therm. Anal.* **5**, 271 (1973).

80. Somarjai, G. A., and J. E. Lester, *Prog. Solid State Chem.* **4**, 1 (1968).

81. Stacey, M. H., *Proc. 2nd European Symp. on Thermal Anal.*, D. Dollimore (Ed.), Aberdeen: Heyden, 1981, p. 408.

82. Turkevich, J., S. Larach, and P. N. Yocom, *5th Int. Symp. on Reactivity of Solids*, G. M. Schwab (Ed.), Amsterdam: Elsevier, 1965, p. 115.

83. *Vacuum Microbalance Techniques*, Vols. 1–8, New York: Plenum, 1961–1969.

84. Whitehead, R., D. Dollimore, D. Price, and N. S. Fatemi, *Proc. of the 2nd European Symp. on Thermal Anal.*, D. Dollimore (Ed.), Aberdeen: Heyden, 1981, p. 51.

85. Wischin, A., *Proc. R. London Soc. Ser. A* **172**, 314 (1939).

86. Young, D. A., *Decomposition of Solids*, New York: Pergamon, 1966, p. 209.

87. Zasuko, J., E. Kekedy, and C. Varhelgi, *J. Therm. Anal.* **1**, 339 (1969).

88. Zawadzki, J., and S. Z. Bretsnajder, *Z. Electrochem.* **41**, 721 (1935).

89. Ziman, J. M., *Principles of the Theory of Solids*, UP: Cambridge, 1964, p. 27.

90. Zsako, J., *J. Phys. Chem.* **72**, 2406 (1968).

Chapter 2

THERMOMETRIC TITRATIONS AND ENTHALPIMETRIC ANALYSIS

By Joseph Jordan, *Professor Emeritus of Chemistry,*
The Pennsylvania State University,
University Park, PA,
and John Stahl, *Associate Professor of Chemistry,*
Geneva College,
Beaver Falls, PA

Contents

Treatise on Analytical Chemistry, Part 1, Volume 13,
Second Edition: Thermal Methods, Edited by James D. Winefordner.
ISBN 0-471-80647-1   © 1993 John Wiley & Sons, Inc.

I. INTRODUCTORY SURVEY

A. METHODOLOGICAL PRINCIPLES

Calorimetry is one of the oldest physical measurement techniques. Thermometric titration and related enthalpimetric methods, which use the heat of a chemical reaction for analysis of dilute solutions (and occasionally of gases), are of a more modern vintage. Enthalpimetric methods have historically developed apart from classical calorimetry, as analytical requirements of speed and simplicity mandated a compromise in rigor. Three distinct methodologies are currently employed under the broader label of enthalpimetric analysis: (1) thermometric titrations, (2) direct injection enthalpimetry (DIE), and (3) flow enthalpimetry.

Thermometric titration (often referred to as thermometric enthalpy titration, TET) in its simplest form, involves adding a titrant (usually added continuously at a constant rate) to a sample solution under quasi-adiabatic conditions while monitoring the temperature of the resulting mixture. A typical titration curve is illustrated in Fig. 1a. The endpoint is indicated by a change in slope. Thermometric titration is a linear titration technique, i.e., the observed variable, temperature, approximates a linear function of the amounts reacted. As such, it is similar to conductometric and amperometric titrations and requires a negligible volume change during the titration to

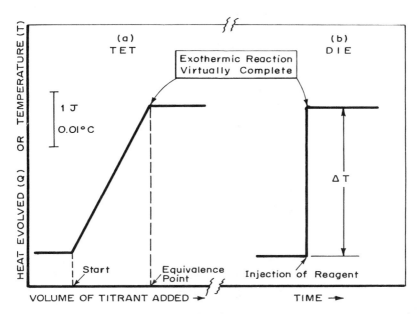

Fig. 1. Thermometric enthalpy titration (TET) and direct injection enthalpimetry (DIE). Idealized readouts.

facilitate endpoint determination. In practice, this is achieved through the use of concentrated titrants and/or by subsequent data reduction.

Conventional thermometric titrations are applicable to virtually any class of reaction provided the enthalpy change is finite ($\Delta H \neq 0$) and the reaction proceeds relatively fast. Novel endpoint determinations (Section III.B.2) permit even use of reactions with $\Delta H = 0$. The shape of the titration curve depends on the heat of reaction and the equilibrium constant. In addition to its utility for endpoint indication, thermometric titration is a powerful tool for evaluating the formal thermodynamic parameters ($\Delta H_r^{\circ\prime}$, $\Delta G_r^{\circ\prime}$, $\Delta S_r^{\circ\prime}$) of a reaction.

Direct injection enthalpimetry (DIE) involves the rapid mixing of a sample with excess reagent solution under effectively adiabatic conditions. The heat effect of the ensuing reaction is monitored versus time, as illustrated in Fig. 1b. DIE is essentially a simplified version of classical batch calorimetry. The total heat produced by the completion of reaction is directly related to sample concentration. DIE has been particularly useful in conjunction with enzyme-catalyzed reactions which combine the advantages of selectivity and reliance on ΔH. For reactions that are slow ($t_{1/2} > 5$ s), the heat versus time curve can yield kinetic information.

Both TET and DIE are essentially batch methods of analysis. For rapid routine analysis, flow systems are becoming increasingly popular. Various enthalpimetric flow methods have been developed, all of which, in essence, combine sample (the analyte) and reagent in a flowing stream and monitor the temperature change. The fundamental principle is, as in DIE, that a direct proportionality exists between the observed heat effect and amount of analyte.

B. HISTORY AND NOMENCLATURE

Thermometric titration was first described in 1913 (12). Its acceptance as a modern instrumental technique was ushered in by two developments in the 1950s: use of thermistors as temperature transducers (80) and motor driven burets for automated, continuous addition of titrant (67). A novel type of catalytic endpoint determination was pioneered in 1965 (129). DIE was developed in the 1960s (132), and flow methods followed in the 1960s and 1970s (25, 54, 101). Enthalpimetric analysis has aroused widespread interest, as is evidenced by several monographs (3, 10, 124, 128) and authoritative reviews (22, 29, 47, 56, 69, 73, 85). Commercial equipment for enthalpimetric analysis is currently offered by manufacturers in several countries, including the United States, the United Kingdom, Japan, Hungary, and Sweden.

More than 25 years ago the consensus at a Thermoanalytical Titrimetry Symposium of the American Chemical Society was that the term "thermometric titration" be used to denote adiabatic titrations yielding plots of temperature versus volume of titrant (61). However, because the technique relies upon enthalpy changes, the term "thermometric enthalpy titration" (TET) has also been used. For the same reason, the generic, but loosely defined, designation "enthalpimetric analysis" has gained widespread acceptance in recent years (47, 69).

C. SPECIAL FEATURES OF THERMOMETRIC METHODS

The measurement of heat effects or temperature changes is not prone to many of the interferences that plague other physical measurements, such as spectrophotometry. For example, turbid solutions, such as whole blood, serum, and slurries, are readily analyzed without prior cleanup.

Because enthalpy change is a universal property of nearly all chemical reactions, the methods of enthalpimetric analysis are broadly applicable. Diverse inorganic, organic, and biochemical reactions have been used, including acid–base, precipitation, complexation, oxidation–reduction, phosphorylation, and immunological processes. The principal drawback is lack of specificity, which is due to the fact that heats of reaction are omnipresent. Consequently, specificity has to be achieved chemically, i.e., by judicious selection of reagents, catalysts, etc. Indeed, enzymatic and immunological reactions, which are highly specific by their very nature, have provided some preferred applications of enthalpimetric analysis.

Thermometric titration, being a linear titration method based on enthalpy change, is generally less likely to fail due to equilibrium considerations than

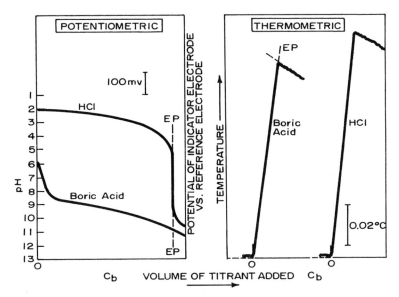

Fig. 2. Comparison of potentiometric and thermometric titration curves: 0.01 M acids titrated with standard sodium hydroxide in aqueous solutions at 25°C. EP, equivalence point.

are potentiometric titration methods. A well-known example is the titration of boric acid with sodium hydroxide in aqueous solution, as is shown in Fig. 2 (66). Because boric acid is a very weak Brønsted acid ($pK_a = 9.2$), the potentiometric (pH) titration curve displays an ill-defined endpoint. However, the thermometric endpoint is well defined, being nearly as good as that of the strong acid HCl. The reason for this can be understood by consideration of the data in Table 1. Neutralization of an acid with a strong base can be understood as the summation of two steps:

$$\text{Dissociation: HA} \rightleftharpoons \text{H}^+ + \text{A}^- \qquad K_a = (\text{H}^+)(\text{A}^-)/(\text{HA}) \qquad (1)$$

$$\text{Neutralization: H}^+ + \text{OH}^- \rightleftharpoons \text{H}_2\text{O} \quad K = [(\text{H}^+)(\text{OH}^-)]^{-1} = K_w^{-1} \quad (2)$$

$$(1) + (2): \text{HA} + \text{OH}^- = \text{A}^- + \text{H}_2\text{O} \quad K = K_a/K_w \qquad (3)$$

where the parentheses denote thermodynamic activities.

TABLE 1

Ionization Constants and Heats of Neutralization in Aqueous Solution at 25°C

Acid	K_a	Heat of Neutralization (kJ/mol)
HCl	∞	-55.9
H_3BO_3	5.8×10^{-10}	-42.2

In the potentiometric titration, the signal is proportional to log (H^+), which is determined by pK_a and pK_w (and thus by the free energy of dissociation of boric acid). Even though the neturalization is nearly complete (98% in 0.01 M boric acid) at the endpoint, the pH does not change dramatically.

In contradistinction, the thermometric titration succeeds because the signal, ΔT, is linearly related to the overall heat of neutralization and the amounts reacted. Boric acid has a relatively large heat of neutralization. Under the prevailing experimental conditions (see Fig. 2), reaction 3 was 98% complete at the equivalence point, yielding a well-defined break at the endpoint. The situation illustrated in Fig. 2 is fairly typical. Thermometric titrations can, in many other instances, be successfully performed when equilibrium constants are so unfavorable that potentiometric titrations fail.

D. THERMODYNAMIC CONSIDERATIONS

The basic concept of all of thermochemistry and enthalpimetry is that enthalpy is a state function. For a general reaction of the type,

$$a\mathrm{A} + b\mathrm{B} = p\mathrm{P} \qquad (4)$$

the total heat effect is correlated by Eq. 5 with the number of moles of product n_p formed.

$$Q = -n_p \, \Delta H_{r4} \qquad (5)$$

where ΔH_{r4} is the enthalpy of reaction 4 expressed in joules per mole of P, and Q is the integral heat evolved (in joules). The enthalpy of reaction ΔH_{r4} is the sum of the heats of formation of the products minus that of the reactants:

$$\Delta H_{r3} = \Delta H_{fP} - (a/p \, \Delta H_{fA} + b/p \, \Delta H_{fB}) \qquad (6)$$

The appropriate ΔH_f assignments depend on the prevailing temperature and other conditions (e.g., solvent and ionic strength).

The amount of product formed $n_p = [\mathrm{P}]v$ (where n denotes the number of moles, the brackets denote concentration in moles per liter, and v is the volume of the solution expressed in liters) is limited by the equilibrium relationship:

$$K_3 = \frac{\gamma_p[\mathrm{P}]^p}{\gamma_\mathrm{A}[\mathrm{A}]^a \gamma_\mathrm{B}[\mathrm{B}]^b} \qquad (7)$$

In the dilute solutions often employed in enthalpimetric analysis, activity coefficients γ are generally near unity. They will be subsequently ignored in this writeup, although situations occur where this approximation is not justified.

For a given combination of A and B, with no P initially present, the total heat effect can be calculated by combining Eqs. 5 and 7. In the case of $a = b = p = 1$, the result is

$$Q = -\Delta H_r v \left\{ \frac{([A]_0 + [B]_0 + 1/K) \pm \sqrt{([A]_0 + [B]_0 + 1/K)^2 - 4[A]_0[B]_0}}{2} \right\} \quad (8)$$

where $[A]_0$ and $[B]_0$ are the respective initial concentrations. Notice that in the limit of $K \to 0$, $Q = 0$. Also, as $K \to \infty$, $Q = -\Delta H_r v [A]_0$ (or if A is in excess, $Q = -\Delta H_r v [B]_0$). An intermediate value of some interest is that for $[A] = [B] = 10^{-3}$ M and $K = 10^9$, $Q = -(0.999)(-\Delta H_r v [A]_0)$ corresponding to 99.9% completion of reaction. Reactions of other stoichiometries yield expressions similar to Eq. 8, but somewhat more complicated.

The preceding discussion assumed that chemical equilibrium was instantaneously attained. Real reactions reach equilibrium at a finite rate, which can limit the applicability of a particular reaction for use in enthalpimetric analysis. A discussion of rate limitations is presented later.

Under adiabatic conditions, the heat produced by a reaction is associated with a temperature change ΔT,

$$\Delta T = \frac{Q}{\kappa} \quad (9)$$

where κ is the heat capacity of the calorimetric system (J/°C). Combining Eqs. 5 and 9 and using the approximation that $\kappa \simeq \bar{\kappa} v$, where $\bar{\kappa}$ is the specific heat of the solvent (per unit volume), yields

$$\Delta T = \frac{n_p}{v} \frac{\Delta H_r}{\bar{\kappa}} = \left(\frac{b}{p}\right) [B]_0 \frac{\Delta H_r}{\bar{\kappa}} \quad (10)$$

where $[B]_0$ is the concentration of B (sample) in moles/liter. The sensitivity of enthalpimetric analysis is thus dependent on both sample concentration $[B]_0$ and heat of reaction ΔH_r. With modern temperature measuring devices (such as thermistors), temperature changes of $10^{-4}°C$ and up can be measured accurately. Enthalpimetric analysis is thus applicable when

$$[B]_0 \Delta H_r > 4 \times 10^{-4} \text{ kJ/L} \qquad (11)$$

For example, acid–base reactions typically have an enthalpy of neutralization of $\simeq 40$ kJ/mol, thereby having a lower limit of detection of 10^{-5} M. Routine enthalpimetric analysis is most conveniently done in the concentration range between 10^{-2} and 10^{-4} M.

II. INSTRUMENTATION FOR THERMOMETRIC TITRATION AND DIRECT INJECTION ENTHALPIMETRY

Thermometric enthalpy titration and direct injection enthalpimetry have primarily been developed as a variation of adiabatic calorimetry for reasons of speed and simplicity. Increased calorimetric precision is available by other methods, such as classical adiabatic calorimetry, heat flow (Tian–Calvet) calorimetry (42), and heat compensation isothermal calorimetry (27, 32). However, these methods are more time-consuming. The interested reader is referred to the pertinent literature for further details.

For both TET and DIE, which are batch methods, similar equipment can be used, which consists of (1) an insulated cell of volume 1–100 mL; (2) a thermostated environment; (3) a temperature sensor and recording system; (4) a reagent delivery system; and (5) an electrical heating calibration system. Figure 3 depicts schematically a typical instrument for batch enthalpimetric analysis.

Insulated (quasi-adiabatic) cells vary from crude polystyrene cups to specially designed thin-walled Dewar flasks. Analytically acceptable titration endpoints can be obtained in quite simple containers. However, for good calorimetric accuracy, the cell must be highly adiabatic and be of a geometry such that the heat leak path is reproducible. An additional requirement is that the calorimeter and its contents attain rapid (< 10 min) thermal equilibrium. Since no real cell can achieve perfect adiabaticity, a finite heat leak will occur whenever the temperature T within the cell is not the same as the temperature of the environment T_E. When all portions of the calorimetric cell have a constant thermal gradient (between T and T_E), the system is said to be in thermal equilibrium. This will be evidenced by a stable and linear temperature baseline. Sunner and Wadsö (122) comparatively studied various cell designs and found the following characteristics to be desirable. The portion of the vessel in direct contact with the sample solution should be of low thermal mass (small heat capacity). The junction of this inner vessel with the surroundings should have a well-designed and reproducible boundary.

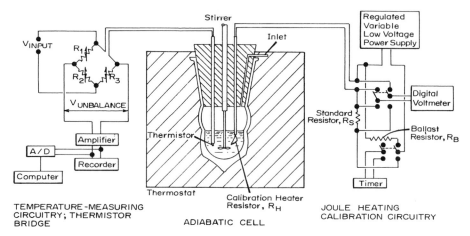

TEMPERATURE-MEASURING
CIRCUITRY; THERMISTOR
BRIDGE

ADIABATIC CELL

JOULE HEATING
CALIBRATION CIRCUITRY

Fig. 3. Diagram of apparatus and circuitry used in enthalpimetric analysis. Redrawn with permission from Henry (60). Copyright 1976, American Chemical Society.

The design of Christensen et al. (30), shown in Fig. 4, meets the desired requirements. They used a Dewar flask with a very thin (0.6 mm) round-bottomed inner wall. Adiabaticity was improved by interrupting the silvering at the top of the evacuated space, as is evident in Fig. 4. The remarkably fast thermal equilibration time for this vessel was 3 s. The basic principles of this design have been widely adopted.

The cell is attached to a well-fitted head, which holds the various inserts (thermistor, heater, titrant delivery tube, etc.). If the calorimetric vessel is to be completely submerged in a liquid thermostat bath, the seal must be leakproof. A good seal is also required for work with air-sensitive compounds or volatile (nonaqueous) solvents. Rubber or cork stoppers are convenient for analytical work. Ground glass or machined Teflon stoppers provide a more reproducible seal. Christensen et al. (32) found that a better defined thermal boundary was achieved by use of an O-ring seal instead of a ground glass stopper. In the cell, provision is made for stirring, either by a magnetic stirring bar or via an overhead motor and shaft. Well-regulated stirring is important for rapid mixing of the reactants and also to maintain thermal homogeneity in the solution. Best results are obtained if the direction of stirring is such that the titrant delivery tube is immediately "downstream" from the thermistor.

The room temperature can serve as a thermostat for rough analytical work. However, to attain a high degree of calorimetric accuracy and reproducibility, the usual practice is to immerse the adiabatic cell in a precisely (± 0.003°C) thermostated water bath. The thermostat serves several functions. It allows the temperature of the reagent and sample to be reproducibly matched. It also eliminates random thermal fluctuations. This configuration

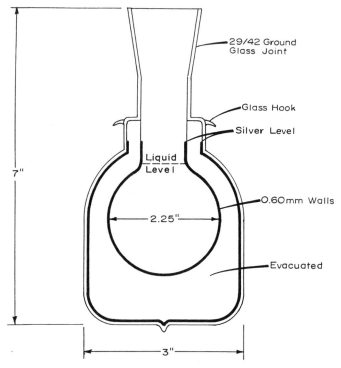

Fig. 4. Thermometric titration calorimeter design of Christensen et al. (30). Copyright 1965, American Institute of Physics.

of a quasi-adiabatic cell in a precisely thermostated environment has been termed an "isoperibol" calorimeter (77). Air thermostat baths have also been used as a compromise between convenience and precision (22, 102).

The temperature sensor is the heart of the instrument. The earliest thermometric titrations used a mercury-in-glass thermometer. Thermistors are unmatched as transducers for measuring small changes in temperature and are now universally used in enthalpimetric analysis. Thermistors are semiconductors, made up of a mixture of sintered oxides (Fe, Mn, Ni, . . .), in which electrons are thermally excited into the conduction band. Their resistance is typically rather large (10^3–10^6 Ω) and they have a large negative temperature coefficient of resistance, about 4% per degree. R_T is accurately described by Eq. 12, first given in Bosson et al. (15):

$$\log R_T = \alpha + \frac{\beta}{T + \theta} \tag{12}$$

For small changes in T, R_T approximates a linear function. For typical values of the constants α, β, and θ in Eq. 12, near 25°C, resistance is a linear function of temperature over a range of 0.1°C with a maximum nonlinearity amounting to 0.2% (22). Thermistors are generally fabricated as small beads $\simeq 0.1$ mm in diameter. Appropriately mounted, they have a low thermal mass and a correspondingly fast response ($\simeq 0.1$ s) to temperature changes in the solution in which they are immersed. Modern thermistors are quite stable, i.e., they have a well-reproducible resistance at any given temperature, provided that (1) reasonable protection is provided from mechanical shock and intense light, and (2) a small current is continuously maintained through the thermistor.

The resistance of the thermistor is generally monitored by wiring it as one arm of a Wheatstone bridge, as depicted in Fig. 3. The unbalance potential of the bridge is given by

$$V_{un} = V_{input}\left(\frac{R_3}{R_T + R_3} - \frac{R_2}{R_1 + R_2}\right) \tag{13}$$

By setting $R = R_1 = R_2 = R_3 \simeq R_T$ and $R - R_T = \Delta R$,

$$V_{un} \simeq V_{input}\frac{\Delta R}{4R} \tag{14}$$

Equation 14 is an approximation valid only for very small changes in R_T. In the usual arrangement (Fig. 3) the nonlinearities in Eqs. 12 and 13 are in opposite directions and partially compensate for one another. Over a temperature interval of 0.1°C, the net nonlinearity of a typical thermistor bridge combination amounts to only 0.03% (22), which is less than other experimental uncertainties in enthalpimetric analysis. For larger temperature changes, more complicated bridge configurations have been suggested which have a greater range of linear response (40).

Equation 14 suggests that the sensitivity of the bridge can be increased by increasing V_{input}. There is, in fact, a practical limit to such improvement. As the bridge voltage is increased, the power dissipated in the thermistor also increases. This not only creates a more steeply sloping baseline for the titration curve, it also creates more noise in the output signal. At some point as V_{input} is increased, the thermistor can no longer conduct its internal heat to the surrounding solution in a uniform and consistent manner. The internal temperature, and thus the resistance, of the thermistor experiences random fluctuations ("thermal noise"). The best bridge voltage to apply is a compromise between sensitivity and noise, and depends on thermistor resistance, thermistor heat dissipation constant, and stirring conditions. Generally, a power dissipation of less than 50 μW in the thermistor is desirable.

An alternating current (ac) bridge with lock-in-amplifier detection has been used to advantage (117). However, the simple dc bridge remains popular. The unbalance potential can be directly measured with a high input impedance strip chart recorder or digital voltmeter. Of course, dc measurements require careful shielding and grounding.

Possibilities exist for other types of temperature transducers, although none have been widely exploited. Multijunction thermocouples of comparable sensitivity to thermistors are bulky, slow, and inconvenient. A quartz crystal thermometer has a similar temperature resolution, but a larger thermal mass, a slower response time, and a much greater cost. Barium titanate ceramic devices, doped to adjust the Curie point to the temperature range of interest, can also function as resistance temperature-change transducers with a *positive* temperature coefficient of about $+10\%$ to 20% resistance change per degree (25, 81).

In thermometric titration, the titrant is generally delivered continuously at a constant rate. Usually a constant speed motor is used to slowly depress the plunger of a syringe containing the titrant. To ensure temperature control, the titrant is thermostated, often by passing through a long tube or a reservoir immersed in the thermostat bath. Very simple and inexpensive delivery systems have been described (59) as well as sophisticated stepping motors. Commercial drives are also available. In any case, excellent reproducibility is required ($< 0.5\%$ variation) for accurate titrations. Such automatic burets are usually calibrated gravimetrically.

In direct injection enthalpimetry the reagent and sample are mixed quickly in the Dewar flask. In cases where the excess reagent is injected, precise volume control is not required and a manually operated syringe is adequate. DIE can also be performed by the injection of a small volume of sample into a large excess of reagent. Accurate control of the sample volume is then needed. Syringes equipped with appropriate adapters (e.g., the so-called Chaney adapter) can be used to deliver highly reproducible volumes, which are calibrated gravimetrically. To minimize any reagent sample temperature mismatch, the syringe is generally immersed in the thermostat. Other types of immersion pipets have also been devised (112).

To correlate the observed bridge unbalance potential for a given chemical reaction to the heat of that reaction, it is necessary to know the heat capacity of the adiabatic cell, including its contents (solution) and all inserts. This is most conveniently obtained by electrical calibration via a circuit, such as that shown in Fig. 3. A constant current is passed through a heater resistor in the cell and a standard resistor R_s for a time t. The heat dissipated in the calorimeter, assuming negligible lead resistances, is given by

$$Q = \frac{V_s V_H t}{R_s} \tag{15}$$

where V_s and V_H are the voltages measured across the standard and heater resistors, respectively. (The heater resistor is also conveniently used to bring the cell contents up to the thermostat temperature if necessary.) To guard against systematic errors, it is advisable to occasionally check the heat capacity measurement with a reaction of accurately known enthalpy. The neutralization of perchloric acid with sodium hydroxide ($\Delta H° = -13.34$ kcal/mol, -55.81 kJ/mol) has been suggested for this purpose (55).

In addition to the basic equipment described, other devices have been variously employed. Tyson et al. (125) employed a differential apparatus with two thermistors, one of which was immersed in the actual titrate solution and the other in a corresponding blank. Smith et al. (117) used a linear ramp generator and a summing operational amplifier to cancel out the sloping baseline caused by stirrer and thermistor self-heating to attain very high sensitivities. Various workers (83, 85) have included a potentiometric probe (such as a pH microelectrode) in the calorimetric cell. Considerable information can be obtained in a single titration by this combination of thermometric and potentiometric detection.

The thermistor bridge provides an unbalance potential that is proportional to temperature. Three basic approaches have been used to record the data:

1. The traditional method for recording the data, which is quite versatile, is on a strip chart recorder, often with preamplification of the signal. The data are recorded as ΔT versus time (which is proportional to volume for titrant delivery at a constant rate). First- and second-derivative plots can also be recorded, using appropriate circuitry as first suggested by Zenchelsky and Segatto (140).

2. Alternatively, thermometric titrators for routine use (100, 110, 127) can use the electronically derived first-derivative of the titration curve to automatically shut off the titrant at the endpoint where an abrupt slope change occurs (see Fig 1a). A direct digital readout of volume of titrant delivered is provided.

3. The third method to record the data is to convert it into a series of digital signals that can be read by a computer. The availability of mini- and microcomputers for use in the laboratory has made this approach feasible in recent years (a trend seen in all of analytical instrumentation). A computer-interfaced analytical calorimeter is currently commercially available (Tronac Inc., Orem, UT). While enthalpimetric analyses are not by nature contingent on computerization as are methods such as Fourier transform IR and Fourier transform NMR, two principal advantages can be realized by computerized instrumentation:

    • Routine analysis can be largely automated. By use of appropriate multichannel interfaces, instrumental functions such as titrant addition or injection, sensitivity, and electrical calibration can be monitored and controlled. Appropriate programming can be used to automatically evaluate the data. In addition, a "smart" computer and program can be used to tutor an inexperienced operator.

- Tedious data corrections and calculations are accelerated manyfold for fundamental research. Corrections for thermal background effects (Section III.B) and the evaluation of thermodynamic and kinetic parameters (Section VI.A) are more readily and accurately implemented by computerized calculation.

III. THERMOMETRIC TITRATIONS

A. THEORY

The shape of a thermometric titration curve is determined by two distinct classes of considerations:

1. Background effects, including sources of heat loss or gain inherent in the apparatus and heats of dilution
2. chemical factors, i.e., the thermodynamics and kinetics of the reactions occurring during the titration.

Consider first an idealized titration reaction that is both thermodynamically complete ($K = \infty$) and kinetically fast ($k_f = \infty$). In the absence of any background effects, the ideal titration curve consists of straight line segments and is described by the following equations:

$$T = T_0 \quad \text{for } t \leq 0 \tag{16}$$

$$T = T_0 - \Delta H_r r C_A t / \kappa_0 \quad \text{for } 0 \leq t \leq t_{EP} \tag{17}$$

$$T = T_0 - \Delta H_r r C_A \frac{t_{EP}}{\kappa_0} \quad \text{for } t > t_{EP} \tag{18}$$

where t is the time from the start of titrant addition, t_{EP} is the time when a stoichiometrically equivalent amount of titrant has been added, r is the rate of titrant addition (mL/min), and C_A is the titrant concentration (mol/L). Such an idealized curve was shown in Fig. 1a.

1. Background Effects

The thermal factors that cause deviations from ideality are listed in Table 2. Figure 5 illustrates the regions of a titration curve in which the various thermal factors are effective. Carr (22) has analyzed these factors in detail and points out that factors 1, 2, 3, 4, and 5 are essentially constant throughout the titration. Provided their magnitude is not so large as to swamp out the heat of reaction of interest, these factors are "innocuous," i.e., they do not affect the linearity or the analytical usefulness of the titration curve. Factors 6, 7 and 8 (in Table 2) cause nonlinear deviations (curvature) which

TABLE 2
Background Effects Contributing to Nonideal Thermometric Titration Curves

Process	Mathematical Representation for Heat Flux of Process	Equation Number	Regions (Fig. 5)	Comments
1. Stirring	$\dfrac{dQ}{dt} = w_1$	19	All	Exothermic
2. Thermistor heating	$\dfrac{dQ}{dt} = w_2$	20	All	Exothermic
3. Evaporation	$\dfrac{dQ}{dt} = w_3$	21	All	Endothermic, depends on atmosphere above solution
4. Heat of titrant dilution[a]	$\dfrac{dQ}{dt} = -\Delta H_D r c_A$	22	II, III	May be exo- or endothermic
5. Side reactions	$\dfrac{dQ}{dt} = -\Delta H_s r c_A$	23	II, III	Usually eliminated through choice of reagents
6. Temperature mismatch of titrant	$\dfrac{dQ}{dt} = -\kappa_t r(T - T_t)$	24	II, III	Assumes additivity of of heat capacities
7. Heat transfer (imperfect adiabacity)	$\dfrac{dQ}{dt} = k(T - T_E)$	25	All	k may be time and volume dependent
8. Increase in heat capacity	$\kappa = \kappa_0 + \dfrac{d\kappa}{dv} rt$	26	II, III	Not a heat loss per se, but causes nonlinearity in observed variable, ΔT

[a] Because the titrate undergoes a relatively small volume change during the titration, the dilution of the titrate species can *normally* be ignored.

Note. c_A = titrant concentration; κ_t = specific heat capacity of titrant: T_t = temperature of titrant (usually equal to T_E); T_E = temperature of calorimeter environment; k = Newtonian cooling constant.
SOURCE: Derived from Carr (22).

are detrimental to precision and accuracy. These are referred to as parasitic effects and must be minimized or corrected. The four regions of the titration curve in Fig. 5 can be rationalized in terms of the appropriate heat flux processes.

Pretitration Baseline (I). Factors 1, 2, and 3 are invariant throughout the entire titration curve and can be combined into a single constant:

$$w = w_1 + w_2 + w_3 \tag{27}$$

Heat transfer also occurs in all regions, at a rate roughly proportional to the temperature difference between the calorimeter's contents and its environment. Equation 25 (Table 2), Newton's law of cooling, is an approximate representation. An exact representation of the heat loss results from the

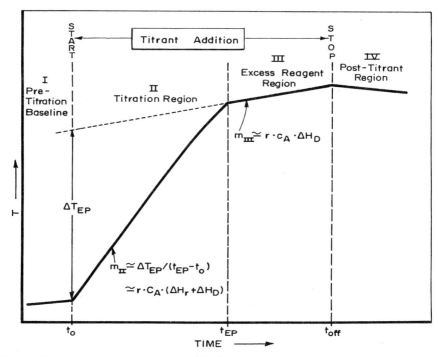

Fig. 5. Thermometric enthalpy titration (TET) curve illustrating various regions of the curve (see Table 2). Time is proportional to volume of titrant (A) added between t_0 and t_{off} because titrant is added at a constant rate.

solution of Fourier's law:

$$\nabla^2 T(x,t) = \frac{\bar{\kappa}}{\lambda} \frac{\partial T(x,t)}{\partial t} \tag{28}$$

In Eq. 28, λ is the thermal conductivity of the appropriate calorimeter material. Explicit solution of Eq. 28 is practically impossible for a real calorimeter because the boundary conditions are unknown and several different materials are used in the construction. Nonetheless, the results for a simplified case (10, p. 163) demonstrate the time dependence of the thermal gradient in a conducting material. Martin et al. (86) have used a two-hole heat leak model that successfully accounts for the time dependence of heat leak characteristics. In practice the time dependence of k in Eq. 25 is minimized by careful design of the calorimeter.

In the description of the pretitration baseline, Eq. 25 can be used if the calorimeter has been allowed to attain a thermal steady state (i.e., thermal gradients in all calorimeter materials have reached constant values). The slope of region I will thus be

$$\frac{dT}{dt} = \frac{dQ}{dt}\frac{1}{\kappa_0} = \frac{w + k(T - T_E)}{\kappa_0} \quad (29)$$

For a small thermal lead $(T - T_E)$ and over short time intervals, Eq. 29 is essentially a linear relationship.

Titration Region (II). As titrant is added and the chemical reaction of interest proceeds, all of the factors in Table 2 become consequential. The rate of temperature change at any time t is given by

$$\frac{dT}{dt} = \frac{-(\Delta H_r + \Delta H_D + \Delta H_s)C_A r + w - k(T - T_E) - \bar{\kappa}_t r(T - T_t)}{\kappa_0 + (d\kappa/dv)rt} \quad (30)$$

A more precise description can be obtained for small vessels (< 5 mL) if the heat loss constant k in Eq. 30 is given as a function of volume, as done by Hansen et al. (57):

$$k = k_0 + \frac{dk}{dv}rt \quad (31)$$

The initial heat loss constant k_0 and its volume dependence dk/dv are determined by independent experiments for a given calorimeter.

Equation 30 has been integrated to yield an expression for the temperature at any given time (c.f. Carr (22)) for a hypothetical titration curve. Conversely, Eq. 30 alone is the sufficient theoretical basis for correcting an actual thermometric titration curve to obtain the desired net (chemical) heat of reaction (ΔH_r). Inspection of Eq. 30 is also illuminating in regard to minimizing the "parasitic" effects that cause nonlinear thermograms. Linearity is enhanced by high adiabaticity k, concentrated titrants (so that $(d\kappa/dv)rt \ll \kappa_0$), and small overall temperature changes ($T - T_E < 0.1°C$).

Excess Reagent Region (III). If the equilibrium constant $k \to \infty$ and all relevant rate constants are large, then the reaction attains virtual completion at the equivalence point t_{EP}. However, as titrant continues to be added, all background effects continue to play a role. The rate of temperature change at any time t is

$$\frac{dT}{dt} = \frac{-(\Delta H_D + \Delta H_s)c_A r + w - k(T - T_E) - \bar{\kappa}_t r(T - T_E)}{\kappa_0 + (d\kappa/dv)rt} \quad (32)$$

Integration will yield an explicit expression for the temperature at any time.

If the nonlinear thermal factors have been kept to a minimum (high adiabaticity, concentrated titrant), the heat of reaction can be conveniently estimated graphically by back-extrapolation of region III to the start of the titration, as has been recommended by Jordan and Alleman (67); this process is shown in Fig. 14.5, where $Q \simeq \Delta T_{EP}/\kappa_0$. This simplification is often not sufficiently accurate, and a more rigorous treatment of the thermal factors is required. In that case, region III can be used to evaluate the heat of dilution ΔH_D and the heat of any side reactions ΔH_s.

Post-titration Baseline (IV).   The same extraneous processes are effective here as in region I, so that the slope of region IV will be

$$\frac{dT}{dt} = \frac{w - k(T - T_E)}{\kappa_0 + (d\kappa/dv)rt_{off}} \tag{33}$$

where t_{off} is the elapsed time when the titrant is turned off. For small vessels, the heat loss constant k must also be corrected for volume change according to Eq. 31.

2.   Chemical Factors

The success of a thermometric titration, like any titration, ultimately depends on the chemistry of the titration reaction. Having eliminated or corrected for the background effects, the shape of the titration curve depends on the heat(s), equilibrium constant(s), and rate(s) of the chemical reaction(s) occurring in the calorimeter. Either thermodynamic incompleteness or kinetic slowness can cause rounding in the region of the endpoint. These two types of curvature are illustrated for an exothermic reaction in Fig. 6.

In the case of a fast, but incomplete reaction, the stoichiometric endpoint can nonetheless be accurately located by extrapolation of the linear regions of the curve as shown by Fig. 6a. Rosenthal et al. (108) has generally analyzed the effect of equilibrium curvature on the precision and accuracy of linear titration methods, and Carr (22) has specifically extended those concepts to TET. Generally, the endpoint can be located with a precision and accuracy of 1% if the reaction is 80% complete at the stoichiometric endpoint. For the reaction given in Eq. 4 (for $a = b = p = 1$), at a sample concentration of 10^{-3} M, 80% completion at the endpoint corresponds to $K_4 = 2 \times 10^4$ M^{-1} (calculated from Eq. 8).

For a reaction whose kinetics are slow compared to the rate of titrant delivery, the extrapolated endpoint does not correspond to the stoichiometric equivalence point (see Fig. 6b). This poses a particularly serious limitation to the application of TET to certain classes of reactions. The continuous addition of titrant is a rate process, which competes with the rate of chemical reaction. If these two rate processes are of comparable magnitude, the attainment of chemical equilibrium will lag behind the theoretical equivalence point. For reliable results, the rate of chemical reaction (under the

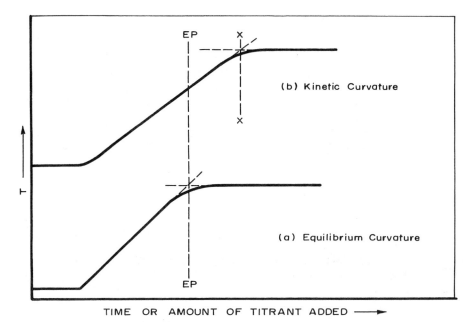

Fig. 6. Endpoint rounding of thermometric titration curves caused by chemical factors. Titrant delivered at a constant rate. EP, equivalence point; X, spurious endpoint.

prevailing conditions) must be at least 100-fold greater than the rate of titrant addition. In TET, the useful range of titrant addition rates is 10^{-4} to 10^{-7} mol/s; therefore, the reaction rate must be faster than about 10^{-4} mol/s. Most ionic processes in aqueous solution (such as acid–base, precipitation, and complexation reactions) satisfy this requirement. Organic functional group reactions often do not.

A convenient check for the presence of sluggish kinetics is to vary the titrant addition rate by at least a factor of 3. Kinetic errors will be indicated by a change in the apparent stoichiometry of the extrapolated endpoint [X in Fig. 6b]. Carr and Jordan (24) have more thoroughly developed the theoretical treatment of kinetic effects in constant rate titrations, including equations for the calculation of a meaningful endpoint from the spurious endpoint X. Catalysts may sometimes be used to enhance the rate of reaction.

B. PRACTICE

1. Endpoint Detection

The practice of thermometric titration for analytical purposes requires the accurate and precise determination of meaningful endpoints. For an endpoint to be meaningful, interfering side reactions must be obviated. Reagents must be pure, i.e., free from reacting contaminants. Interferences in the sample must also be considered. Naturally, titrants must be standardized with appropriate accuracy. To minimize the problems of volume change during the titration, it is generally desirable to use titrants at least 50 times more concentrated than the sample. Finally, an endpoint is meaningful only if the reaction is relatively fast (as previously outlined).

In general, the necessary condition for obtaining a TET endpoint is a change in slope at the equivalence point. The larger and more abrupt the change, the better is the endpoint. The sharpness of the endpoint is dictated by the equilibrium constant of the reaction as has previously been discussed. The magnitude of the slope m change is given by $|m_{II} - m_{III}|$, where $m_{II} = rc_A(\Delta H_r + \Delta H_D)$ and $m_{III} = rc_A H_D$ (see Fig. 5). Consequently, the heat of reaction must differ significantly from the heat of dilution.

The overall heat of reaction (and thus m_{II}) can be increased by coupling the primary reaction between sample and titrant with a secondary reaction. An example is a method for the determination of sulfur in steel (112) in which the heat of oxidation of sodium sulfide is coupled to the neutralization of protons with a strong base as follows:

Primary reaction

$$S^{2-} + 8MnO_4^- + 4H_2O \rightleftharpoons SO_4^{2-} + 8MnO_4^{2-} + 8H^+$$

$$\Delta H \simeq +20 \text{ kJ/mol} \qquad (34)$$

Coupled reaction

$$8H^+ + 8OH^- \rightleftharpoons 8H_2O \qquad \Delta H \simeq -440 \text{ kJ/mol} \qquad (35)$$

Actual overall reaction

$$S^{2-} + 8MnO_4^- + 8OH^- \rightleftharpoons SO_4^{2-} + 8MnO_4^{2-} + 4H_2O$$

$$\Delta H \simeq -420 \text{ kJ/mol} \qquad (36)$$

Reaction 36 has a huge heat of $\simeq 420$ kJ/mol. Such chemical amplification of the reaction heat can sometimes be manipulated by judicious selection of buffering media. The concept of buffer amplification is equally applicable to enhance sensitivity in TET, DIE, and flow enthalpimetry. Secondary reactions other than proton transfer may also be employed.

Enhancement of the heat of reaction is one way to facilitate endpoint determination in TET. However, since it is the *change* in slope that actually defines the endpoint, increased precision can also be obtained by adjusting the slope of the region after the endpoint (m_{III} in Fig. 5). If the experiment is designed so that a very exo- or endothermic endpoint indicator reaction will occur after the stoichiometric equivalence point, TET can be extended to reactions with small heats. One approach is the direct reaction of titrant and

TABLE 3

Thermodynamic Data Explaining the Use of Sulfate as a Direct Thermochemical
Indicator in the Titration of Carboxylate Salts by Strong Acid

Reaction	K	ΔH (kJ/mol)
$H^+ + RCO_2^- \rightleftharpoons RCO_2H$	10^3–10^5	0 to -5
$H^+ + SO_4^{2-} \rightleftharpoons HSO_4^-$	1.2×10^2	$+19.7$

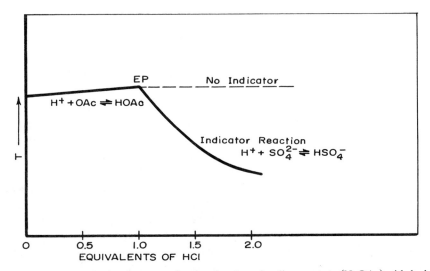

Fig. 7. Thermometric titration curve for the titration of sodium acetate (NaOAc) with hydrochloric acid using sodium sulfate as a direct thermochemical indicator. EP, endpoint. Reprinted
from Hansen et al (56, p. 22), by courtesy of Marcel Dekker, Inc.

indicator. Hansen et al. (56) have used excess of sulfate as an indicator in the
titration of salts of carboxylic acid (RCO_2^-) by strong acid. Relevant data are
contained in Table 3. The protonation of most carboxylic acids is weakly
exothermic, and as a result, endpoints are difficult to locate. Sulfate is a
much weaker base than RCO_2^- and has a large endothermic heat of protonation. Figure 7 illustrates the use of sulfate as an endpoint indicator in the
TET of sodium acetate. The slope change at the acetate endpoint is increased fivefold by this method.

2. Catalytic Thermometric Titration

While an endpoint indicator that reacts directly with the titrant is quite
useful, the lower limit of concentrations accessible is about 10^{-3} M. To
titrate a sample that is more dilute requires a relatively dilute titrant for good
volumetric precision. Correspondingly, the heat evolved by the indicator
reaction will be smaller, making the endpoint less perceptible. A novel

approach that overcomes this limitation is to use an endpoint indicator reaction that is *catalyzed* by the first excess of titrant added upon virtual completion of the primary "determinative" reaction. The general reaction scheme is

Determinative reaction

$$\underset{\text{Titrant}\quad\text{Sample}}{A + B} \rightleftharpoons \underset{\text{Products}}{P} \tag{37}$$

Indicator reaction

$$\underset{\text{Indicator reactants}}{D + E} \xrightarrow{\text{A(catalyst)}} \underset{\text{Products}}{F} \tag{38}$$

where, ideally, reaction 38 does not commence until reaction 37 has reached stoichiometric equivalence. The catalyzed indicator reaction proceeds rapidly and yields a steep slope. The slope can, in principle, be endothermic or exothermic, depending on the indicator reaction. However, all indicator reactions used to date have been exothermic. The indicator reactants D and E can be present in large amounts, so that the total temperature change may be as much as several degrees. Because the temperature change is so large, adequate endpoint location can be obtained even with simple manual equipment (44). Catalytic thermometric titrations are amenable to microanalysis, having been used on samples as dilute as 10^{-6} M (19). A remarkable feature of catalytic thermometric titrations is that they are feasible in situations where the heat of the determinative reaction (reaction 37) is negligible.

The applicability of catalytic thermometric titration is limited by the availability of suitable indicator reaction/titrant–catalyst combinations. Table 4 summarizes the representative reaction types that have been developed to date. These can be divided into two broad categories: (1) the determination of acidic or basic organic compounds in nonaqueous solvents, and (2) endpoint indication, which is by dimerization, polymerization, or hydrolysis of a component of the solvent system, catalyzed by excess strong acid or base.

An example of this is the pioneering work of Vaughan and Swithenbank (129) in 1965. They used acetone as a solvent in the titration of weak organic acids (such as phenols) with alcoholic potassium hydroxide. As shown in Fig. 8, a sharp temperature increase occurs at the expected endpoint, due to the base-catalyzed dimerization of acetone. Their work found application in the determination of phenolic groups in coal (131).

Greenhow and Spencer (44, 46) have extensively evaluated the use of ionic polymerization reactions (which are acid or base catalyzed) of vinyl monomers (such as acrylonitrile) as catalytic endpoint indicators. Various titrant–solvent–vinyl monomer combinations have been studied. In addition to traditional acid–base titrants such as perchloric acid and alcoholic potassium hydroxide, Greenhow and Spencer also report on the use of Lewis acid and

TABLE 4
Catalytic Thermometric Titrations

Sample Species	Titrant (Catalyst)	Solvent	Indicator Reaction	Reference
Weak organic acids	Alcoholic KOH	Acetone	$2CH_3COCH_3 \xrightarrow{OH^-} CH_3COCH_2C(CH_3)_2OH$	129
Weak organic acids (sulfanilamides)	Alcoholic KOH or tetra-n-butyl ammonium hydroxide	Acrylonitrile	$n\text{-}CH_2{:}CHCN \xrightarrow{OH^-} HO(CH_2CHCN)_nH$	44
Weak bases (amines, alkaloids)	$HClO_4$	Toluene and 2-phenyl-propene	$n\text{-}CH_2{:}C\phi CH_3 \xrightarrow{H^+} H(CH_2C\phi CH_3)_n^+$	44
Weak bases (pyridine, caffeine)	$HClO_4$ or coulometrically generated H^+	Acetic acid acetic anhydride and trace water	$(CH_3CO)_2O + H_2O \xrightarrow{H^+} 2CH_3COOH$	126
Hg^{2+}, Ag^+, Pd^{2+}	KI	Water	$As(III) + 2Ce(IV) \xrightarrow{I^-} As(V) + 2Ce(III)$	19, 137
Hg^{2+}, Ag^+, Pd^{2+}	KI	Water	$As(III) + 2Mn(III) \xrightarrow{I^-} As(V) + 2Mn(II)$	75
EDTA	$Mn(NO_3)_2$	Water	$H_2O_2 \xrightarrow{Mn^{2+}} H_2O + 1/2O_2$	136
EDTA	Cu^{2+}	Water	$2H_2O_2 + N_2H_4 \xrightarrow{Cu^{2+}} 4H_2O + N_2$	138

bases such as boron trifluoride and n-butyl lithium as titrants. Using polymerization indicators, drugs with acidic functional groups, such as sulfanilamides (46), catecholamines (45), and barbituates (4) have been successfully assayed.

Catalytic thermometric titrations have also been successfully applied to trace metal analysis. In this instance, the classical iodide catalyzed Sandell–Kolthoff reaction,

$$As(III) + 2Ce(IV) \xrightarrow{I^-} As(V) + 2Ce(III) \qquad (39)$$

has been used as endpoint indicator in the precipitation titration of various cations with standard potassium iodide. Thus, mercury, silver, and palladium were titrated with KI. A back-titration procedure of mercury, silver, or palladium added in excess was used for the determination of anions $(Cl^-, Br^-, S^{2-}, S_2O_3^{2-})$, which form complexes or insoluble salts with these cations. Another variation termed *substitution* titration has been developed (76) which further extends the capabilities of the method. In these applications, the indicator reactants and an inactive form for the catalyst (such as HgI_2) are added to the sample solution. Potassium thiocyanate is used as a titrant for mercury, silver, and palladium. The first excess of thiocyanate displaces free iodide from the HgI_2, which then catalyzes the indicator reaction 39.

The Sandell–Kolthoff reaction 39 cannot be used in the presence of phosphate due to the precipitation of ceric and cerous phosphates. Instead,

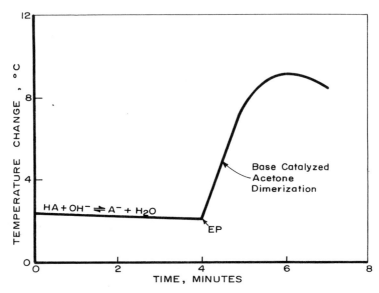

Fig. 8. Titration of an acidic substance with KOH using acetone as solvent and catalytic thermochemical indicator. Time is proportional to volume because titrant is added at a constant rate. From Vaughan and Swithenbank (129).

the iodide catalyzed reaction of Mn(III) and As(III)

$$As(III) + 2Mn(III) \xrightarrow{I^-} As(V) + 2Mn(II) \qquad (40)$$

can be used as an indicator when phosphate is present (75).

Weisz and Kiss (136) have also used oxidation and decomposition reactions of hydrogen peroxide as endpoint indicators. The decomposition of H_2O_2 (Eq. 41) and the reaction of H_2O_2 with resorcinol (1,3-dihydroxybenzene) (Eq. 42) are catalyzed by manganese(II):

$$H_2O_2 \xrightarrow{Mn^{2+}} H_2O + \tfrac{1}{2}O_2 \qquad (41)$$

$$H_2O_2 + C_6H_4(OH)_2 \xrightarrow{Mn^{2+}} \text{red-colored oxidation products} \qquad (42)$$

Thus, Mn(II) can be determined by reverse titration of the sample into an EDTA solution of known concentration. Other cations that are strongly complexed by EDTA can be determined by back titration of excess EDTA with standard manganese(II) nitrate. Alternatively, the following type of substitution titration can be performed to determine cations that are more strongly complexed than manganese (e.g., zinc) (76). A trace of Mn^{2+}/EDTA complex is added to a standard EDTA solution, along with the indicator reactants (H_2O_2 and resorcinol), and titrated with the zinc, which is complexed by the free EDTA. The indicator reaction is activated when the first excess of zinc abstracts EDTA from the Mn^{2+}/EDTA complex, liberating catalytic amounts of Mn^{2+}.

The reaction of hydrogen peroxide with hydrazine is catalyzed by copper(II) and can also be used as an indicator in compleximetric titrations (138):

$$2H_2O_2 + N_2H_4 \xrightarrow{Cu^{2+}} N_2 + 4H_2O \tag{43}$$

It yields a large heat effect and thereby an increased sensitivity.

A thorough discussion of catalytic thermometric titration and its applications can be found in a comprehensive review article by Greenhow (43). Catalytic endpoint indication methods extend the dynamic range of conventional TET to concentrations as low as 10^{-6} M. A disadvantage is the inability to obtain multiple endpoints for multicomponent mixtures. Also, the scope of application is limited by the handful of indicator reactions and catalysts presently available. However, the increased sensitivity and the possibility of using simpler equipment may make this a desirable option when applicable.

C. TITRATION OF MIXTURES

In practice a sample may contain a mixture of components that will react with a given titrant. The ability to differentiate between these components by TET depends on the thermodynamics of the relevant reactions. A TET curve will yield discrete, sequential endpoints for each reacting component if the following conditions are met:

$$\frac{K_i}{K_{i+1}} \geq 50 \tag{44}$$

$$\frac{|\Delta H_{i+1} - \Delta H_i|}{\Delta H_i} \geq 0.2 \tag{45}$$

where K_i denotes the equilibrium constant and ΔH_i the heat for the reaction of the ith component with the titrant. The requirement of Eq. 44 is less stringent than that for potentiometric titration of mixtures, which is

$$\frac{K_i}{K_{i+1}} \geq 10^4 \tag{46}$$

An example of a TET curve with sequential endpoints is shown in Fig. 9. That figure represents the titration of a mixture of calcium and magnesium

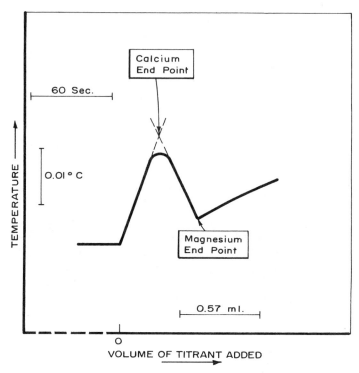

Fig. 9. Titration curve of a mixture of Ca^{2+} and Mg^{2+} with ethylenediaminetetraacetate. Reprinted with permission from Jordan and Alleman (67). Copyright 1957, American Chemical Society.

with EDTA (Y^{4-}). The relevant reactions are

$$Ca^{2+}_{aq} + Y^{4-}_{al} \rightleftharpoons CaY^{2-}_{aq} \tag{47}$$

$$Mg^{2+}_{aq} + Y^{4-}_{aq} \rightleftharpoons MgY^{2-}_{aq} \tag{48}$$

where Y^{4-} denotes the quadrivalent ethylenediamine tetraacetate anion. For calcium (reaction 47), $K = 10^{+10.6}$ and $\Delta H = -25.2$ kJ/mol, while for magnesium (reaction 48), $K = 10^{8.7}$ and $\Delta H = 16.8$ kJ/mol. The difference in heats for these reactions has been rationalized as due to ordering of the hydration sphere of the aqueous magnesium ion (67).

Even when the equilibrium constants for two components in a mixture are nearly the same, if the heats are different and known, differentiation of the mixture is possible on the following considerations. For the reactions

$$A + B = P \qquad K_B, \Delta H_B$$
$$A + C = R \qquad K_C, \Delta H_C$$

where $K_B = K_C$ and $\Delta H_B \neq \Delta H_C$

only one TET endpoint will be observed. However, two pieces of information are available from the titration curve: the endpoint n_A corresponding to the sum of B + C, and the total heat evolved by both reactions Q_T. The following equations are applicable:

$$n_A = n_B + n_C \tag{49}$$

$$Q_T = n_B \, \Delta H_B + n_C \, \Delta H_C \tag{50}$$

If the heats ΔH_B and ΔH_C are known from prior experiments, Eqs. 49 and 50 can be simultaneously solved for n_B and n_C. Hansen and Lewis (58) demonstrated that analysis of binary mixtures by this procedure is practicable with a precision of 5%.

IV.   DIRECT INJECTION ENTHALPIMETRY

Direct injection enthalpimetry may be viewed as a quasi-instantaneous thermometric titration. The volume axis is compressed to a single point (the injection point) and the analytical information is contained within the temperature change alone (see Fig. 1b). Because DIE analysis is based solely on measurement of heats evolved (or absorbed) and not on volumetric endpoints, DIE and TET are analogous to spectrophotometry and spectrophotometric titration, respectively. In contrast to spectrophotometry, DIE is a "one-shot" destructive method.

A.   THEORY

DIE can be performed in either of two modes. A small volume of excess concentrated reagent can be injected into the sample solution, or the sample solution can be injected into excess reagent (inverse DIE). In either case for a chemical reaction that is instantaneous ($k_f = \infty$) and complete ($K = \infty$) and an ideally adiabatic system, a sharp temperature "step" results, whose height is a direct measure of sample concentration (as shown in Fig. 10). In a real isoperibol calorimeter, all of the background effects listed in Table 2 will influence the height and shape of the DIE step, as discussed below.

1.   Background Effects

(1), (2), (3)   (numbered as in Table 2) Stirring, thermistor self-heating, and evaporative cooling.   To within a first approximation, these effects can be considered constant and equal before, during, and after the injection. The small volume increase upon injection causes a negligible increase in the heat of stirring (the interface area between solution and cell walls and stirrer shaft increases with solution volume). The temperature change engendered by the chemical reaction between sample and reagent causes a concomitant change in thermistor resistance (and therefore power dissipation) and also affects the

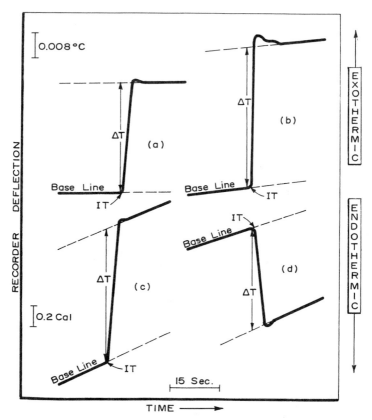

Fig. 10. Typical direct injection enthalpograms for exothermic and endothermic reactions: (*a*) HCl + NaOH, (*b*) H_3BO_3 + NaOH, (*c*) Pb^{2+} + EDTA, (*d*) Mg^{2+} + EDTA. IT, Injection time. Reprinted with permission from Wasilewski et al. (132). Copyright 1964, American Chemical Society.

rate of solvent evaporation. However, for small temperature changes, these effects are inconsequential.

(4), (5)   Heat of dilution and side reactions.   Within a given DIE experiment, these effects are generally indistinguishable from the heat of reaction. They must be eliminated or accounted for as discussed in Section IV.B.

(6)   Temperature mismatch of the injected solution.   If the injected solution is of a different temperature than the contents of the calorimetric cell, the magnitude of the measured temperature change will be affected. Assuming additive heat capacities, the temperature change $T_f - T_1$ upon mixing nonisothermal solutions is

$$T_f - T_1 = \frac{(T_2 - T_1)v_2}{v_1 + v_2} \tag{51}$$

where the subscripts 1, 2, and f denote the injected solution, the solution initially in the calorimeter, and the mixture of the two, respectively. In classical solution calorimetry, the mismatch $(T_2 - T_1)$ is made zero by placing the reagent in a thin glass vial which is immersed in the calorimetric cell. Mixing is accomplished by breaking the vial. In DIE, a more convenient, though less precise arrangement is used, the external thermostating of the reagent. The mismatch error is also minimized by making reagent volume v_2 small compared to v_1, the volume in the Dewar. In the original publication on DIE (132), it was pointed out that linear calibration plots (concentration versus T) are obtained even in the presence of a *constant* temperature mismatch.

(7) Heat transfer due to imperfect adiabaticity. The same considerations apply here as in TET (see Section III.A.1). Whenever the calorimeter's internal temperature differs from the external temperature, finite heat transfer will occur. The Newtonian cooling expression, Eq. 25, can be used to provide an approximate correction. However, for very rapid chemical reactions, the thermal equilibrium time of the calorimetric cell becomes important. This is apparent from Fig. 10, which illustrates typical DIE curves associated with virtually instantaneous exothermic and endothermic reactions: all curves display a temperature "overshoot," due to the slow thermal equilibration of the cell walls. The quantity ΔT is nevertheless correctly measured by extrapolating from the linear region beyond the overshoot, as in Fig. 10. The shape of the DIE curve, plotted as Q versus time, however, is in these cases determined primarily by thermal characteristics of the vessel, and not by chemical considerations. DIE can be used to measure rate constants and for differential kinetic analysis (see below), but this kinetic information will be obscured for reactions that occur quickly. With a good cell design, reactions occurring in as little as 3–5 s can be followed and kinetically analyzed. For very slow reactions the accuracy of heat loss corrections becomes increasingly dubious, so that about 30 min is the practical upper limit for a DIE experiment.

The isothermal calorimeter developed by Christensen et al. (27) is useful over a time scale of many hours, but its response time is quite slow (several minutes).

(8) Increase in heat capacity. In DIE the heat capacity changes in one sudden step upon injection of reagent or sample. The final heat capacity can be measured by electrical calibration (Fig. 3) and the T axis directly converted to heat evolved Q. In determinations performed using similar volumes, the final heat capacity will be constant. In that case, the linear relationship between ΔT and amounts reacted n_P (Eq. 52) is maintained:

$$\Delta T = \frac{n_P \Delta H_r}{\kappa} \tag{52}$$

and electrical calibration is not necessary for routine analyses. Instead, the ΔT (or Q) scale can be calibrated with the aid of a known sample.

2. Chemical Effects

Equation 52 for the height of a DIE step (ΔT in Fig. 10) is useful only if the sample reacts totally, i.e., if conversion to the product P (Eq. 4) is virtually complete. The actual extent of reaction is limited by equilibrium considerations according to Eq. 8. The equilibrium can be driven toward completion by using coupled reactions which consume a primary product. Also, DIE allows the use of a manyfold excess of reagent which will both shift the equilibrium toward completion and increase the forward rate of reaction.

Many analytically useful reactions occur too slowly to be used as titration reactions, but are amenable to DIE. Originally DIE was used only with fast reactions (those complete within a few seconds). New and exciting applications of DIE have been with slower reactions, and also those in which the analytical information is contained in the rate of reaction. For a general reaction of the type

$$A + B \longrightarrow P + R \tag{53}$$

an often encountered rate law in solution chemistry is

$$\frac{d[P]}{dt} = k[A][B] \tag{54}$$

Equation 54 may be rewritten in terms of heat evolved Q:

$$\frac{d(Q/\Delta H_{53}v)}{dt} = k[A][B] \tag{55}$$

The integrated form of Eq. 55, with initial conditions $[A] = [A]_0$, $[B] = [B]_0$, is

$$\frac{1}{[A]_0 - [B]_0} \ln \frac{[B]_0([A]_0 - Q/\Delta H_{53}v)}{[A]_0([B]_0 - Q/\Delta H_{53}v)} = kt \tag{56}$$

In the special case when $[A]_0 = [B]_0$,

$$\frac{1}{[A]_0 - Q/\Delta Hv} = kt + \frac{1}{[A]_0} \tag{57}$$

In the presence of a large excess of A (reagent), Eq. 54 becomes pseudo-first order and upon integration yields

$$Q = \Delta Hv[B]_0[1 - \exp(-k[A]_0 t)] \tag{58}$$

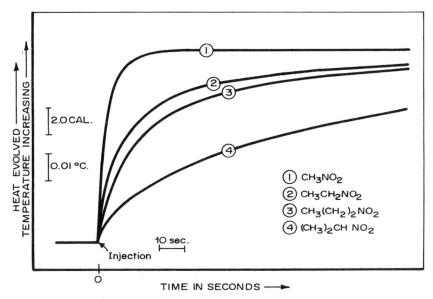

Fig. 11.   Idealized plots of heat evolved versus time for the reaction

$$RCH_2NO_2 + OH^- \text{ (excess)} \rightleftharpoons RCHNO_2^- + H_2O$$

RCH_2NO_2 at approximately millimolar concentrations. Reprinted from Henry (60).

Figure 11 displays a series of pseudo-first-order DIE curves for reactions having different rate constants.

Enzyme-catalyzed reactions have been a preferred application of DIE. Enzymatic reaction rates are generally successfully described by a classical two-step mechanism:

$$S + E \underset{k_{-1}}{\overset{k_1}{\rightleftharpoons}} E \cdot S \tag{59}$$

$$E \cdot S \overset{k_2}{\longrightarrow} E + P \tag{60}$$

where E is enzyme, S is substrate, E · S is an enzyme–substrate complex, and P is product. The rate law for this mechanism is the Michaelis–Menton equation, which has a broad applicability:

$$\nu_0 \text{ (initial reaction velocity)} = \frac{d[P]}{dt} = \frac{k_2[E]_0[S]}{[S] + (k_{-1} + k_2)/k_1} \tag{61}$$

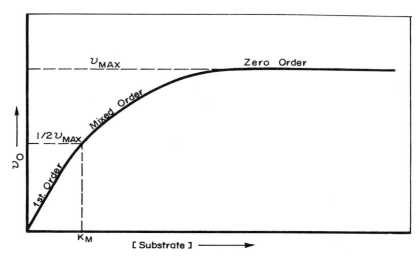

Fig. 12. Plot of the initial reaction velocity (ν_0) of an enzymatic reaction as a function of the initial substrate concentration. K_M denotes the Michaelis constant.

The following parameters are normally substituted into Eq. 61:

$$\nu_{max} = k_2[E]_0 \tag{62}$$

$$K_m \ (\text{Michaelis constant}) = \frac{k_{-1} + k_2}{k_1} \tag{63}$$

Figure 12 is the well-known plot of initial rate ν_0 versus substrate concentration [S] according to Eq. 61. As the enzyme becomes saturated at high substrate concentrations, ν_0 attains a limiting value ν_{max} and zero-order kinetics prevail. Under such conditions, and if [S] remains effectively constant throughout the experiment, a DIE plot will be a straight line, with slope, dQ/dt, corresponding to ν_{max} or the enzymatic activity (EA) present. EA is usually reported in moles of substrate converted per unit time,

$$\text{EA} = \frac{dn_s}{dt} \tag{64}$$

from which it follows that

$$\text{EA} = -\left(\frac{1}{\Delta H_{59+60}}\right)\left(\frac{dQ}{dt}\right) \tag{65}$$

This is the basis of a novel enthalpimetric enzyme assay (95).

B. PRACTICE

1. Experimental Implementation

Instrumentation for DIE is identical to that for TET (see Fig. 3) with a modification for rapid injection. As previously mentioned, DIE can be implemented in either of two modes: reagent injected or sample injected (for enzyme-catalyzed reactions, the reaction can also be initiated by injecting enzyme solution into the reagent/sample mixture). In any case, upon rapid mixing the adiabatic cell will contain n_B moles of sample and n_A moles of reagent in a volume $(v_1 + v_2)$. The temperature step is proportional to $n_B/(v_1 + v_2)$. When the sample solution is injected (inverse DIE), a sample of sufficient concentration is required, since a small volume of sample v_2 must be used to minimize the temperature mismatch error (see Eq. 51). Despite this requirement of higher concentrations, inverse DIE can be performed to advantage in a seriatim fashion. Samples can be injected repeatedly into a large excess of reagent.

The temperature step ΔT produced upon injection can be correlated to sample concentration in two ways. One is the dependable empirical approach of preparing a calibration curve by measuring ΔT for a series of standards of known concentration. Linear calibration curves are obtained if identical volumes are maintained and reasonable precautions are taken to ensure constant and minimal temperature mismatch. Another approach is to directly utilize the heat of reaction ΔH_r as a standard of reference by independently determining the heat capacity after injection (i.e., by electrical calibration). By this approach, all heat effects must be known. Heats of reaction must be well characterized, temperature mismatch must be negligible, and heats of dilution ΔH_D must be compensated. The heat of dilution of a concentrated reagent is totally unrelated to the desired analytical information, viz., sample concentration. However, ΔH_D for dilution of a concentrated sample (inverse DIE) will appear as though it is a component of ΔH_r in Eq. 52. Heats of dilution can be minimized by careful matching of ionic strength, estimated from tabulated data and/or measured by a blank experiment.

The precision and accuracy of DIE is generally 2–5%. In the case of rapid reactions, simple graphical extrapolation (Fig. 10) yields good quantitative results. Computerized data handling facilitates corrections for heat losses and data analysis for slower reactions.

2. Heterogeneous Reactants

An oft-quoted virtue of enthalpimetric methods is their relative immunity to deleterious matrix effects (viscosity, turbidity, etc.). This advantage can be taken one step further by using heterogeneous systems for analysis. For example, a method for the assay of water in organic solvents involves the injection of the sample into a slurry of dry solvent and zeolite molecular sieves (106). The adsorption of water into the sieves creates a heat effect quantitatively related to water concentration.

Gaseous reagents or samples can also be used in DIE, with some adaptation of the equipment (139). The gaseous reagent (or sample) is slowly bubbled into the sample (or reagent) solution. An advantage of gases is their negligible heat capacity compared to liquids, which minimizes temperature mismatch effects.

3. Analysis of Mixtures

Conventional DIE yields only one data point (ΔT, the temperature or heat step) per sample. Therefore, resolution of multicomponent mixtures by a single experiment is not normally possible. Obviously, if a specific reagent for each component can be found, the mixture's composition can be determined by a series of experiments. Even if the reagents are nonselective, provided they react with the various components of the mixture with a different ΔH_r, a y-component mixture can be resolved by y DIE experiments without prior separation. For each reagent added in excess the heat evolved will be

$$-Q_{A1} = n_{B1} \Delta H_{B1-A1} + n_{B2} \Delta H_{B2-A1} + \cdots + n_{By} \Delta H_{By-A1} \quad (66)$$

where ΔH_{B1-A1} is the heat of reaction of sample component one (B1) with reagent one (A1) under the experimental conditions. These heats must be known from prior experiments. A system of y equations in y unknowns is thus set up and solved. An example of this type of serial DIE analysis is the determination of mixture of sulfur-containing anions (7). Sulfide, sulfate, sulfite, and thiosulfate were determined at the millimolar level with about 2% precision and accuracy. The reagents used were Cd^{2+}, Ba^{2+}, I_2, and Ag^+. Other applications of serial DIE have also been reported (5, 6).

DIE has also been applied to the analysis of mixtures by differential reaction kinetics. Differential kinetic methods of analysis have been widely investigated since first proposed by Lee and Kolthoff (79), and particular application has been found in the functional group analysis of mixtures of similar organic molecules. The use of quasiadiabatic enthalpimetry to monitor reaction progress in kinetic analysis was first reported by Papoff and Zambonin (98).

Consider a two component mixture, B_1 and B_2, each of which reacts with reagent A:

$$A + B_1 \xrightarrow{k_1} P_1 \qquad \Delta H_1 \quad (67)$$

$$A + B_2 \xrightarrow{k_2} P_2 \qquad \Delta H_2 \quad (68)$$

When pseudo-first-order conditions prevail (large excess A), Eq. 69 represents the total heat evolved (or absorbed):

$$-\frac{\Delta I(t)}{\kappa} = -Q(t) = v \Delta H_1 [B_1]_0 \{1 - \exp(-k_1' t)\}$$
$$+ v \Delta H_2 [B_2]_0 \{1 - \exp(-k_2' t)\} \qquad (69)$$

In Eq. 69, k_1' and k_2' are the conditional pseudo-first-order rate constants. If these parameters, plus the heats ΔH_1 and ΔH_2 are known, Eq. 69 can be solved for the concentrations of B_1 and B_2 by picking two points off the plot of ΔT versus t. Logarithmic extrapolation has also been used to analyze the data (98, 60). Binary mixtures can thus be resolved by a single experiment.

4. Enzymatic Reactions

The increased availability of pure enzymes in the past decade has opened the door for a new subfield of enthalpimetry. The universal property, enthalpy change, is advantageously combined with the inherent selectivity of enzyme reactions in biochemical and clinical applications. The use of kinetic DIE as a method of enzyme assay was discussed in Section IV.A.2 (95). A fixed-time integral approach to enzyme assay has recently been reported (50). In this approach, the unknown amount of enzyme is combined with a known amount of substrate and the reaction allowed to proceed for a fixed time (30 min) under conditions of zero-order kinetics. The remaining substrate is then determined by injecting a large amount of enzyme and monitoring the magnitude of the ensuing temperature change. From the change in substrate concentration over the fixed time (30 min), the original enzyme activity is readily calculated. This indirect method attains better sensitivity than direct kinetic DIE. However, for routine clinical enzyme assays, enthalpimetry evidently cannot compete with spectrophotometry on considerations of sensitivity and sample throughput.

A number of applications of enzyme catalyzed reactions for the determination of substrates (e.g., glucose, glycerol, cholesterol, etc.) have been developed in conjunction with DIE (Section VI.B). An example is the determination of glucose in serum or whole blood by the phosphorylation reaction:

$$\text{glucose} + \text{Mg(ATP)}^{2-} \underset{\text{hexokinase}}{\rightleftharpoons} \text{glucose-6-phosphate}^{-2} + \text{Mg(ADP)}^{-} + \text{H}^{+}$$
$$\Delta H = -27.6 \text{ kJ/mol} \qquad (70)$$

The reaction is allowed to go to virtual completion and the total heat evolved is correlated to glucose concentration. Enzyme activities are, of course, pH dependent, so that buffered solutions are required. McGlothlin and Jordan (94) used tris(hydroxymethyl)aminomethane (THAM) as a buffer (pH 8.0) for reaction 70. The release of protons by that reaction was coupled to the protonation of the buffer,

$$C(CH_2OH)_3NH_2 + H^+ \rightleftharpoons C(CH_2OH)_3NH_3^+$$

$$\Delta H = -47.3 \text{ kJ/mol} \qquad (71)$$

which nearly tripled the overall heat of reaction and is an excellent example of buffer amplification. The coupled buffer reaction also helps drive equilibrium 70 toward completion. Buffer amplification or use of other types of coupled reactions is generally applicable and advantageous in enzymatic DIE.

An important consideration when using enzymes is their inhibition by various contaminants. This shortcoming has been turned into virtue by some very sensitive methods for the determination of the inhibitors themselves (11, 51).

V. FLOW ENTHALPIMETRY

Leonard Skeggs has been called the Henry Ford of analysis because of his revolutionary concept of continuous flow analysis (116). Classical chemical analysis uses flasks, test tubes, cuvettes, etc., to carry out reactions and subsequent measurements (spectral absorbance, potential, etc.) on one sample at a time. Skeggs introduced an assembly-line approach to such analysis. Sample and reagent are combined in flowing streams and the physical measurement is made in a flow-through type cell. Thus, both sample handling and measurement are automated. A considerable increase in sample throughput is achieved with good reliability and accuracy. The Technicon Auto-Analyser and other commercial instruments incorporate these principles and have been widely accepted in clinical and industrial laboratories.

Analysis by measurement of heat of reaction is equally suitable for flow-type sample handling, with a corresponding decrease in analysis time. This represents a significant advance in the development of enthalpimetry. Conceptually, flow enthalpimetry methods are little more than a dynamic, automated form of DIE. Sample is combined with excess reagent and the corresponding temperature change measured in the flowing stream. A single instrumental output ΔT is obtained which is correlated to instantaneous sample concentration.

A. CONTINUOUS FLOW

In 1965 Priestley et al. (101) introduced the method termed continuous-flow enthalpimetry (CFE). A block diagram of the apparatus is shown in Fig. 13a. The sample and reagent streams were pumped at identical rates by a dual

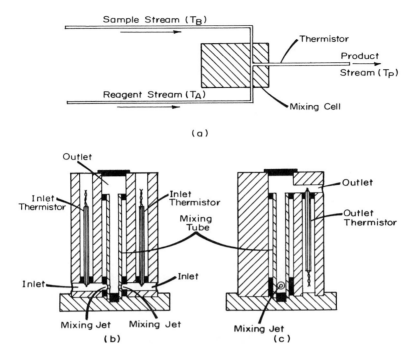

Fig. 13. Apparatus for continuous flow enthalpimetry: (a) block diagram of flow system; (b) front cross section of mixing cell; (c) side cross section of mixing cell. Reprinted from Priestley et al. (101).

channel peristaltic pump. A critical factor in the implementation of flow enthalpimetry is provision for rapid mixing of reagent and sample, immediately followed by temperature measurement of the product stream. This is necessary to maintain nearly adiabatic conditions (heat losses to the environment are negligible over a short time scale). The design of the mixing cell is thus very important. Cross sections of Priestley's reaction vessel, made of a suitable plastic material, are shown in Fig. 13b and c. Rapid tubulent mixing is ensured by introduction of sample and reagent in opposing jets.

Assuming negligible heat losses to the environment, the following heat balance equation is readily derived:

$$\phi_B c_B \, \Delta H_r + \bar{\kappa} \phi_A T_A + \bar{\kappa} \phi_B T_B = \bar{\kappa} (\phi_A + \phi_B) T_P \qquad (72)$$

where the subscripts A, B, and P denote reagent, sample, and product, respectively, ϕ is the flow rate, and c is the molar concentration. Equation 72 also assumes an excess of reagent, i.e.,

$$\phi_B c_B < \frac{a}{b} \phi_A c_A \qquad (73)$$

One additional assumption implicit in Eq. 72 is that the specific heats $\bar{\kappa}$ of sample, reagent, and product streams are identical. Equation 72 is rearranged to yield

$$T_P - \frac{\phi_A T_A + \phi_B T_B}{\phi_A + \phi_B} = \frac{\phi_B}{\phi_A + \phi_B} \frac{\Delta H}{\bar{\kappa}} c_B \qquad (74)$$

This working equation is analogous to Eqs. 51 and 52 for DIE with the substitution of flow rates for volumes. If all temperature mismatch is eliminated so that $T_A = T_B = T_i$, then

$$T_P - T_i = \frac{\phi_A}{\phi_A + \phi_B} \frac{\Delta H}{\bar{\kappa}} c_B \qquad (75)$$

In Preistley's arrangement for CFE, the flow streams were not thermostatted, but the temperatures T_A and T_B were monitored by placing thermistors in reagent and sample streams, respectively. When the flow rates are equal ($\phi_A = \phi_B$),

$$T_P - \frac{T_A + T_B}{2} = \frac{1}{2} \frac{\Delta H}{\bar{\kappa}} c_A \qquad (76)$$

A three-thermistor differential bridge can be used to directly monitor the left side of Eq. 76.

Other workers (33) increased the sensitivity of CFE by thermostatting the reagent and sample streams ($T_A - T_B \leq 0.002°C$). By doing so, down to 1 ppm of water in hydrocarbons could be detected via the reaction of water with triethyl aluminum, $\Delta H_r \cong 420$ kJ/mol. Approximately 10 mL of sample is required in CFE to obtain a steady-state temperature reading. An analysis can be performed in less than a minute and the method is suitable for intermittent or continuous on-line monitoring of industrial processes (33, 121).

B.   SEGMENTED FLOW

Preistley's version of continuous flow enthalpimetry is not to be confused with the general method developed by Skeggs, which has also been called continuous flow analysis. In reality, Skegg's method and subsequent commercial instruments utilize a segmented flow. Snyder et al. (119) have reviewed segmented flow methodology in general. The salient distinctive feature is that sample and reagent solutions are moved through the flow system as discrete plugs, segmented by air bubbles.

Hagedorn et al. (54) and Weber et al. (135) have developed an automated thermometric analyzer that relies on the principles of segmented flow

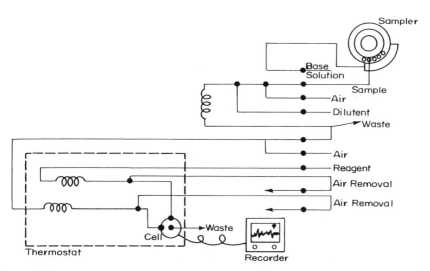

Fig. 14. Schematic diagram of a segmented flow automatic thermometric analyzer. Reprinted from Hagedorn et al. (54).

methodology. A schematic diagram of the instrument, which is commercially available, is given in Fig. 14. Precise amounts of sample and a wash fluid are intermittently aspirated by a peristaltic pump at the sampler and the resulting stream is segmented with air bubbles. The segmentation serves to restrict axial dispersion and carryover of one sample to the next. If needed, the sample is diluted and made to flow through a helical coil to ensure mixing of sample and diluent. The sample stream and a separate reagent stream are thermostatted to within 0.005°C, debubbled, and then combined in a special thermometric flow cell, pictured in Fig. 15. The reagent and sample streams are vigorously mixed by the large-bladed stirrer at 1000 rpm and the resulting temperature change is sensed by a single thermistor affixed to the stainless steel cap of the inner wall of the cell. The temperature change is measured as a Wheatstone bridge unbalance potential in the usual manner and recorded on a strip chart recorder, as shown in Fig. 16. Each sample yields a peak whose height is empirically correlated to sample concentration by linear calibration plots.

Ideally, the peak heights are described by Eq. 75. However, the thermometric flow cell (Fig. 15) is designed for fast and reproducible response, not adiabaticity. Therefore, the so-called steady-state temperature change recorded for a given sample may not truly represent the heat of reaction. Also, a kinetically slow reaction may not reach equilibrium and completion during an average residence time in the thermometric flow cell ($\cong 20$ s). However, if experimental conditions such as sample size and flow rate are constant, the reaction will proceed to an equal and reproducible extent for samples and calibration standards alike, and linearity will be maintained.

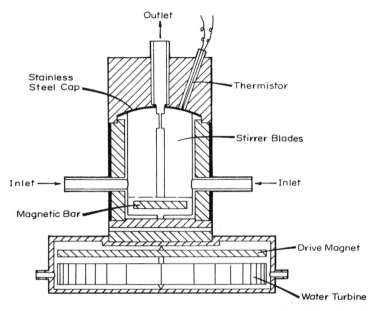

Fig. 15.   Flow cell for segmented flow thermometric analyzer. Reprinted from Hagedorn et al. (54).

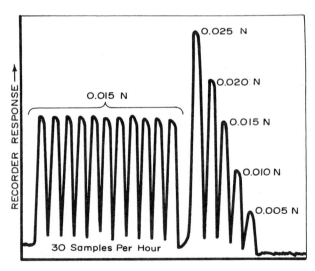

Fig. 16.   Readouts from segmented flow thermometric analyzer for the determination of HCl (concentrations as labeled) using 0.5 M NaOH as reagent. Reprinted from Hagedorn et al. (54).

TABLE 5
Selected Examples of Determinations with the Segmented Flow
Automatic Thermometric Analyzer

Determined Moiety	Reaction Employed	Reagent	Nature of Analyzed Material	Dynamic Range	Precision
Ammonia	Conversion to hexamethylenetetramine	Formaldehyde	Fertilizer	5–7% N_2	0.02%
Magnesium	Precipitation of ammonium magnesium phosphate	$(NH_4)_3PO_4 + NH_4OH$	Fertilizer	0–8%	0.04%
K_2O	Precipitation of potassium perchlorate	Perchloric acid	Fertilizer	46–63%	0.2%
Chloride	Precipitation of silver chloride	Silver nitrate	Kieserite $(MgSO_4 \cdot H_2O$ concentrates	0–4%	0.3%
Sulfate	Precipitation of barium sulfate	Barium chloride	Fertilizer	0–9%	0.1%
Hydrogen peroxide	Reduction of H_2O_2 to H_2O	Iodide	Aqueous solution	0.5–5%	0.02%
Sodium hydroxide	Neutralization with a strong acid	HCl	Aqueous solution	10–30 g/L	0.1 g/L

SOURCE: Based on Hagedorn et al. (54).

The design of the thermometric flow cell requires a minimum sample size of 1 mL. The emphasis for this instrument was on industrial applications where relatively high concentrations are often found. The sensitivity is rather low. For a precision of 1%, 1.2 kJ/L is required (i.e., for $\Delta H_r = 40$ kJ/mol, $c_A > 0.03$ M; see Eq. 11). However, as many as 40 samples per hour can be analyzed automatically. Any type of reaction can be used that has a sufficient heat. Due to the design of the flow cell, precipitation reactions cause no clogging and carryover effects are minimal. A synopsis of representative determinations in a wide variety of materials is given in Table 5, which documents the versatility of the segmented flow thermometric analyzer.

C. FLOW INJECTION

In the mid 1970s, several groups of workers pointed out that the segmentation of the flow streams with air bubbles was not necessary to perform flow-based analyses. Ruzicka and Hansen (109) and Stewart et al. (120) utilized the direct injection of a sample "plug" into a nonsegmented carrier stream. The general theory of flow injection analysis (FIA) and a discussion of some types of detectors and applications have been reviewed (13). FIA appears to have advantages of increased speed, smaller sample size, and simplicity in operation over conventional segmented flow methods. In addition, the controlled dispersion of the sample plug, based on fundamental principles, permits some types of analyses not otherwise possible (103).

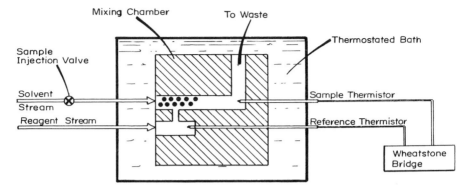

Fig. 17.   Schematic of mixing chamber for peak enthalpimetry. Reprinted from Waugh (133).

Enthalpimetric detection on flow injection systems has been implemented in several configurations. Censullo et al. (25), Jordan et al. (73), and Waugh (133) were the first to report on the method that they called peak enthalpimetry (due to the peak-shaped response curve). Two thermostatted flow streams, a carrier solvent and a reagent, converge in a vibrating mixing chamber packed with glass beads, as shown in Fig. 17. A valve in the solvent stream allows injection (actually intercalcation) of a sample plug, approximately 100 μL in volume. Because of diffusional spreading, the initially cylindrical plug has a Gaussian profile by the time it reaches the reaction chamber. The two streams are combined rapidly and thoroughly in the mixing chamber, which vibrates at 60 Hz. The resulting heat of reaction is sensed as a temperature change by a differential thermistor bridge, which measures the temperature difference between the effluent stream and the thermostatted entering reagent stream. If the reaction proceeds virtually to completion, the measured temperature change at any time is proportional to the instantaneous concentration of the sample. Because the temperature sensor is located immediately after the mixing area, heat losses to the environment are negligible. The corresponding working equation for peak enthalpimetry is

$$\int \Delta T(t)\, dt = \left(\frac{\phi_B}{\phi_A + \phi_B} \right) \left(\frac{-\Delta H}{\bar{\kappa} v} \right) n_B \tag{77}$$

Integration of the temperature–time profile yields the analytically significant information, viz., moles of sample, n_B, provided the other factors in Eq. 77 are constant.

The analytical potential of peak enthalpimetry has been documented by the determination of chloride in human serum (25, 133). The classical

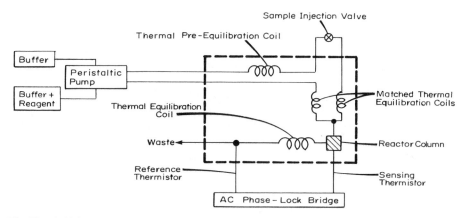

Fig. 18. A high-sensitivity flow injection thermochemical analyzer. Reactor column and thermal equilibration coils are immersed in a thermostatted water bath. Reprinted with permission from Schrifreen et al. Copyright 1979, American Chemical Society.

reaction of chloride with mercuric nitrate was used:

$$2Cl^- + Hg^{2+} = HgCl_2 \qquad \Delta H_r^\circ = -56.1 \ kJ/mol \qquad (78)$$

A one-molar aqueous solution of mercuric nitrate was employed as the reagent stream. The solvent stream was pure water. Chloride was successfully determined in the range of 50 to 180 meq/L with a precision and accuracy of 1%. This more than encompasses the clinically significant range of chloride concentrations in serum. The pertinent reference (25) concludes that "by recycling a sufficiently concentrated mercuric nitrate reagent stream the seriatim performance of several hundred chloride analyses is evidently feasible. The technology for automatic injection and digital printout is readily available."

Schifreen et al. (113) made refinements on the methodological principles of peak enthalpimetry to develop a high-sensitivity flow injection thermochemical analyzer. Fig. 18 displays a block diagram of the flow system. The main differences compared to the peak enthalpimeter are in the reaction cell and the temperature-sensing system. An insulated packed column was used to mix the reagent and sample streams. An ac phase-lock Wheatstone bridge recorded the differential temperature between a sensing thermistor positioned in the adiabatic reactor column near the outlet and a reference thermistor positioned downstream from the reactor in the effluent stream. The effluent stream was reequilibrated to the water bath temperature prior to measurement. This arrangement was used because the response of a thermistor was found to be dependent on the net flow around the sensing thermistor bead. Sensing and reference thermistors are exposed to exactly the same flow rate, which was necessary for a stable baseline.

Three well-characterized chemical reactions were tested in this flow enthalpimeter, which documents its versatility: the protonation of THAM [tris(hydroxymethyl)aminomethane], the complexation of calcium by EGTA [ethylene bis(oxyethylenenitrilo)tetraacetic acid], and the oxidation of nitrite by sulfamic acid. Precipitation reactions are not feasible in this analyzer due to clogging of the mixing column. The readouts consisted of peaks, whose height and integral both were linearly related to sample concentration over a wide range. The peaks did exhibit appreciable tailing and non-Gaussian shapes. The reported sensitivity is quite good, being about 4 J/L (10^{-5} M nitrite for the nitrite/sulfamic acid reaction, which has a $\Delta H = 400$ kJ/mol; see Eqs. 10 and 11).

Elvecrog and Carr (39) have also adapted this same flow enthalpimeter to determine traces of iodide. A cerium(IV) reagent stream was mixed with an arsenic(III) carrier stream into which samples of iodide (the catalyst in the Sandell–Kolthoff reaction 39) had been injected. After the streams are combined, the reaction proceeds for a precisely reproducible time in the reactor column before temperature sensing. Under proper experimental conditions, peak heights were linearly related to iodide (the catalyst) concentration over a range of 10^{-8} to 10^{-5} M, because the rate of reaction 39 was proportional to $[I^-]_0$. This is a remarkable sensitivity for an enthalpimetric determination and represents an ingenious combination of catalytic and flow methodologies.

D.   RELATED ENZYMATIC METHODS

The usefulness of enzyme specificity for analysis has already been discussed. One drawback to implementation of enzyme-catalyzed reactions for routine analysis is the costly consumption of enzyme. Much interest has focused on technology for immobilizing or entrapping enzymes on solid supports and on subsequent analytical methods utilizing immobilized enzymes. The combination of immobilized enzymes with flow injection enthalpimetric analysis (17) results in a means for rapid, selective analysis, which is not prone to matrix interferences.

Such a system has been described by Canning and Carr (20) for the determination of urea (Fig. 19). It actually preceded the high-sensitivity flow injection thermochemical analyzer (39). Reaction 79 proceeds very slowly in the absence of the specific catalyst urease, so that the sample could be injected directly into the reagent stream:

$$NH_2CONH_2 + H_2O \xrightarrow{\text{urease}} CO_2 + 2NH_3 \qquad \Delta H^0 = -55.10 \text{ kJ/mol}$$

$$(79)$$

The single thermostatted stream passed through a column, packed with controlled porosity glass on which urease had been immobilized. Enough enzyme was used to ensure virtually total hydrolysis of urea in the largest sample expected. A single thermistor, positioned near the column outlet, was

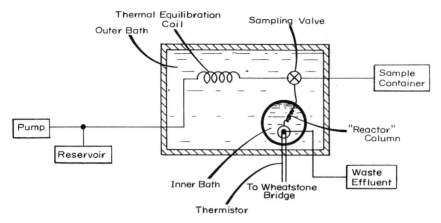

Fig. 19.   Flow injection analysis for enzymatic reactions. The reactor column contains immobilized enzyme. Outer bath is thermostatted. Reprinted from Canning and Carr (20).

used to detect the heat evolved by the reaction in the column. Peak heights
were proportional to urea concentration over a range of 1–100 mM.

A similar device has been developed by Mosbach and Danielsson (96, 97).
The final version of their "enzyme thermistor" (97) was virtually identical in
principle and design to the device developed by Canning and Carr. A
refinement added later was the split-flow enzyme thermistor (91). Two
columns were used, one with enzyme, one without, and the stream containing
the as yet unreacted sample was evenly split between them. The differential
temperature at the outlets of the two columns was used as the measure of
substrate concentration. The advantage of this arrangement was to cancel out
extraneous effects caused by the sample plug alone (for example, a change in
viscosity will alter the flow pattern around the thermistor and thus affect its
temperature response characteristics).

Another novel idea, developed by Mattiasson (87) was to utilize an
antibody–antigen immunological interaction to immobilize interchangeable
enzymes in the flow column. The concept is similar to that employed in
thermometric enzyme-linked immunosorbant assay (TELISA), Section
VI.B.3.d. The column contains an immobilized antibody (antihuman serum
albumin). An enzyme that is cross-linked with the relevant antigen is perco-
lated through the column where they are bound to the antibody. The column
is then ready for use to determine a specific substrate by flow injection
enzymatic enthalpimetry. The enzyme/antigen pair can be removed from the
column by a glycine wash and replaced with a different enzyme/antigen pair.

The heat produced by immobilized enzyme reactions has had analytical
applications in configurations other than flow systems (134, 123, 107). Weaver
et al. (134) pioneered in the development of a thermal enzyme probe (TEP).
The enzyme was immobilized directly onto the thermistor bead using a

special glue. This measuring thermistor and a similarly treated reference thermistor (without enzyme) were incorporated in two arms of a differential Wheatstone bridge. Both thermistors were immersed in a well-stirred solution containing the appropriate reagents and buffers. After the establishment of a steady baseline, the substrate (sample) was introduced. The ensuing reaction at the TEP surface engendered a measurable temperature change.

A simplified treatment of TEP theory (134) is instructive in defining relevant experimental parameters. The reaction velocity v at the TEP is given by a modified Michealis–Menton equation:

$$v = \frac{v_{max}C_s'}{K_m + C_s'} \tag{80}$$

where v and v_{max} are expressed as rate per unit area and C_s' is the substrate concentration at the TEP. If $C_s' \ll K_m$, first-order kinetics will prevail at the TEP (see Fig. 12) and Eq. 80 simplifies to

$$v = \left(\frac{v_{max}}{K_m}\right)C_s' \tag{81}$$

The flux of substrate toward the TEP, J_s, is given by

$$J_s = \frac{D_s(C_s - C_s')}{l} \tag{82}$$

where D_s is the diffusion coefficient and l is the thickness of a one-dimensional diffusion layer. The heat flux J_Q away from the TEP is similarly given by

$$J_Q = \frac{\lambda_w(\Delta T)}{l} \tag{83}$$

where λ_w is the thermal conductivity of water. The parameter l is governed by the particular stirring conditions prevailing at the TEP. When these various rate processes attain a steady state, the following will be true:

$$v = J_s \tag{84}$$

$$v|\Delta H| = J_Q \tag{85}$$

Combining Eqs. 81–85 and solving for ΔT_{ss} (the steady-state temperature difference) yields

$$\Delta T_{ss} = \frac{(D_s/l)(v_{max}/K_m)}{(D_s/l + v_{max}/K_m)}\left(\frac{l|\Delta H|}{\lambda_w}\right)C_s \tag{86}$$

If $\nu_{max}/K_m \gg D_s/l$, then diffusion becomes the limiting factor, and Eq. 86 simplifies to

$$\Delta T_{ss} = (D_s|\Delta H|/\lambda_w)C_s \qquad (87)$$

which is independent of small changes in stirring parameters and reaction rate. Substituting typical values of $D_s = 10^{-5}$ cm^2/s, $\Delta H = 50$ kJ/mol, and $\lambda_w = 6$ mW/cm K into Eq. 87 indicates that 1.2 mM substrate is necessary to cause a ΔT_{ss} of 10^{-4} °C between bulk solution and thermistor surface. A direct proportionality between ΔT and D_s is indicated by Eq. 87. Weaver et al. (134) used such a TEP coated with glucose oxidase and also one coated with hexokinase for the determination of glucose.

Another type of enzyme bound thermistor (123) is pictured in Fig. 20a. In this case, the enzyme was cross-linked together with albumin around the thermistor bead using glutaraldehyde. A perforated glass shield was affixed around the thermistors to minimize the convective heat loss. Various immobilized enzyme/substrate combinations were successfully employed, e.g., glucose oxidase for glucose, catalase for hydrogen peroxide, and urease for urea. Concentrations down to 3 mM could be determined with a precision of 5% (with better precision at higher concentrations).

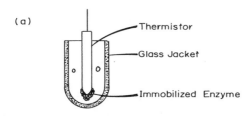

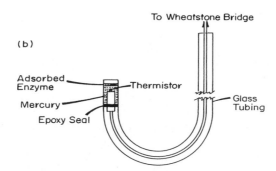

Fig. 20. Design features of enzymes immobilized on thermistors. (a) Enzyme bound thermistor (EBT) with enzyme immobilized by cross-linked coating. The glass jacket shown was perforated to permit access to ambient solutions. (b) Thermistor enzyme probe (TEP) with enzyme immobilized by adsorption on mercury. Reprinted with permission from Tran-Minh and Vallin (123) and Rich et al. (107), respectively. Copyrights 1978 and 1979, American Chemical Society.

One other novel and radically different thermistor enzyme probe has been reported (107). It consisted of an encapsulated thermistor sealed in a U-shaped glass tube, as shown in Fig. 20b. A thin layer of mercury covered the thermistor. The sensing thermistor was first immersed in a solution of urease, which adsorbed onto the mercury surface. (Urease maintains its enzymatic activity while so adsorbed.) A reference thermistor was constructed in the same way, but without adsorbed enzyme. The pair of thermistors were used to record the temperature difference in an unstirred solution containing urea. Because of the high thermal conductivity of mercury (13.6 times that of water) most of the heat of reaction remained localized at the thermistor. The measured ΔT yielded a linear measure of urea concentration. While this adsorbed enzyme configuration is easy to construct, it has the disadvantage of a short lifetime ($\simeq$ 4–8 measurements).

All enzyme-coated thermistor configurations differ from other types of enthalpimetry in that they are essentially a nondestructive technique. In the time required to attain a steady-state temperature difference, only a very small fraction of the substrate actually reacts. Continuing interest in these types of devices is to be expected.

VI. APPLICATIONS OF ENTHALPIMETRIC ANALYSIS

A. DETERMINATION OF STOICHIOMETRIES AND THERMODYNAMIC AND KINETIC PARAMETERS

1. General Considerations

There is no sharp dividing line between enthalpimetric analysis and conventional solution calorimetry. Classical calorimetry has traditionally utilized state-of-the-art instrumentation for very accurate measurement of reaction heats and for the determination of other thermodynamic parameters. Enthalpimetric analysis makes use of the heats of known reactions for determining concentrations. Obviously, these two functions are complementary, and the practicing analytical chemist can have an interest in both. The equipment and methodology described in this chapter have been primarily analytical in nature. However, TET and DIE can be used to acquire thermodynamic and kinetic data. The accuracy, obviously, is dependent on the instrumental sophistication. Even quite simple equipment can be used to obtain good estimates of the heat, equilibrium constant, and rate. In addition, reaction stoichiometries can often be found from the breaks on a TET curve.

2. Determination of Thermodynamic Parameters

The heat of reaction ΔH_r is the parameter most directly accessible. DIE can be used to determine ΔH_r as in classical adiabatic calorimetry. The observed temperature change ΔT (see Fig. 1b) is correlated to the observed

TABLE 6
Comparison of Methods for Evaluating Thermodynamic Parameters
from Thermometric Titration Curves

	ΔH_r	K	Advantages
Initial slope method	Fairly good estimate	No	Rapid and simple; can be used to evaluate multiple-stage titration reactions
Graphical extrapolation method	Good estimate	Rough estimate	Rapid and simple
Simultaneous equations or least-squares methods	Precise evaluation	Precise evaluation	Good accuracy when properly used

heat effect Q by electrical or chemical calibration (Eq. 9). ΔH_r can be evaluated from Q provided the amount of product formed is known independently (i.e., the reagent must be of known concentration and the reaction virtually complete under the experimental conditions).

Every TET experiment is per se a titration calorimetry experiment as well. TET curves can be used to evaluate the stoichiometry ΔH_r and the equilibrium constant K. Basically, three different methods have been used to evaluate thermodynamic data from thermometric titration curves. The methods are listed in Table 6, along with their salient advantages. Details are outlined below.

a. INITIAL SLOPE METHOD

The heat of reaction ΔH_r may be evaluated from the initial slope of the titration region $m_{II} = dT/dt$ of a recorded titration curve (see Fig. 5). Heat of reaction is related to slope via Eq. 88:

$$\Delta H_r = m_{II} \frac{\kappa}{r c_A} \tag{88}$$

By measuring the slope soon after the start of the titration, the nonlinearizing effects of changing heat capacity, titrant temperature mismatch, and cooling heat losses converge toward zero. If the titrant reacts with several titrate species, these nonlinearizing effects are negligible for only the first reaction. Nonetheless, an estimate of ΔH to within a few percent may be obtained from the slopes of subsequent reaction regions on the TET curve. If the TET curve is first corrected for the nonlinearizing background thermal effects, then the slopes provide a quantitative measure of ΔH_r in all regions of the TET curve where the reaction is occurring.

b. GRAPHICAL EXTRAPOLATION METHOD

The excess reagent region (III) of the titration curve may be graphically extrapolated back to the start of the titration t_0 (see Figs. 5 and 21) to determine ΔT and the ΔH_r. This procedure has been extensively analyzed by

Carr (23) and is surprisingly accurate if moderate care is given to experimental conditions. The heat of titrant dilution is approximately subtracted by this extrapolation. To convert ΔT into Q (which is required to calculate ΔH_r via Eq. 5) an electrical calibration curve is needed which is comparable to the TET curve. If the postheating portion of that curve is likewise extrapolated to time zero (when the heater is first turned on in the calibration experiment), heat exchange losses with the environment will be corrected for to a good approximation. The extrapolation method does not account for the change in heat capacity during the titration or for titrant temperature mismatch. When both of these problems are minimized through the use of concentrated titrants, the heat of reaction can be determined with an accuracy of 0.2%. These procedures for the determination of ΔH_r are in practice limited to reactions that attain virtual completion with a twofold excess of titrant.

Both the initial slope and graphical extrapolation methods yield approximate values for ΔH_r and have as major advantages simplicity and speed. These two graphical methods do not provide for the accurate evaluation of the equilibrium constant K. However, the extrapolation method can allow an estimation of K within a factor of two or better. Jordan and Billingham (68) used this method to rapidly accumulate extensive thermodynamic data for precipitation reactions in molten salt media. The procedure is illustrated in Fig. 21. Tangents are drawn to the initial reaction region (II) and the final part of the excess reagent region (III). The implicit assumptions are that the first titrant added reacts completely to form product, due to the excess of titrate, and that the titrate is completely reacted with a reasonable excess of

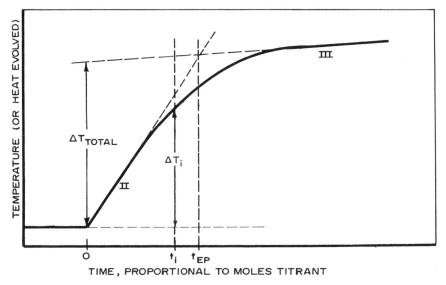

Fig. 21. Thermometric titration curve illustrating the graphical extrapolation method for estimating the equilibrium constant of a reaction.

titrant. By extrapolating from these regions of the curve where the reaction is nearly going to completion, the tangents represent the ideal case of $K = \infty$. The actual equilibrium constant K is evaluated by measuring the actual temperature change ΔT_i at selected points, and applying the following equations (which are derived for the reaction, titrant (A) + titrate (B) $\rightleftharpoons$ P):

$$\Delta H_r = \frac{\Delta T_{\text{total}} \kappa}{n_A} \tag{89}$$

$$[P]_i = \frac{n_{pi}}{v_i + v_0} \frac{\Delta T_i \kappa}{\Delta H_r (rt_i + v_0)} \tag{90}$$

$$[A]_i = \frac{n_{Ai} - n_{pi}}{(rt_i + v_0)} = \frac{t_i r C_A - n_{pi}}{(rt_i + v_0)} \tag{91}$$

$$[B]_i = \frac{n_{B0} - n_{pi}}{rt_i + v_0} \tag{92}$$

$$K = \frac{[P]_i}{[A]_i [B]_i} \tag{93}$$

In these equations, v_0 is the initial volume of titrate and n_{B0} is the initial number of moles of titrate.

c. Simultaneous Equations or Least-Squares Methods

The extrapolation method becomes less and less accurate as the curvature becomes more pronounced (i.e., small K value). To apply the third method for accurate evaluation of both ΔH_r and K, a rigorous mathematical treatment of the data is necessary. Of course, sophisticated data treatment is useless if the data are not of high quality. All background thermal effects must be carefully controlled and appropriate corrections applied to yield a plot of Q versus time (or volume or moles of titrant). The theoretical foundations for such corrections were given in Section III.A. Their implementation has been described in detail (26). The general approach is to break the titration curve down into a series of discrete points and apply a point-by-point correction to the data. Because the corrections and subsequent thermodynamic calculations are rather tedious, computerized data acquisition and treatment are almost essential.

For the rigorous evaluation of corrected thermometric titration data, Eqs. 90–93 are applicable. Equation 89 does not apply because no assumptions are made about regions where the reaction is complete. These basic equations may be combined into Eq. 94, which describes any point i on the curve:

$$K = \frac{(Q_i/\Delta H_r)(v_0 + rt_i)}{(n_{B0} - Q_i/\Delta H_r)(c_A rt_i - Q_i/\Delta H_r)} \tag{94}$$

In this equation, which is analogous to Eq. 8, n_{B0}, v_0, r, and c_A are known constants, (Q_i, t_i) are the data points (variables), and K and ΔH_r are the unknown parameters. By selecting two points from the Q versus t_i curve, (Q_1, t_1) and (Q_2, t_2), two equations in two unknowns can be solved for K and ΔH_r, using an appropriately rearranged form of Eq. 94. Christensen et al. (31) were among the first to use simultaneous solution of equations (SSE) for the interpretation of a continuous thermometric titration curve. By selecting several pairs of data points and averaging the resulting K and ΔH_r values, a greater portion of the titration curve is utilized and some smoothing of random noise is gained.

A more efficient use of the data and greater precision may be gained by a least-squares approach to solving for the parameters K and ΔH_r. Such an approach has been adopted by Christensen et al. (26) in lieu of SSE and has also been used by Sillen et al. (115) and Arnek (1). The general principle is to minimize the error function $U(K, \Delta H_r)$, which is the sum of the squares of the differences between the experimental points and the calculated points resulting from the selected values of the parameters:

$$U(K, \Delta H) = \sum_{i=1}^{m} w_i (Q_i - Q_{calc})^2 \qquad (95)$$

where m is the number of data points and w_i is a weighting function. Slightly different numerical methods have been used to arrive at the best values of K and ΔH_r, and the interested reader is referred to the relevant literature for details.

3. Determination of Kinetic Parameters

Direct injection enthalpimetry often provides a simple direct way to evaluate reaction rates. Application of DIE is particularly straightforward, because each point (T, t) is a direct measure of amounts reacted. The experimental readout T versus t is readily converted to Q versus t and replotted as $\log Q$ versus t. The slope of the latter yields the rate constant k_f for a first-order (or pseudo-first-order) reaction. This use of DIE is discussed in Henry (60) and Papoff and Zambonin (98). DIE can be used to study processes that have half-lives of 3 s to 30 min. Biochemical rate parameters, such as Michaelis constants, that have been determined by DIE and other calorimetric methods have been reviewed (47, 85). Calorimetric methods are also amenable to qualitative study of processes occurring in cells, tissues, bacteria, or complex matrices.

Although little exploited, titration calorimetry also can be used for measuring rate parameters. Carr and Jordan (24) have developed the foundations of kinetic thermometric titration and studied several systems. In kinetic TET, the continuous titrant addition serves as a competing rate process with the rate of reaction. The rate constant is evaluated from the shift in the apparent endpoint (X in Fig. 6). The methodology apparently has capabilities comparable to stopped-flow kinetic measurements.

B. ANALYTICAL USES

In the 30 years since thermometric titration became a viable analytical technique, numerous researchers have explored the application of the methodology to countless reactions, both classical and novel. Use of the heat of reaction simplifies many classical procedures and has made possible many new ones.

A comprehensive listing of all analytical uses of enthalpimetric analysis is beyond the scope of this chapter. The monograph by Vaughan (128) provides a thorough listing of inorganic and organic applications through the early 1970s. Jordan et al. (69) have tabulated some newer applications through 1975. Some selected applications more recent than that are included in the remainder of this chapter. However, the purpose of this section is not to be a comprehensive updating, but to survey selected examples that illustrate the range of application of enthalpimetric analysis. Biochemical and clinical assays are well covered in recent reviews by Martin and Marini (85) and Grime (47).

1. Inorganic Analysis

a. BRØNSTED ACIDS AND BASES

Acid–base reactions have been in the forefront of titrimetric methods since their inception. The neutralization of a strong acid with a strong base has an exothermic heat of approximately 56 kJ/mol. This fairly large heat allows TET to be nearly competitive with potentiometric titrations for accuracy and sensitivity. It is, however, in the titration of weak acids or weak bases that thermometric titration holds a distinct advantage, being a linear titration method. This advantage has already been discussed in Section I.C using the example of boric acid. TET also will, in many cases, provide separate endpoints for the various components of a mixture of weak acids or bases, or for polyprotic species.

Theoretical aspects of acid–base thermometric titration have been discussed by Barthel (10). Specific analyses are reviewed by Vaughan (128). A valuable compilation of proton transfer thermodynamic data is available (28).

b. LEWIS ACIDS AND BASES

The adduct-forming reactions of Lewis acids and bases have found extensive use in thermometric titrations in nonaqueous solvents. For example, pyridine and di-n-butyl ether are Lewis bases (electron donors) toward trialkyl aluminum compounds, AlR_3 (R-alkyl). Dialkyl aluminum hydrides, AlR_2H, do not react or interfere. This is the bais for a determination of commercial aluminum alkyls (41).

Complexation reactions in aqueous solutions can also be regarded as Lewis acid–base interactions. Ethylenediaminetetraacetic acid (EDTA) has been extensively studied and used as a complexation reagent for the analysis of metals in all forms of volumetric analysis, including thermometric titration. Table 7 adapted from Vaughan (128) (with the addition of a new value in column vii) provides a compilation of thermodynamic data for EDTA/metal ion reactions. The feasibility of resolving a mixture of calcium and magnesium by thermometric titration with EDTA was outlined in Section III.B.3. Ligand substitution methods have been used to improve selectivity (37). Other complexation reagents for metal ions have been used, including di(2-aminoethoxy)ethane tetraacetic acid (EGTA) (18) and tartrate (64).

Of considerable interest and importance are the macrocyclic ligands, crown ethers and cryptands. A group at Brigham Young University has extensively studied the complexation thermodynamics of these ligands (62). Thermometric titration with Ba^{2+} has been found to be the best way to standardize solutions of crown ethers (78).

In general complexation reactions in aqueous solution suffer from the drawback of having somewhat small heats (usually less than 30 kJ/mol).

c. DETERMINATIONS BY PRECIPITATION REACTIONS

Enthalpimetric analysis provides a rapid, convenient, and direct means to use precipitation reactions for chemical analysis, since the presence of the precipitate does not interfere in the temperature measurement (as it does, e.g., in spectrophotometric titrations). The classical reactions of silver nitrate with halides (and the pseudo-halides cyanide and thiocyanate) have been used in both TET and DIE, as has the classical precipitation of barium sulfate. Reactions 96 and 97 have been used for the determination of sodium (111) and potassium (21), respectively:

$$Na^+ + K_2AlF_6^- \rightleftharpoons NaK_2AlF_{6(s)} \tag{96}$$

$$K^+ + B(C_6H_5)_4^- \rightleftharpoons KB(C_6H_5)_{4(s)} \tag{97}$$

Because the nucleation of precipitates in dilute, but supersaturated, solutions is often a slow process, the TET methodology is usually limited to higher concentrations (> 5 mM) for precipitation reactions. Mixed solvents, such as 50% ethanol/water, and seeding techniques (adding some precipitate moiety before the determinative reaction) have been used to speed nucleation. Alternatively, the DIE methodology has been used, which is less affected by sluggish kinetics. The precipitation reactions of metal sulfides have been successfully used in enthalpimetric procedures. H_2S has been employed as a gaseous reagent for the determination of metals by DIE (9).

Pioneering work with precipitation reactions in molten salts at temperatures of 105–200°C illustrates the range of application of enthalpimetric analysis (63, 71).

TABLE 7
Thermodynamic Data for Metal Ion–EDTA Chelate Formation
$M^{n+} + EDTA^{4-} \rightleftharpoons [M\ EDTA]^{(4-n)-}$

Cation	Stability Constant log K	$\Delta H^{0\prime}$ Values, kJ/mol at 20–25°C						
		(i)	(ii)	(iii)	(iv)	(v)	(vi)	(vii)
Fe^{3+}	25.1	—	—	—	—	—	—	−11.5
In^{3+}	25.0	—	—	—	—	−30.3	—	
Hg^{2+}	21.8	—	—	—	—	—	−79	
Sn^{2+}	—	(−)nq	—	—	—	—	—	
Cu^{2+}	18.8	−35	−34	−36	−34	−36.3	−34	
Ni^{2+}	18.5	−30	−32	−35	−31	−34.9	−32	
Y^{3+}	18.1	—	—	—	—	−1.3	—	
Pb^{2+}	18.0	−53	−55	−59	−54	−55.2	—	
Gd^{3+}	17.0	—	—	—	—	−4.6	—	
Cd^{2+}	16.5	−41	−38	−42	−38	−42.2	−38	
Zn^{2+}	16.5	−19	−19	−23	−19	−23.5	−21	
Nd^{3+}	16.5	—	—	—	—	−12.5	—	
Al^{3+}	16.1	+46	—	—	—	+52.6	—	
Co^{2+}	16.1	−15	−17	—	−18	−18.4	−18	
Ce^{3+}	16.0	—	—	—	—	−10.2	—	
Fe^{2+}	14.4	—	—	—	—	—	−17	
UO_2^{2+}	—	(+)nq	—	—	—	—	—	
Mn^{2+}	13.8	−27	−22	—	—	−22.8	−19	
Ca^{2+}	10.7	−23	−24	−27	−24	−27.0	—	
Be^{2+}	—	+9.6	—	—	—	—	—	
Mg^{2+}	8.7	+20	+13	+13	+23	+13.1	—	
Cr^{3+}	—	+31	—	—	—	—	—	
Sr^{2+}	8.6	−20	−18	−17	—	−17.2	—	
Ba^{2+}	7.8	−19	−21	−20	—	−20.2	—	
Ag^+	7.3	—	—	—	—	—	—	
NH_4^+	—	+20	—	—	—	—	—	
Li^+	2.8	+9.2	+0.4	—	—	—	—	
K^+	—	(+)nq	—	—	—	—	—	
La^{3+}	—	−16	—	—	—	—	—	
Na^+	—	—	−5.8	—	—	—	—	

Note. (i) Priestley, P. T., et al., *Proc. Soc. Anal. Chem.* **3**, 17 (1966);
(ii) Charles, R. G., *JACS* **76**, 5854 (1954);
(iii) Care, R. A., and L. A. Stonely, *J. Chem. Soc.* 1956, 4571;
(iv) Jordan, J., and T. G. Alleman, *Anal. Chem.* **29**, 9 (1957);
(v) Sillen, L. G., and A. E. Martell, Stability Constants, The Chemical Society, London, Special Pub. 17 (1964);
(vi) Brunetti, et al., *J. Am. Chem. Soc.* **91**, 4680 (1969);
(vii) Doi, K., *Talanta* **25**, 97 (1978);
nq value not quoted, only sign.
SOURCE. Adapted from Vaughan (128).

d. REDOX METHODS

Oxidation–reduction reactions are generally quite exothermic, sometimes as much as several hundred kilojoules per mole. This makes them very desirable for use in enthalpimetric analysis on considerations of sensitivity. Many such reactions, however, are relatively slow to reach equilibrium, making them unsuitable for thermometric titration. DIE has been successfully applied in such instances. Many common oxidants have been used, such as cerium(IV), dichromate, permanganate, hydrogen peroxide, and N-bromosuccinimide (8).

2. Organic Functional Group Analysis

a. ACIDS

Thermometric titrations of organic acids with a strong base have been widely carried out and a considerable body of data is available on these reactions in both aqueous and nonaqueous media (128, 28). Amino acids, phenols, nucleotides and nucleosides, and sulfonamides, as well as carboxylic acids, have been determined by thermometric titration. Titrations using neutralization reactions in nonaqueous solvents can be carried out with normal or catalytic (see below) endpoint indication.

b. AMINES

Compounds containing a basic nitrogen can be titrated with strong acid to a thermometric endpoint. A mixture of aliphatic and aromatic amines (e.g., pyridine) yields separate endpoints in a TET experiment, due to the considerable difference in K and ΔH_r values for the two classes of amines. This was the basis of an analysis of the basic components of tar products, an otherwise difficult determination (130). Other reactions, such as diazotization with nitrous acid can be used for the determination of the amine functional group (60).

c. ALCOHOLS

Terminal OH groups in polymers of industrial importance have been determined by DIE (74, 14). The reaction utilized was esterification with acetic anhydride in a dry organic solvent.

Alcohols and other oxygen-containing impurities in hydrocarbons have been determined by reaction with triethyl aluminum in a continuous flow enthalpimeter (33).

d. THIOLS

A method for the determination of cysteine and other thiols by DIE has been developed (99). The reaction used was the selective addition of the thiol

(RSH) to the activated double bond of N-ethylmaleimide (NEM):

$$RSH + CH\colon CHCONCOC_2H_5 \rightleftharpoons RSCHCH_2CONCOC_2H_5 \quad (98)$$

e. Surfactants

The wide scope of applicability of enthalpimetric analysis is documented (72) by a method for the assay of alkyl benzene sulfonate (ABS) detergents. Thermometric titration curves with a well-defined endpoint and an apparent heat of -24 kJ/mol were obtained by titrating an aqueous solution of 2-dodecylbenzenesulfonate, $CH_3(CH_2)_{11}\phi SO_3^-$ ($=$ an), with benzyldimethyl-(octylphenoxyethoxyethyl)ammonium, $[CH_3C(CH_3)_2CH_2C$ $(CH_3)_2\phi O(CH_2)_2O(CH_2)_2N(CH_3)_2CH_2\phi]^+$ ($=$ cat). In a range of concentrations between 0.001 and 0.2 M, ABS was determined with a precision of 0.5% and an accuracy of 1%, which is comparable to other methods. The thermometric endpoint corresponds to a theoretical stoichiometric ratio of $1\colon 1$ between ABS and titrant. The titration reaction is

$$\frac{x}{n}(AN)^{n-}_{(aq)} + \frac{x}{m}(CAT)^{m+}_{(aq)} = (an)_x(cat)_x(s) \quad (99)$$

where the symbols $(AN)^{n-}$ and $(CAT)^{m+}$ denote aquated anionic and cationic surface-active micelle ions. The product is a colloidal precipitate or inactive complex.

3. Biochemical and Clinical Applications

a. Protein Analysis

Thermometric titration has been used in the study of the prototropic groups of proteins (65, 70). Additional resolution can be obtained by titrating a protein solution with strong base and *simultaneously* monitoring both temperature and pH (83–85). The binding of ligands other than H^+, such as calcium and magnesium, to proteins has been studied calorimetrically (85, 38).

The determination of total serum protein, an important clinical procedure, has been investigated by TET using 12-phosphotungstic acid ($H_3PW_{12}O_{40}$) as titrant (118). Phosphotungstic acid reacts stoichiometrically with N^+ sites on the protein.

b. Enzymatic Assays

The use of DIE as an assay of enzymatic activity was outlined in Section IV.A.2. Enzymes assayed by enthalpimetry include hexokinase (95), serum cholinesterase (53), and horseradish peroxidase (48). The inhibition of en-

TABLE 8
Applications of Enthalpimetric Immobilized Enzyme Flow Reactors

Analyte	Reaction	Enzyme	Reference
Glucose	Glucose + $\frac{1}{2}O_2 \rightleftharpoons$ glucose-δ-lactone + H_2O	Glucose oxidase, catalase	35, 82, 90, 92, 97, 114
Glucose	Glucose + ATP $\rightleftharpoons$ glucose-6-phosphate + ADP	Hexakinase	17
Cholesterol	Cholesterol + $O_2 \rightleftharpoons$ 4-cholesten-3-one + H_2O_2	Cholesterol oxidase	90, 92
Uric acid	Uric acid + $O_2 \rightleftharpoons$ allantoin + CO_2	Uricase	90, 104
Urea	$CO(NH_2)_2 + H_2O \rightleftharpoons CO_2 + 2NH_3$	Urease	16, 20, 34, 36, 82, 97, 114
Penicillin G	Penicillin + $H_2O \rightleftharpoons$ penicilloic acid	penicillinase	97
Lactose	Lactose + $H_2O \rightleftharpoons$ galactose + glucose	Lactase	90, 92
Amylose	Amylose + n-$H_2O \rightleftharpoons$ n-glucose	Amylase	36
Benzoyl-L-arginine ethyl ester ("T")	"T" + $H_2O \rightleftharpoons$ benzoyl-L-arginine + ethanol	Tryspin	97
p-nitrophenyl phosphate	$O_2N\phi OPO_3H^- + H_2O \rightleftharpoons O_2N\phi OH + H_2PO_4^-$	Alkaline phosphatase	36
Oxalic acid	$HOOCCOOH \rightleftharpoons HOOCH + CO_2$	Oxalate decarboxylase	92
Hydrogen peroxide	$H_2O_2 \rightleftharpoons H_2O + \frac{1}{2}O_2$	Catalase	36
Pyruvate	Pyruvate + NADH + $H^+ \rightleftharpoons$ lactate + NAD^+	Lactate dehydrogenase	104
Cyanide	$CN^- + S_2O_3^{2-} \rightleftharpoons SCN^- + SO_3^{2-}$	Rhodanase	93
Cyanide	$CN^- +$ L-cysteine $\rightleftharpoons HS^- + \beta$-cyanoalanine	Injectase	93
Hg(II), Cu(II)	Inhibition of urea hydrolysis	Urease	89

SOURCE: Adapted from Grime (47).

zyme activity by various substances can be used as an assay of the inhibitors. Silver ions (2) and various alkaloids (51) have been determined by their inhibitory effects.

c. Enzyme Substrate Determinations

A considerable number of clinically important substances may be determined by their selective enzyme-catalyzed reaction monitored by DIE or flow enthalpimetry. The general method was outlined for the determination of glucose (94) in Section IV.B.4. Other substrates determined in this manner include cholesterol and cholesterol ester (105), iodide (49), and penicillin (52). Immobilized enzyme methods have also been widely developed. Table 8 taken from Grime (47), provides a comprehensive listing of enthalpimetric flow analysis methods using immobilized enzymes.

d. Immunological Reactions

The feasibility of using the inherent selectivity of antibody–antigen reactions in enthalpimetric analysis was demonstrated by Jespersen (63, 70) in a thermodynamic study of immunological reactions. The analytical potential of these reactions has been further exploited with the development of thermometric enzyme-linked immunosorbent assay (TELISA) (88). The technique uses a packed column flow enthalpimeter similar to that shown in Fig. 19. The column contains immobilized human serum albumin (HSA) antibody. The experimental procedure is analogous to radioimmunoassay. A sample of HSA antigen is mixed with a known amount of labeled HSA antigen. The labeling is accomplished by chemically bonding (linking) an enzyme such as catalase to the antigen (using glutaraldehyde). The mixture is then pumped through the column where competitive binding of the sample HSA and the enzyme-linked HSA occurs with the immobilized antibody. Subsequently, a hydrogen peroxide solution is passed through the column, which is catalytically decomposed by the now immobilized catalase. The temperature pulse recorded is proportional to catalase concentration (because of the fixed reaction time in the column), and thus inversely proportional to the amount of HSA antigen in the sample. The detection limit is 10^{-10} M HSA antigen. While not ultimately as sensitive as radioimmunoassay, considerations of cost and safety may make TELISA an attractive alternative.

ACKNOWLEDGMENT

The authors' relevant research and development work was supported by the U.S. Department of Energy, under Grant DE-FG22-81PC40783.

REFERENCES

1. Arnek, R., *Ark. Kemi* **32**, 81 (1970).
2. Baldridge, J. N., and N. D. Jespersen, *Anal. Lett.* **8**, 683 (1975).
3. Bark, L. S., and S. M. Bark, *Thermometric Titrimetry*, Oxford, UK: Pergamon, 1969.
4. Bark, L. S., and O. Ladipo, *Analyst* **101**, 203 (1976).
5. Bark, L. S., and A. E. Nya, *Anal. Chim. Acta* **87**, 473 (1976).
6. Bark, L. S., and A. E. Nya, *J. Therm. Anal.* **12**, 277 (1977).
7. Bark, L. S., and A. E. Nya, *Thermochim. Acta* **23**, 321 (1978).
8. Bark, L. S., and P. Prachuabpaibul, *Anal. Chim. Acta* **87**, 505 (1976).
9. Bark, L. S., and P. Prachuabpaibul, *Fresenius' Z. Anal. Chem.* **283**, 293 (1977).
10. Barthel, J., *Thermometric Titrations*, New York: Wiley, 1975.
11. Beezer, A. E., and C. D. Stubbs, *Talanta* **20**, 27 (1973).
12. Bell, J. M., and C. F. Cowell, *J. Am. Chem. Soc.* **35**, 49 (1913).
13. Betteridge, D., *Anal. Chem.* **50**, 832A (1978).
14. Bieber, O., G. Degler, A. Pfeffer, H. Schnecko, and W. Weigelt, *J. Appl. Polym. Sci.* **23**, 1043 (1979).
15. Bosson, G., F. Gutmann, and L. M. Simmons, *J. Appl. Phys.* **21**, 1267 (1950).
16. Bowers, L. D., L. M. Canning, C. N. Sayers, and P. W. Carr, *Clin. Chem.* **22**, 1314 (1976).
17. Bowers, L. D., and P. W. Carr, *Clin. Chem.* **22**, 1427 (1976).
18. Boyd, S., A. Bryson, G. H. Nancollas, and K. Torrance, *J. Chem. Soc.* 7353 (1965).
19. Burton, K. C., and H. M. N. H. Irving, *Anal. Chim. Acta* **52**, 441 (1970).
20. Canning, L. M., and P. W. Carr, *Anal. Lett.* **8**, 359 (1975).
21. Carr, P. W., *Anal. Chem.* **43**, 756 (1971).
22. Carr, P. W., *Crit. Rev. Anal. Chem.* **2**, 491 (1972).
23. Carr, P. W., *Thermochim. Acta* **3**, 427 (1972).
24. Carr, P. W., and J. Jordan, *Anal. Chem.* **45**, 634 (1973).
25. Censullo, A. C., J. A. Lynch, D. H. Waugh, and J. Jordan, in J. F. Johnson and R. S. Porter, eds., *Analytical Calorimetry*, Vol. 3, New York: Plenum, 1974, p. 217.
26. Christensen, J. J., D. J. Eatough, and R. M. Izatt, *Thermochim. Acta* **3**, 219 (1972).
27. Christensen, J. J., J. W. Gardner, D. J. Eatough, R. M. Izatt, P. Watts, and R. M. Hart, *Rev. Sci. Instrum.* **44**, 481 (1973).
28. Christensen, J. J., L. D. Hansen, R. M. Izatt, *Handbook of Proton Ionization Heats*, New York: Wiley, 1976.
29. Christensen, J. J., and R. M. Izatt, Thermochemistry in inorganic solution chemistry, in H. A. O. Hill and P. Day, eds., *Physical Methods of Advanced Inorganic Chemistry*, New York: Interscience, 1968, p. 538.
30. Christensen, J. J., R. M. Izatt, and L. D. Hansen, *Rev. Sci. Instrum.* **36**, 779 (1965).
31. Christensen, J. J., R. M. Izatt, L. D. Hansen, and J. A. Partridge, *J. Phys. Chem.* **70**, 2003 (1966).
32. Christensen, J. J., H. D. Johnston, and R. M. Izatt, *Rev. Sci. Instrum.* **39**, 1356 (1968).
33. Crompton, T. R., and B. Cope, *Anal. Chem.* **40**, 274 (1968).
34. Danielsson, B., K. Gadd, B. Mattiasson, and K. Mosbach, *Anal. Lett.* **9**, 987 (1976).
35. Danielsson, B., K. Gadd, B. Mattiasson, and K. Mosbach, *Clin. Chim. Acta* **81**, 163 (1977).
36. Danielsson, B., and K. Mosbach, *FEBS Lett.* **101**, 47 (1979).
37. Doi, K., *Anal. Chim. Acta* **74**, 357 (1975).

38. Eatough, D. J., T. E. Jensen, L. D. Hansen, H. F. Loken, and S. J. Rehfeld, *Thermochim. Acta* **25**, 289 (1978).

39. Elvecrog, J. M., and P. W. Carr, *Anal. Chim. Acta* **21**, 121 (1980).

40. Everson, W. L., *Anal. Chem.* **39**, 1894 (1967).

41. Everson, W. L., and E. M. Ramirez, *Anal. Chem.* **37**, 806 (1965).

42. Goldberg, R. N., E. J. Prosen, B. R. Staples, R. N. Boyd, and G. T. Armstrong, in H. Kambe and P. D. Garn, eds., *Thermoanalytical Investigations by New Techniques*, New York: Wiley, 1975.

43. Greenhow, E. J., *Chem. Rev.* **77**, 835 (1977).

44. Greenhow, E. J., and L. E. Spencer, *Analyst* **98**, 90 (1973).

45. Greenhow, E. J., and L. E. Spencer, *Analyst* **98**, 485 (1973).

46. Greenhow, E. J., and L. E. Spencer, *Anal. Chem.* **47**, 1384 (1975).

47. Grime, J. K. *Anal. Chim. Acta* **118**, 191 (1980).

48. Grime, J. K., and K. R. Lockhart, *Anal. Chim. Acta* **106**, 251 (1979).

49. Grime, J. K., and K. R. Lockhard, *Anal. Chim. Acta* **108**, 363 (1979).

50. Grime, J. K., and E. D. Sexton, *Anal. Chim. Acta* **121**, 125 (1980).

51. Grime, J. K., and B. Tan, *Anal. Chim. Acta* **106**, 39 (1979).

52. Grime, J. K., and B. Tan, *Anal. Chim. Acta* **107**, 319 (1979).

53. Grime, J. K., B. Tan, and J. Jordan, *Anal. Chim. Acta* **109**, 393 (1979).

54. Hagedorn, F., G. Peuschel, and R. Weber, *Analyst* **100** 810 (1975).

55. Hale, J. D., R. M. Izatt, and J. J. Christensen, *J. Phys. Chem.* **67**, 2605 (1963).

56. Hansen, L. D., R. M. Izatt, and J. J. Christensen, Applications of thermometric titrimetry to analytical chemistry, in J. Jordan, ed., *New Developments in Titrimetry*, New York: Dekker, 1974.

57. Hansen, L. D., T. E. Jensen, S. Mayne, D. J. Eatough, R. M. Izatt, and J. J. Christensen, *J. Chem. Thermodyn.* **7**, 919 (1975).

58. Hansen, L. D., and E. A. Lewis, *Anal. Chem.* **43**, 1393 (1971).

59. Hansen, L. D., W. M. Litchman, E. A. Lewis, and R. E. Allred, *J. Chem. Educ.* **46**, 876 (1969).

60. Henry, R. A., Dissertation, Pennsylvania State University, University Park, PA, 1967.

61. Hume, D. N., and J. Jordan, *Anal. Chem.* **30**, 2064 (1958).

62. Izatt, R. M., R. E. Terry, B. L. Haymore, L. D. Hansen, N. K. Dalley, A. G. Avondet, and J. J. Christensen, *J. Am. Chem. Soc.* **98**, 7620 (1976).

63. Jespersen, N. D., Dissertation, Pennsylvania State University, University Park, PA, 1971.

64. Jespersen, N. D., *Anal. Lett.* **5**, 497 (1972).

65. Jespersen, N. D., and J. Jordan, *Anal. Lett.* **3**, 323 (1970).

66. Jordan, J., *Chimia* **17**, 101 (1963).

67. Jordan, J., and T. G. Alleman, *Anal. Chem.* **29**, 9 (1957).

68. Jordan, J., and E. J. Billingham, Chemical and thermodynamic properties at high temperatures, IUPAC, XVIII Congress, Montreal, 1961, pp. 144–148.

69. Jordan, J., J. K. Grime, D. H. Waugh, C. D. Miller, H. M. Cullis, and D. Lohr, *Anal. Chem.* **48**, 427A (1976).

70. Jordan, J., and N. D. Jespersen, in Thermochimie, *Colloq. Int. Cent. Rech. Sci.* **201**, 59 (1972).

71. Jordan, J., J. Meier, E. J. Billingham, and J. Pendergrast, *Anal. Chem.* **32**, 651 (1960).

72. Jordan, J., P. T. Pei, and R. A. Javick, *Anal. Chem.* **35**, 1534 (1963).

73. Jordan, J., J. D. Stutts, and W. J. Brattlie, *Proceedings of the Workshop on the State-of-the-Art of Thermal Analysis*, NBS-SP 580, 149 (1980).

74. Kaduji, I. I., and J. H. Rees, *Analyst* **99**, 435 (1974).

75. Kiba, N., and M. Furosawa, *Anal. Chim. Acta* **98**, 343 (1978).

76. Kiss, T., *Fresenius' Z. Anal. Chem.*, **252**, 12 (1970).

77. Kubaschewski, O., and R. Hultgren, in H. A. Skinner, ed., *Experimental Thermochemistry*, Vol. 2, New York: Wiley Interscience, 1962, p. 351.

78. Lamb, J. D., J. E. King, J. J. Christensen, and R. M. Izatt, *Anal. Chem.* **53**, 2127 (1981).

79. Lee, T. S., and I. M. Kolthoff, *Ann. N.Y. Acad. Sci.* **53**, 1093 (1951).

80. Linde, H. W., L. B. Rogers, and D. N. Hume, *Anal. Chem.* **25**, 404 (1953).

81. Lynch, J. A., Dissertation, Pennsylvania State University, University Park, PA, 1976.

82. Marconi, W., F. Bartoli, F. Morisi, and F. Pittalis, *Int. J. Artif. Organs* **2**, 159 (1979).

83. Marini, M. A., W. J. Evans, and C. J. Martin, *Anal. Lett.* **14**, 707 (1981).

84. Marini, M. A., C. J. Martin, R. L. Berger, and L. Farlani, in R. S. Porter and J. F. Johnson, eds., *Analytical Calorimetry*, Vol. 3, New York: Plenum, 1974, p. 407.

85. Martin, C. J., and M. A. Marini, *Crit. Rev. Anal. Chem.* **8**, 221 (1979).

86. Martin, C. J., B. R. Sreenathan, and M. A. Marini, *Biopolymers* **19**, 2047 (1980).

87. Mattiasson, B., *FEBS Lett.* **77**, 107 (1977).

88. Mattiasson, B., C. Barrebaeck, B. Sanfridson, and K. Mosbach, *Biochim. Biophys. Acta* **483**, 221 (1977).

89. Mattiasson, B., B. Danielsson, C. Hermansson, and K. Mosbach, *FEBS Lett.* **85**, 203 (1978).

90. Mattiasson, B., B. Danielsson, and K. Mosbach, *Anal. Lett.* **9**, 217 (1976).

91. Mattiasson, B., B. Danielsson, and K. Mosbach, *Anal. Lett.* **9**, 867 (1976).

92. Mattiasson, B., B. Danielsson, and K. Mosback, in L. B. Wingard and E. K. Pye, eds., *Enzyme Engineering*, Vol. 3, New York: Plenum, 1978, p. 453.

93. Mattiasson, B., K. Mosbach, and A. Svenson, *Biotechnol. Bioeng.* **14**, 1643 (1977).

94. McGlothlin, C. D., and J. Jordan, *Anal. Chem.* **47**, 786 (1975).

95. McGlothlin, C. D., and J. Jordan, *Anal. Chem.* **47**, 1479 (1975).

96. Mosbach, K., and B. Danielsson, *Biochim. Biophys. Acta* **364**, 140 (1974).

97. Mosbach, K., B. Danielsson, A. Borgerud, and M. Scott, *Biochim. Biophys. Acta* **403**, 256 (1975).

98. Papoff, P., and P. G. Zambonin, *Talanta* **14**, 581 (1967).

99. Pau, C. P., Dissertation, Pennsylvania State University, University Park, PA, (1984).

100. Priestley, P. T., *Analyst* **88**, 194 (1963).

101. Priestley, P. T., W. S. Sebborn, and R. F. W. Selman, *Analyst* **90**, 589 (1965).

102. Raffa, R. J., M. Stern, and L. Malspeis, *Anal. Chem.* **40**, 70 (1968).

103. Ramsing, A. U., J. Ruzicka, and E. H. Hansen, *Anal. Chim. Acta* **129**, 1 (1981).

104. Rehak, N. N., J. Everse, N. O. Kaplan, and R. L. Berger, *Anal. Biochem.* **70**, 381 (1976).

105. Rehak, N. N., and D. S. Young, *Clin. Chem.* **23**, 1153 (1977).

106. Reynolds, C. A., and S. M. J. Harris, *Anal. Chem.* **41**, 348 (1969).

107. Rich, S., R. M. Ianniello, and N. D. Jespersen, *Anal. Chem.* **51**, 204 (1979).

108. Rosenthal, D., G. L. Jones, and R. Megargle, *Anal. Chim. Acta* **53**, 141, (1971).

109. Ruzicka, J., and E. H. Hansen, *Anal. Chim. Acta* **78**, 145 (1975).

110. Sadtler, P., and T. Sadtler, *Am. Lab.* **14**, 86 (1982).

111. Sajo, I., *Mag. Kem. Foly.* **75**, 1 (1968).

112. Sajo, I., and B. Sipos, *Mikrochim. Acta* 248 (1967).

113. Schifreen, R. S., C. S. Miller, and P. W. Carr, *Anal. Chem.* **51**, 278 (1979).

114. Schmidt, H. L., G. Krisam, and G. Grenner, *Biochim. Biophys. Acta* **429**, 283 (1976).

115. Sillen, L. G., and B. Warnquist, *Ark. Kemi* **31**, 315 (1969).

116. Skeggs, L. T., *Am. J. Clin. Pathol.* **28**, 311 (1957).

117. Smith, E. B., C. S. Barnes, and P. W. Carr, *Anal. Chem.* **44**, 1663 (1972).

118. Smith, E. B., and P. W. Carr, *Anal. Chem.* **45**, 1688 (1973).

119. Snyder, L., J. Levine, R. Stoy, and A. Conetta, *Anal. Chem.* **48**, 942A (1976).

120. Stewart, K. K., G. R. Beecher, and P. E. Hare, *Anal. Biochem.* **70**, 167.

121. Strafelda, F., and J. Kroftova, *Collect. Czech. Chem. Commun.* **33**, 3694 (1968).

122. Sunner, S., and I. Wadsö, *Acta Chem. Scand.* **13**, 97 (1959).

123. Tran-Minh, C., and D. Vallin, *Anal. Chem.* **50**, 1874 (1978).

124. Tyřell, H. J. V., and A. E. Beezer, *Thermometric Titrimetry*, London: Chapman & Hall, 1968.

125. Tyson, B. C., W. H. McCurdy, and C. E. Bricker, *Anal. Chem.* **33**, 1641 (1961).

126. Vajgand, V. J., F. F. Gaal, and S. S. Brusin, *Talanta* **17**, 415 (1970).

127. Van Dalen, E., and L. G. Ward, *Anal. Chem.* **45**, 2248 (1973).

128. Vaughan, G. A., *Thermometric and Enthalpimetric Titrimetry*, London: Van Nostrand Reinhold, 1973.

129. Vaughan, G. A., and J. J. Swithenbank, *Analyst* **90**, 594 (1965).

130. Vaughan, G. A., and J. J. Swithenbank, *Analyst* **92**, 364 (1967).

131. Vaughan, G. A., and J. J. Swithenbank, *Analyst* **95**, 890 (1970).

132. Wasilewski, J. C., P. T-S. Pei, and J. Jordan, *Anal. Chem.* **36**, 2131 (1964).

133. Waugh, D. J., Dissertation, Pennsylvania State University, University Park, PA, 1978.

134. Weaver, J. C., C. L. Cooney, S. P. Fulton, P. Schuler, and S. R. Tannenbaum, *Biochim. Biophys. Acta* **452**, 285 (1976).

135. Weber, R., G. Blanc, G. Peuschel, and F. Hagedorn, *Anal. Chim. Acta* **86**, 79 (1976).

136. Weisz, H., and T. Kiss, *Fresenius' Z. Anal. Chem.* **249**, 302 (1970).

137. Weisz, H., T. Kiss, and D. Klochow, *Fresenius' Z. Anal. Chem.* **247**, 248 (1969).

138. Weisz, H., and S. Pantel, *Anal. Chem. Acta* **62**, 361 (1972).

139. Zambonin, P. G. and J. Jordan, *Anal. Chem.* **41**, 437 (1969).

140. Zenchelsky, S. T., and P. R. Segatto, *Anal. Chem.* **29**, 1856 (1957).

Chapter 3

THERMOGRAVIMETRY

By Jeffrey G. Dunn, *School of Applied Chemistry, Curtin University of Technology, Perth, Western Australia* and John H. Sharp, *Department of Engineering Materials, University of Sheffield, Sheffield, UK*

Contents

Treatise on Analytical Chemistry, Part 1, Volume 13,
Second Edition: Thermal Methods, Edited by James D. Winefordner.
ISBN 0-471-80647-1 © 1993 John Wiley & Sons, Inc.

I. INTRODUCTION

A. DEFINITIONS

Early in the educative process a chemist learns that many compounds, such as hydroxides and carbonates, although stable for indefinite periods at room temperature decompose on heating with an associated loss in mass. Conversely, the oxidation of a metal at elevated temperatures is accompanied by a gain in mass. The change in mass may occur in several steps dependent on the temperature. When calcium oxalate monohydrate is heated, three reactions occur which all involve loss in mass:

$$Ca(COO)_2 \cdot H_2O \longrightarrow Ca(COO)_2 + H_2O\uparrow \longrightarrow$$
$$CaCO_3 + CO\uparrow \longrightarrow CaO + CO_2\uparrow$$

When the mineral siderite is heated in air two reactions are observed, one involving loss in mass and the other gain in mass:

$$2FeCO_3 \longrightarrow 2FeO + CO_2\uparrow + \xrightarrow{\frac{1}{2}O_2} Fe_2O_3$$

In gravimetric analysis, the anaytical chemist usually heats a sample to bring about some reaction involving a change in mass and then cools the sample to room temperature and subsequently reweighs the reacted sample to determine this change in mass. If, however, the change in mass is recorded while the sample is heated (or cooled) under controlled conditions, the technique is known as thermogravimetry. The Nomenclature Committee of the International Confederation for Thermal Analysis (ICTA) has defined *thermogravimetry* (233, 234) as a technique in which the mass of a substance is measured as a function of temperature while the substance is subjected to a controlled temperature program. The term thermogravimetry and the abbreviation TG are to be preferred to the older thermogravimetric analysis and TGA (233, 234).

Strictly, TG is concerned with changes in mass rather than with changes in weight, but the distinction between these terms is rarely, if ever, made in practice and it is common to refer to the TG curve as the plot of change in weight against temperature, even though the local value for the acceleration due to gravity has not been taken into account (241).

It should be emphasized at the outset that thermogravimetry, unlike the complementary techniques differential thermal analysis (DTA) and differential scanning calorimetry (DSC), cannot give any information about reactions that do not involve mass change, such as polymorphic transformations and double decomposition reactions. TG is not, therefore, particularly useful as a technique for identification of a substance or mixture of substances. On the other hand, when a positive identification has been made by X-ray powder diffraction, DTA, or some other method, TG by its very nature is a quantitative technique and can frequently be used to estimate the amount of a particular substance present in a mixture or the purity of a single substance.

To obtain a thermogravimetric curve an apparatus known as a *thermobalance* is used. By means of a thermobalance the temperature range over which a reaction involving weight change occurs may be determined; this range depends not only on thermodynamic and kinetic factors, but also on procedural variables, such as the heating rate, crucible geometry, atmosphere, sample weight, and sample preparation.

The origins of thermogravimetry have been discussed by Keattch and Dollimore (214). The first worker to use the term thermobalance was Honda (191) in 1915. Subsequent development of apparatus has been particularly rapid since 1950 and thermobalances that simultaneously provide reliable temperature control and measurement with accurate recording of weight change have become commonplace. For precise work, however, even a sophisticated thermobalance should be calibrated to allow for the buoyant effect of the atmosphere on the object weighed, convection within the furnace tube, and the effect of heat on the balance mechanism.

B.   DERIVATIVE THERMOGRAVIMETRY AND RELATED TECHNIQUES

Thermogravimetric data can be presented either as the TG curve, which is a plot of the mass against time or temperature, with the mass loss on the ordinate plotted downward (233, 234), or as the *derivative thermogravimetric* (*DTG*) curve, which is a plot of the rate of change of mass with respect to time or temperature against time or temperature. DTG mass losses should also be plotted downward and gains upward, but in much of the literature this convention has not been followed. Schematic TG and DTG curves are shown in Fig. 1. In practice, the weight losses are not nearly so well resolved since the reactions are temperature-dependent rate processes which require time to reach completion. Hence, the first reaction may not be complete before the second reaction commences, and the base line between the reactions may be a slope rather than a plateau region. When a horizontal plateau is obtained in a TG curve, a minimum at which $dm/dt = 0$ is observed in the DTG curve (Fig. 1). A maximum in the DTG curve corresponds to a point of inflection in the TG curve at which mass is lost (or gained) most rapidly. Minima can also occur in a DTG curve at which

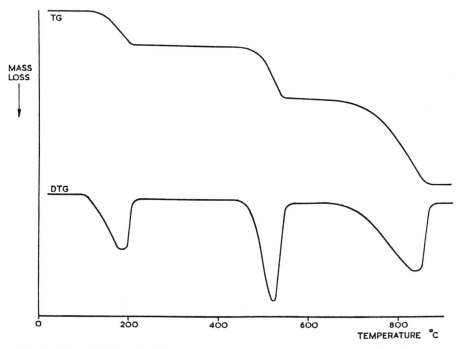

Fig. 1.   TG and DTG curves for calcium oxalate, 8.2 mg, 30°C/min, in argon at 25 mL/min.

$dm/dt > 0$, corresponding also to a point of inflection in the TG curve, rather than to a horizontal plateau (88, 100, 214).

A relatively minor technique, known as differential thermogravimetry, should not be confused with derivative thermogravimetry. Differential thermogravimetry was developed by de Keyser (102) who heated two identical samples suspended on either arm of a beam, using two furnaces in which one sample was maintained at a temperature approximately 5°C higher than the other. The technique is principally of historical interest as the forerunner of DTG.

Another technique that is closely related to TG but has not become established as of major importance is fractional thermogravimetry, developed by Waters (376). Volatile reaction products are selectively condensed and weighed as well as the sample. The method is particularly useful in the case of overlapping reactions, as in the thermal decomposition of coal. Fractional thermogravimetry is rarely used nowadays because of the development of combined thermogravimetry and evolved gas analysis (TG–EGA, discussed in Section X.B) in which the volatile products are analyzed usually by means of mass spectrometry or gas chromatography.

Other techniques placed by the ICTA Nomenclature Committee (233, 234) within the branch of the family of thermal analysis methods based on changes of mass are evolved gas detection, evolved gas analysis, emanation thermal analysis, and thermoparticulate analysis. Emanation thermal analysis involves the release of radioactive emanation from a substance, which involves change in mass if, for example, α particles are emitted. The method was developed by Balek, who has reviewed it on several occasions (11–13, 131). Thermoparticulate analysis involves the release of particulate matter which can be detected as condensation nuclei when introduced into a cloud chamber. The method was devised by Van Luik and Rippere (372) and further developed by Murphy and Doyle (264).

C. RELATIONSHIP TO GRAVIMETRIC ANALYSIS

The relationship between thermogravimetry and gravimetric analysis is clearly of interest to an analytical chemist. The traditional instructions to an analyst to "dry the sample" or "to ignite the precipitate to a dull red heat" are hopelessly vague to the thermal analyst and must have caused many errors and much confusion. The systematic work of Duval (126) was the first major attempt to rectify this situation, but caused further confusion, since he used only one heating rate and failed to emphasize the dependence of TG on procedural variables. The second edition of his book (127) and other articles (128) partly meet these criticisms.

It is important to distinguish clearly between thermogravimetry, isobaric mass-change determination and gravimetric analysis. In thermogravimetry the mass of a substance is measured as a function of temperature while the substance is being heated and the mass is determined at elevated temperatures. In isobaric mass-change determination the *equilibrium* mass of a substance at constant partial pressure of the volatile product(s) is measured as a function of temperature; i.e., the mass is determined at elevated temperatures after a period of isothermal heating to constant weight. In gravimetric analysis the sample is heated, held under isothermal conditions for a period, then cooled and weighed at room temperature. It is essential, therefore, that the validity of a temperature recommended for heating an analytical precipitate on the basis of a TG curve should always be checked under isothermal conditions and it should also be confirmed that the reaction product does not absorb moisture or react chemically on cooling to room temperature.

With these provisos the application of TG during the development of a gravimetric method (see Section VI.D) can provide detailed information on drying and ignition temperatures for specific precipitates. It is ironic that the advent of alternative instrumental techniques has led to a decline in the use of gravimetric procedures at a time when they could be developed in a more rigorous manner than hitherto. Even so, the reagents used as calibrants in instrumental methods, such as in atomic absorption spectroscopy, are frequently analyzed by gravimetric methods to provide accurate calibration standards.

D. LITERATURE

In addition to Duval's book (127) on *Inorganic Thermogravimetric Analysis*, several books on thermal methods contain major sections on thermogravimetry, including those written by Garn (162), Wendlandt (382), Blazek (41), and Todor (365). An excellent introductory text has been written by Daniels (100).

Keattch and Dollimore (214) have provided a good introductory book specifically on TG. Several important papers and reviews on TG appear in books edited by Kambe and Garn (211) and Wendlandt and Collins (384), while Liptay (222) has collected together DTA–TG–DTG curves obtained from the derivatograph, an instrument that is extensively used in Eastern Europe.

Murphy (265) has provided a thorough coverage of the literature concerning TG in the biennial reviews that are published in *Analytical Chemistry* in even-numbered years. References to TG can also be found in the Application Reviews, which appear in the same journal in odd-numbered years. Of particular value is the section on high polymers written for several editions by Mitchell and Chiu (260) and more recently by Cobbler and Chou (90). An extensive coverage of the literature of interest to thermal analysts is provided by *Thermal Analysis Abstracts* (335), which was first published in 1972 and appears bimonthly. Short review articles have appeared in this journal since 1985 (45, 160).

II. PHYSICAL BASIS OF THE TECHNIQUE

The shapes of TG and DTG curves depend on procedural, thermodynamic, and kinetic factors. A theoretical TG curve may be calculated if the kinetic mechanism and parameters are known, on the assumption that heat transfer is instantaneous and no temperature gradient exists within the sample. Thus, the kinetics of most reactions under isothermal conditions can be summarized by the general equation:

$$\frac{d\alpha}{dt} = kf(\alpha) \tag{1}$$

where α is the fraction reacted in time t, k is the rate of reaction, and $f(\alpha)$ is some function of α. The various forms adopted by the function $f(\alpha)$ have been discussed elsewhere (139, 319, 329, 332, 334).

The temperature dependence of the rate constant usually follows the Arrhenius equation

$$k = Ae^{-E/RT} \tag{2}$$

where T is the absolute temperature and A, E, and R are constants known as the pre-exponential factor, the activation energy, and the gas constant, respectively.

If the rate of heating is linear with time, then

$$T = T_0 + \beta t \tag{3}$$

where T_0 is the initial temperature and β is the heating rate.

Equations 1 and 2 can be combined to give

$$\frac{d\alpha}{dt} = A f(\alpha) e^{-E/RT} \tag{4}$$

and substitution for dt using Eq. 3 leads to

$$\frac{d\alpha}{dT} = \frac{A}{\beta} f(\alpha) e^{-E/RT} \tag{5a}$$

or

$$\frac{d\alpha}{f(\alpha)} = \frac{A}{\beta} e^{-E/RT} dT \tag{5b}$$

Equation 5 is the basic equation of the DTG curve, and when integrated of the TG curve. The integration of the left side of the equation is straightforward when the form of function $f(\alpha)$ is known, and leads to the associated function $g(\alpha)$. Doyle (112) showed that the right side can be integrated if E is constant, leading to

$$g(\alpha) = \int \frac{d(\alpha)}{f(\alpha)} = \frac{A}{\beta} \int e^{-E/RT} dT = \frac{AE}{\beta R} p(x) \tag{6}$$

where $p(x) = \dfrac{e^{-x}}{x} - \displaystyle\int_x^\infty \frac{e^{-u}}{u} du$

where $u = \dfrac{E}{RT}$ and $x = \dfrac{E}{RT_\alpha}$

It is useful to note at this stage that it follows (54, 115) from Eq. 6 that

$$\log g(\alpha) - \log p(x) = \log \frac{AF}{\beta R} \tag{7}$$

where B is a constant that is independent of temperature.

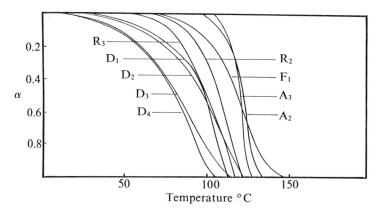

Fig. 2. Calculated TG curves for reactions controlled by various rate-controlling processes (Table 1) with $E = 83.7$ kJ/mol, $A = 10^9$, and $B = 1°C$/min. After Satava and Skvara (319).

Although the kinetic parameters E and A markedly influence the temperature range over which the TG curve is observed, they do not influence the shape of the curve too greatly (319). The kinetic mechanism, i.e., the form of $f(\alpha)$ or the related function $g(\alpha)$, however, determines the shape of the curve, a shown in Fig. 2. The different rate-controlling processes, identified by a letter followed by a number (e.g., A2) are listed in Table 1 (319, 332).

It can be seen from Eq. 6 that as the heating rate β increases, the magnitude of the function $g(\alpha)$ decreases, at any particular temperature. Since a decrease in $g(\alpha)$ indicates a decrease in α, the fraction reacted decreases as the heating rate increases. Consequently, the temperature at which a change in mass is first detected, called the *initial temperature* T_i (see Fig. 3) (also referred to as the onset temperature or procedural decomposition temperature) increases as the heating rate increases. The initial temperature is not sufficiently well defined to be used as a satisfactory reference point since it is dependent on a number of procedural variables and the rate at which the initial reaction occurs. This is illustrated in Fig. 4, which shows the decomposition of calcium carbonate in nitrogen. The slow initial loss of carbon dioxide leads to uncertainly in the determination of T_i, especially if a buoyancy correction is required.

A more satisfactory approach is to use the extrapolated onset temperature T_e, which yields consistent values even when measured by different operators (see Fig. 3). If, however, the decomposition extends over a wide temperature range and becomes rapid only in its final stages, the extrapolated onset temperature will differ considerably from the onset temperature. Thus, in the decomposition of $CaCO_3$ as depicted in Fig. 4, the T_i value is 512°C and the T_e value 772°C, with a difference of 260°C.

Probably the most satisfactory method is to measure the temperature at which a fractional weight loss α has occurred (T_α) (see Fig. 5). Clearly, the

TABLE 1
Integral Forms of Kinetic Equations

Function	$g(\alpha)$	Rate-Controlling Process
$D_1(\alpha)$	α^2	One-dimensional diffusion (parabolic law)
$D_2(\alpha)$	$(1 - \alpha)\ln(1 - \alpha) + \alpha$	Two-dimensional radial diffusion into a disk or cylinder
$D_3(\alpha)$	$\left[1(1 - \alpha)^{1/3}\right]^2$	Three-dimensional diffusion into a sphere (Jander's equation)
$D_4(\alpha)$	$1 - 2\alpha/3 - (1 - \alpha)^{2/3}$	Three-dimensional diffusion into a sphere (Ginstling–Brounshtein equation)
$F_1(\alpha)$	$-\ln(1 - \alpha)$	Random nucleation (first-order kinetics)
$F_0(\alpha)$	α	Linear kinetics (zero order)
$R_2(\alpha)$	$1 - (1 - \alpha)^{1/2}$	Two-dimensional movement of a reaction interface
$R_3(\alpha)$	$1 - (1 - \alpha)^{1/3}$	Three-dimensional movement of a reaction interface
$A_2(\alpha)$	$\left[-\ln(1 - \alpha)\right]^{1/2}$	Random nucleation and growth (Avrami–Erofeev equation)
$A_3(\alpha)$	$\left[-\ln(1 - \alpha)\right]^{1/3}$	Random nucleation and growth (Avrami–Erofeev equation)

SOURCE: After Satava and Skvara (319).

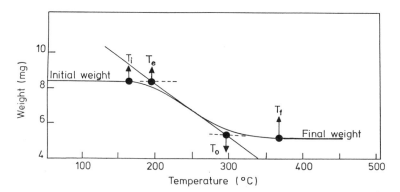

Fig. 3. Schematic diagram indicating the various temperature values used to define a TG weight loss region. T_i = initial, onset, or procedural decomposition temperature; T_e = extrapolated onset temperature; T_o = extrapolated offset temperature; T_f = final temperature.

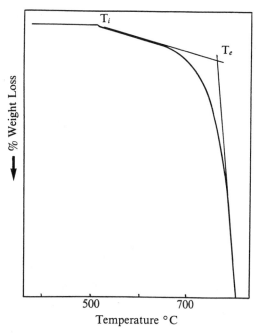

Fig. 4.   TG curve for the decomposition of $CaCO_3$ in N_2 illustrating the differences in measured values of T_i and T_e. 9.5 mg sample heated at 20°C/min in N_2 flowing at 50 mL/min.

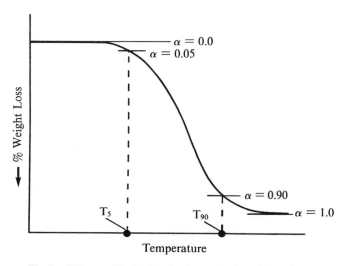

Fig. 5.   TG curve illustrating the determination of T_α values.

TABLE 2
The Effect of Heating Rate and Sample Weight on the Decomposition
of Calcium Carbonate in Nitrogen (238)

Sample Weight (mg)	β (°C/min)	$T_{0.1}$ (°C)	$T_{0.5}$ (°C)	$T_{0.9}$ (°C)	$(T_{0.9} - T_{0.1})$ (°C)
50	1	634	689	712	78
100	1	655	712	740	85
250	1	672	739	773	101
50	7	716	788	818	102
100	7	742	818	855	113
200	7	768	845	890	112
300	7	775	853	902	127
400	7	775	865	915	140

temperature $T_{0.01}$ is close to that at the start of the reaction and $T_{0.99}$ is close to that at the end of the reaction. Then for an endothermic reaction

$$(T_{0.01})_{\beta_1} > (T_{0.01})_{\beta_2}$$

$$(T_{0.99})_{\beta_1} > (T_{0.99})_{\beta_2}$$

and

$$(T_{0.99} - T_{0.01})_{\beta_1} > (T_{0.99} - T_{0.01})_{\beta_2}$$

where $\beta_1 > \beta_2$

These relationships are easily confirmed in practice; typical data for the decomposition of calcium carbonate (238) are shown in Table 2.

The fractional weight loss method has the advantage that using modern digital processing systems the temperature values can be tabulated directly. To define the complete range of reaction, two further temperatures T_o and T_f may be identified as shown in Fig. 3. Similar comments apply to these as to the corresponding temperatures already discussed.

III. APPARATUS

Although there are many thermobalance designs reported in the literature, they all consist of the essential components of a balance and balance controller, a sample chamber to allow experiments to be carried out in a controlled atmosphere, a furnace and furnace controller, and a recorder system. In this section, desirable features of each of these components are discussed before consideration of commercially available and specialist thermobalances.

A. THE BASIC COMPONENTS OF A THEROMOBALANCE

1. The Balance

Nearly all modern thermobalances are based on automatic recording balances. In a comprehensive review, Gordon and Campbell (175) divided these balances into two types: (1) deflection instruments and (2) null-deflection instrument. In deflection instruments the change in weight is monitored by following the displacement of the balance itself. The different balance types available are shown schematically in Fig. 6. In the case of the null-deflection instruments the position of the balance is monitored and a servo-system is used to maintain it in a quasi-equilibrium position, the power supplied to the servo-system being a measure of the weight change. This is shown schematically in Fig. 7.

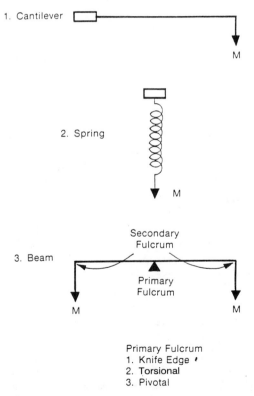

Fig. 6.   The principles of operation of various types of balances.

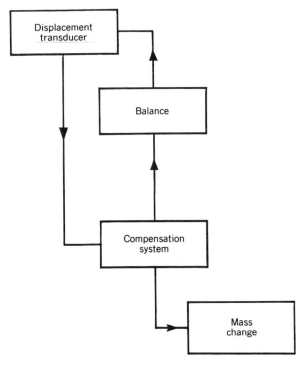

Fig. 7.   Schematic diagram of a null deflection balance.

Automatic recording beam microbalance and ultramicrobalances for use in vacuum and controlled environments have been reviewed by Gast (167) and Czanderna and Wolsky (98), these instruments being capable of detecting weight changes of below 10 and 0.1 μg, respectively. The beam microbalance in the null-deflection mode forms the basis of an increasing number of commercial thermobalances.

In specifying balance performance, a number of the terms shown in Table 3 are often used. It should be noted that sensitivity strictly refers to the balance alone and should be quoted in items of displacement per microgram or milligrams and not as often quoted in micrograms or milligrams alone. Sensibility, which is the more practical term, refers to the complete weighing system, including the displacement transducer and compensation circuitry. When examining specifications of thermobalances it is important to establish whether the data for mass measurement refer to the balance itself or to the complete thermobalance. Thus, a statement such as "10 g capacity balance is used" frequently refers to the total capacity of the balance, and the maximum weight of sample that can be used will be less than this by the weight of the crucible and crucible support system. For many applications the maximum sample weight that can be used will be governed by the volume of the sample

TABLE 3
Definition of Terms Used to Describe Balances

Sensitivity	Magnitude of the reversible or elastic displacement per unit variation in weight or mass.
Response	Reciprocal of the sensitivity.
Capacity	Maximum load that can be suspended and placed on the balance without injury to the balance or its operation.
Precision	Minimum variation in mass that can be observed experimentally in a reproducible manner in the absence of any effects from the balance system.
Sensibility	Minimum variation in mass that can be observed experimentally in a reproducible manner in a complete weighing system.
Range	Maximum variation in mass change that can be measured with the balance at a given load.
Zero-Point Stability	Time-dependent variation of the precision of a balance or the sensibility of a system.

crucible rather than by the balance capacity. Typically this provides a maximum of 50–100 mg sample, depending on the sample density.

In discussing the performance of microbalances the term load to precision ratio (LPR) is often used. Thus a 10 g capacity beam balance with a precision of 2 μg will have an LPR of 5×10^6. The target for the designers of high-performance microbalances is to achieve an LPR of 10^8 or better (98, p. 7). The ultimate sensitivity of an ultramicrobalance of the beam or spring type is considered to be of the order of 10^{-2} to 10^{-4} μg due to the limiting effect of Brownian motion (297, p. 239). For higher sensitivities the quartz crystal oscillator offers the possibility of resolving mass changes of the order of 10^{-6} μg.

The use of load to sensitivity product (LSP) has been recommended as an alternative to the LPR in assessing balance performance and has been used to characterize a wide range of microbalances in a general discussion of their theory and design (325).

2. The Sample Chamber

The function of the sample chamber is to enable work to be carried out in a controlled atmosphere, which may be inert, reactive, or corrosive, at a controlled pressure. In most modern thermobalances the sample chamber forms a sealed system with the balance chamber; this is shown schematically in Fig. 8a. This figure also indicates how a noncorrosive gas can be passed through the balance chamber, entering at A, flowing downward through the sample chamber, and leaving at C. The direction of the flow is important to avoid condensation of reaction products on the hang-down system, with the resultant errors to the weight record, and also to prevent the reaction products, which may be highly corrosive, from entering the balance chamber.

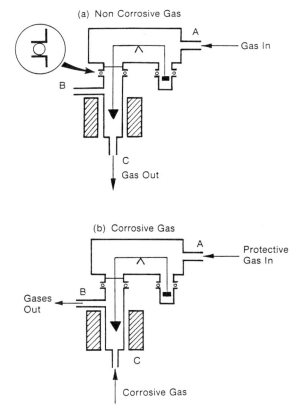

Fig. 8. Schematic of closed-system thermobalances for work in flowing (a) noncorrosive and (b) corrosive atmospheres.

If the chamber has a high volume it will probably be necessary to use high initial purge flow rates to remove air completely from the system, for work in other atmospheres. An alternative approach, particularly useful if a rapid changeover of gases is required, is to introduce the gas at the sidebranch B, splitting the flow so that part of the gas flows through the sample chamber and part through the balance chamber.

For work in corrosive atmospheres the flow system shown schematically in Fig. 8b can be used, allowing work in an undiluted atmosphere of the corrosive gas. The relative flow rates of the corrosive and protective gases will depend on the apparatus being used and on the gases themselves. Initial experiments should use a high flow rate of the protective gas to allow a good safety margin. If condensation of reaction products is a problem, the corrosive gas can be introduced at B and passed in a downward direction, leaving at C. However, in this case it will be diluted with the protective gas, which is typically dry nitrogen or argon. An example of a thermobalance modified for work in a pure sulfur atmosphere has been described (310), and other instructive examples include thermobalances modified (1) to study the reaction between rhenium hexafluoride and hydrogen to produce rhenium metal (16) and (b) to allow work in an atmosphere of sodium vapor (254). The problem of containing corrosive gases is considerably simplified by the magnetic balance designed by Gast (168), where the balance and sample chambers are completely separated (see Section III.B.2).

In a number of designs the furnace tube itself will form the sample chamber, although it may be desirable to protect it by using an inner sheath. Where the balance and sample do not form a closed system, normally because the balance housing is not designed to be gas-tight, work in a corrosive or hazardous atmosphere is generally difficult. Typical flow systems for this type of instrument are shown in Fig. 9. Figure 9a shows the simplest

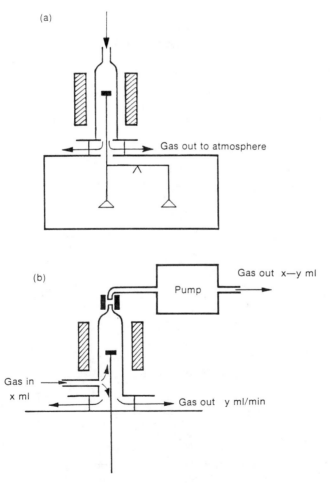

Fig. 9.   Schematic of open-system thermobalances for work in corrosive gases with (a) downward flow of gas and (b) system for condensible or toxic gases.

system with a downward flow of gas and Fig. 9*b* shows the arrangement that
should be adopted when condensible or toxic products are being evolved.
Many recent thermobalances incorporate manometers to measure flow rates
of gases and a valve that permits connection to an external gas source.

If the sample chamber is well sealed, then work in vacuum or under
pressure can be carried out. Many examples of specialist balances produced
to work in high vacuum are reported at the vacuum microbalance conferences held on a worldwide basis. For high-pressure work, a TG apparatus
capable of working up to 300 psi and 350°C has been reported (56), and a
versatile TG has been developed for operation between 10^{-5} torr and
300 bar in the temperature range -200 to 500°C (151).

3. The Furnace

The essential requirement of any thermobalance furnace is that it will
provide a uniform hot zone of sufficient length to contain completely the
sample holder over its entire operating range. In the case of a deflection
instrument, a longer hot zone will be required than with a null-deflection
balance, where the sample is held in a constant position.

The most widely used furnaces are those based on resistance elements in
the form of wires, strips, or rods. A number of element materials and their
operating ranges are given in Table 4. The most frequently used are nichrome,
Kanthal, and platinum–rhodium alloys, often wound noninductively to enable weight changes in magnetic materials to be followed accurately, without
interference from the field generated by the winding. Molybdenum-wound
furnaces, used at temperatures about 1600°C, must be operated in a reducing
atmosphere, e.g., a mixture of 10% hydrogen and 90% nitrogen. Other

TABLE 4
Furnace Resistance Elements

Trade Name	Composition	Form	Operating Range
Nichrome	80% Ni–20% Cr	W	a–1000°C
Kanthal Al	20–23% Cr, 4–5.5% Al, 0.6–1% Co, < 0.1% C Balance Fe	W	a–1300°C
Harwell	SiC	R	a–1500°C
Platinum –rhodium	80% Pt–20% Rh	W	a–1650°C
	60% Pt–40%	W	a–1650 C
Kanthal Super	$MoSi_2$	R	a–1600°C
Molybdenum[a]	Mo	W	a–1800°C
Tungsten[a]	W	W	a–2000°C
Graphite[a]	C	R	a–2800°C

[a] inert or reducing atmosphere required.

Note. a, ambient; W, wire; R, rod.

TABLE 5
Ceramic Materials Used in TG Systems

	Melting Point (°C)	Maximum Operating Temperature (°C)
Al_2O_3	2045	1830–1900
BeO	2550	2000 in vacuum
MgO	2800	2200 in oxidizing atmosphere
		1700 in reducing atmosphere
ZrO_2	2715	
Mullite	1830	1600
(approx. $3Al_2O_3 \cdot 2SiO_2$)		
SiO_2	1713	1500
Aluminous porcelain		1400

furnaces for high-temperature work are based on carbon or tungsten and must also be run in an inert or reducing atmosphere or under a vacuum. Some high-temperature ceramic materials used in TG furnaces are listed in Table 5.

Furnaces have been made or modified that are capable of very high heating rates. Such a furnace has been used for thermogravimetry of textiles at heating rates approaching 3000°C/min (34). These fast-response furnaces enable isothermal temperatures to be achieved rapidly.

For some isothermal work it is advantageous to be able to keep the furnace at the operating temperature between runs and to have the facility of raising it rapidly into place at the start of a run. This is generally easier with the lower mass systems.

Alternative methods of heating include infrared (328) and induction heating, which enable the sample to be heated directly without heating the sample chamber. The former method has been used to obtain heating rates of the order of 6000°C/min for polymer ablation studies (252). The use of induction heating allows very high temperatures to be obtained; a thermobalance for operation up to 3000°C is described by Steinheil (352). In view of the field generated by the system, it was necessary to carry out weighings with the heating system switched off, only 2–3 s being required to reheat to an operating temperature of 2000°C. For the technique to be used the sample must be conducting itself or placed in a conducting sample holder. The method was used for studying the outgassing of space materials up to 500°C at pressures of 10^{-6} torr (27). To eliminate the return of outgassed molecules to the sample surface, the glass walls of the vacuum chamber were cooled with liquid nitrogen. The use of induction heating enabled the sample to be heated directly through the liquid nitrogen-cooled walls. In some cases, particularly for adsorption studies, it may be necessary to carry out thermogravimetric studies below ambient temperature, in which case a programmable cryostat is required.

To minimize experimental errors to the weight record, particularly under reduced pressure, a number of workers have used symmetrical furnaces consisting of two elements, designed to heat the sample and reference materials under identical conditions (see Section V.A).

Furnaces are often wound nonconductively to avoid interaction between magnetic fields and a magnetic sample. In some cases, however, advantage is taken of an inductively wound furnace to calibrate the temperature of the instrument with magnetic samples (see Section V.B).

4. Temperature Measurement

In the majority of thermobalances the sample temperature is measured by using a thermocouple. The most widely used types are chromel–alumel (1000°C), platinum v platinum/(10 or 13% rhodium) (1500°C) and platinum/16% rhodium v platinum/30% rhodium (1750°C). The most satisfactory system is to have the thermocouple directly in contact with the sample crucible. This necessitates the use of fine wires or strips made of the same material as the temperature-sensing thermocouple to enable the measuring circuit to be completed without affecting the balance action. An alternative often adopted with microbalance systems is to place the thermocouple so that it is close to the pan but does not touch it. This system can be used only with null deflection balances. Some typical arrangements are shown in Fig. 10.

The thermocouple output is normally fed by a cold junction consisting of melting ice at 0°C or an electronic system providing automatic compensation. In modern instrumentation the output is frequently linearized electronically

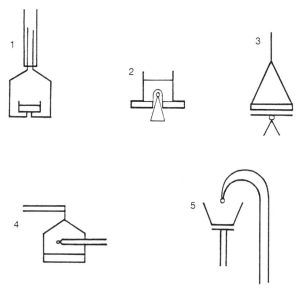

Fig. 10. Diagram showing possible thermocouple locations for temperature measurement in TG.

to provide a direct readout in degrees, thus eliminating the need for conversion tables. It is also possible to feed the signal to a microprocessor containing tabulated data on thermocouple readings, so that the conversion to temperature is made by direct comparison.

Thermocouples do age, especially if exposed to corrosive atmospheres, and the signal output will vary over a period of time. Hence, calibration with an external calibrant is necessary. Problems encountered in calibrating the temperature measuring system are discussed in Section V.

5. The Temperature Programmer

The main functions of a temperature programmer in a thermobalance are to provide a linear and reproducible set of heating rates and to be able to operate at a constant temperature with a stability of $\pm 1°C$ or better. Most modern programmers also allow programmed cooling and cycling between two temperatures. The programmer is usually controlled from a thermocouple in the furnace, located close to the furnace windings, but in some systems the control thermocouple is in direct contact with the sample itself. Three-term control is widely used, whereby the independent corrections are applied (1) *proportional* to the difference between the required and measured value, (2) to the *integral* of this difference, and (3) to the *derivative* of the difference.

A number of programmers offer the facility of executing complex programs where the heating rate is changed several times during a run and isothermal intervals of various duration may be inserted. Increasing use is being made of microprocessors for this application. In addition to executing complex temperature programs, microprocessors also change gas atmospheres at preset temperatures.

It should be noted that not all TG experiments are carried out at linear heating rates or at fixed temperatures. Of particular interest are the interactive programming techniques whereby the temperature program is controlled by the rate of weight loss or gain of the sample, so that this does not rise above a preset value. In the so-called quasi-isothermal technique of Paulik and Paulik (286) the heating rate is controlled from the DTG signal of the balance, initially starting at $10°C/min$ and reducing as the reaction starts, so that reaction takes place slowly over a narrow temperature range. For work at reduced pressures in the range 10^{-5} to 4 torr, Rouquerol and Gauteaume have developed the technique of constant decomposition rate thermal analysis (316), whereby the heating rate is controlled by the pressures of the system. Recently these and related techniques have been grouped together under the term controlled rate thermal analysis, which is discussed in Chapter 1. For polymer studies Flynn and Dickens (141) have advocated a temperature jump method of programming where the sample is examined at a series of temperature plateaus, e.g., $T_1 \rightarrow T_2 \rightarrow T_3 \rightarrow$, and the weight changes are monitored at each successive plateau. Temperature intervals are of the order of 10–15°C and the temperature is raised as rapidly as possible between plateaus. This technique has been automated using a rapid response nichrome furnace under computer control (105). For kinetic studies it has also been suggested that programming the heating rate proportional to the square of the temperature simplifies integral methods of analysis (140).

6. The Derivative Module

At one time derivatization of the TG signal could be carried out only by the laborious task of calculating mass changes over small increments of time or temperature and then plotting the results. There are now two alternative automatic methods available that give much better resolution of the DTG record relative to manual methods, as well as providing the DTG trace simultaneously with the TG record. The first technique involves electronic differentiation and can be either an integral part of the system or an add-on module. Since these devices are based on CR circuits, the time constant needs to be small to provide fast response to changes in the TG signal. The sensitivity of the derivative device is usually expressed as weight per degree per centimeter of chart recorder, and it is useful if more than one selectable range is available.

The other method is to collect the TG data on a computer, which allows other mathematical manipulations, such as smoothing of results, to be carried out in addition to the derivative calculation. This is the most powerful technique for the presentation of DTG curves, but is also the most expensive option.

7. The Recording System

Since most modern thermobalances provide multivolt output that is directly proportional to weight, it is possible to take advantage of the wide range of potentiometric recorders that are commercially available. Two main types are used: (1) the strip chart recorder, where weight and temperature are recorded as a function of time; and (2) the X-Y recorder where weight is recorded directly as a function of temperature. The X-Y recorder enables data to be obtained in the most convenient form, but it relies on a completely linear temperature rise uninterrupted by deviations induced by instrumental or experimental factors. For example, the oxidation of disodium fosfomycin is so exothermic that for a short time the temperature-recording thermocouple is heated above the temperature expected for the linear program (366). This overheating perturbs the TG and DTG records, as illustrated in Fig. 11, and the records do not return to normal until the temperature is once again defined by the programmer. This problem can be overcome on a two-pen X-Y recorder by converting the X axis to a time-based mode and monitoring both mass and temperature changes. A suitable recorder should possess a wide range of switch-selected sensitivities, variable range control useful when expanding or contracting a given sample weight to give full-scale deflection,

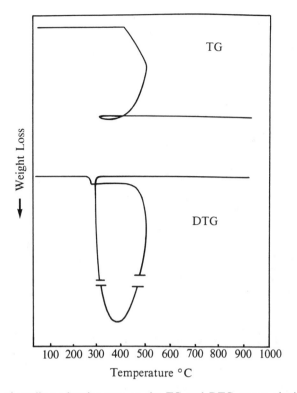

Fig. 11. Effect of nonlinear heating rate on the TG and DTG curves obtained on an *X-Y* recorder (366).

high accuracy (0.3% full-scale deflection or better), and good temperature stability. Temperature stability is important when long-term testing is required. If one is purchasing a recorder other than that supplied by the TG manufacturer, one should ascertain that it is compatible with the instrument, particularly with regard to the DTG output.

Many users with access to computers are now storing the data directly in digital form, particularly where kinetic analysis is to be attempted. Increasing use of microprocessor-based recording systems is also strongly evident. These systems permit ready manipulation of the data in a way that is not possible with conventional potentiometric recorders. In addition the storage of data in digital form allows higher accuracy than can be obtained by readings from a chart recorder, which is rarely better than 0.2%.

B.   COMMERCIALLY AVAILABLE THERMOBALANCES

Since the 1950s, the range of commercial thermobalances available on the market has increased, hence these instruments tend to dominate. Reviews of TG apparatus appear regularly in the literature (see, for example, Wendlandt

and Gallagher (386) and Robens et al. (312)). The manufacturers also produce well-illustrated brochures that describe their products in some detail.
Escoubes et al. (134) list the names and addresses of some 20 manufacturers
of TG equipment. However, reviews become out of date quickly, because
instrument specifications change with some rapidity. Consequently, although
some discussion of the features of commercial thermobalances is pertinent,
this section concentrates more on trends in instrument design and common
features that are considered desirable.

1.   Trends in Instrument Design

Probably the most significant trend in recent years has been toward
miniaturization. The driving force for this has primarily been the adoption of
the microbalance as the weighing device, which has permitted sample sizes of
between 1 and 10 mg to be used, instead of the 100 to 1000 mg sizes
commonly used in beam balances. The smaller sample size is advantageous
experimentally (see Section IX), although care needs to be exercised to
ensure that a representative sample is obtained. Materials should be ground
if possible so that the number of particles selected from a batch sample is
statistically significant, unless there is good evidence that the material is
homogeneous. Mineral samples, for example, frequently consist of inhomogeneous lumps, and it is usually necessary to grind such samples to less than
100 μm, followed by thorough mixing, to provide a representative 10-mg
sample. Plastic or rubber sheeting prepared by compounding at elevated
temperatures, on the other hand, is quite homogeneous, and representative
samples can be obtained by use of a leather punch to provide a disk of
material. Analysis of rubber samples using this sampling technique gave very
reproducible results (78). An obvious precaution to take when dealing with
new samples is to carry out replicate experiments to check reproducibility,
and hence sample homogeneity. Since the smaller sample occupies less space,
a shorter hot zone is required which allows smaller furnaces with lower
thermal mass to be constructed. The lower thermal mass enables the furnace
to respond rapidly to changes in the temperature program. Hence, linearity
of heating rate is achieved quickly at the commencement of a heating
program, and overshooting the temperature required for an isothermal
experiment tends to be small. Finally, the small sample requires only small
volumes of gas for reaction purpose, or evolves small volumes of gas during
reaction, so that the volume of the atmosphere chamber can be small.

Apart from the compactness of modern instruments, another major advantage is that the thermal mass of the TG apparatus is considerably reduced, so
that the instrument can be cooled quickly to carry out more experiments.
This is assisted if the furnace has some additional cooling aid, such as a
forced air current or a water-cooled jacket (76). One parameter quoted
sometimes is the cooling rate of the furnace, and this is useful to know since

it gives an indication of turnaround time. If the cooling rate is slow, then consideration may need to be given to the purchase of an additional furnace. Sometimes, however, changing furnaces is a slow process.

A second major trend is toward modular thermal analysis systems, in which there is a central console that provides the temperature programming and recorder facilities. In more recent models, the console provides microprocessor capabilities and the recorder is replaced by a digital printer plotter. The advantage of this system approach is that the TG apparatus is then one of a series of modules that can be connected to the console, thus saving the cost of purchasing multiple recorders and temperature programmers. The obvious disadvantages are loss of all capability if the console is inactive due to a fault, and nonsimultaneous operation of the modules. The choice of a system as opposed to individual units depends on the specific needs and budgets of each laboratory.

Third, most instrumentation tends to demonstrate greater flexibility in the experimental parameters available. Hence, most recent instruments have a choice of selectable mass ranges and a wider variety of heating rates than previous models.

The most significant recent trend is the addition of microprocessor facilities, for both data recording and manipulation and instrument control. The manufacturers usually provide, at no insignificant cost, software packages that enable various functions to be achieved. One of the most useful features is the ability to produce a report-quality TG record. Data from the experiment are recorded on a floppy disk and then recalled onto a digital printer plotter correctly scaled and with all the required experimental conditions. Various TG record data, such as the initial and final temperatures of the weight losses, can be directly plotted, as can the DTG curve. The weight gains or losses can be expressed as weight or percentage values. More sophisticated software packages permit, for example, the calculation of kinetic parameters such as rate constants and estimates of material lifetimes. The microprocessor also has several control functions and is responsible for the greatly increased range of heating rates and complex heating programs that may involve isothermal intervals of different lengths with different heating rates interspersed. Relay closures can be used to control atmosphere changes. These facilities thus provide semiautomated instrumentation so that once the operator has set up the initial conditions the experiment can be completed without further attention. This is leading to a greater dependence on TG and other thermal techniques as routine quality control methods.

One problem with the applications software is that the necessary algorithms are not usually provided to enable the user to understand the software philosophy; hence, the user has to accept the results provided. Usually, there is no provision either for the user to adapt existing programs to specific needs or to input locally developed programs. While the commercial advantages of locking up both software and instrument intervention can be appreciated, it

TABLE 6
Characteristics of First- and Second-Generation TG Apparatus

	First Generation	Second Generation
Capacity	20 g	1 g
Sensitivity (fsd)	100 mg	1–250 mg
Temperature range	Ambient to 1500°C	
Heating rates (deg/min)	1–10	1–100
Cooling times	Hours	Minutes
Atmosphere required	Modification	In-built
DTG	Manual	Add on facility
Weight record correction	Buoyancy loss	Direct weight
Average number of experiments per day	1 or 2	12–15
Weight of TG apparatus	126 kg	30 kg
Height	2 m	0.5 m

SOURCE: Dunn (118).

is possible to speculate that the long-term effects may be detrimental to thermal analysis, and indeed to the continued adaption of microprocessors in general.

Some of these trends are illustrated in Table 6 (118), which compares the characteristics of first- and second-generation thermobalances. The third-generation differs mainly by the inclusion of microprocessor facilities.

2. Characteristics of Commercial Thermobalances

Some of the essential features of commercially available thermobalances are listed in Table 7. These characteristics have been obtained from an inspection of the manufacturers' literature, and hence represent the total range of values possible. In the balance sensitivity category, for example, the highest quoted sensitivity for all the reviewed thermobalances is 1 μg and the lowest is 50 μg. However, the lower sensitivity balances may also be capable of taking larger sample masses and so the LPR value may not vary significantly across the range. This needs to be checked for each specific balance. The ranges given also tend to be for the basic model, so that TG furnaces commonly operate in the temperature range ambient to 1000°C, but other models may operate from -196°C to 500°C or ambient to 2400°C with steps in between. The choice of balance capability will depend on the particular applications of the user together with economic considerations.

It is evident that the null electronic microbalance is the favored choice as the basis of the instrument. All instruments have selectable mass ranges, which is useful since it enables the correct scale to be chosen for a particular weight gain or loss to maximize sensitivity. Although in many cases it is necessary to counterbalance the mass of the sample crucible and sample with

TABLE 7
Typical Features of a Modern Thermobalance

Balance Characteristics	
Balance	Null deflection electronic beam microbalance
Weighing ranges	0.01 to 1000 mg full-scale deflection in steps
Tare facility	1 g
Sensitivity	1–50 μg
Sample capacity	50–100 mg maximum (but some systems up to 10 g)
Arrangement	Bottom loading > top loading $\gg$ lateral loading
DTG	Usually present, sensitivity in the region of 1 mg/min cm^{-1}
Atmosphere Control	
Flowing gas	200 mL/min, but up to 1 L/min possible. Some instruments have switching valves for gas change, and flow meters for gas flow measurement.
Vacuum	1–760 torr common, but some systems down to 10^{-3} torr
Pressure	Rare for TG systems to have pressure operation capability
Furnace and Temperature Programmer Characteristics	
Temperature range	25–1000°C (but some systems operate down to −196°C or up to 2400°C)
Heating rates	0.1–160°C/min
Cooling time	10 min from 1000 to 100°C (although not many manufacturers quote a value)

weights, there is usually some in-built suppression to make final sensitive adjustment to put the recorder pen on scale, or to zero the balance.

Details of the sensitivity or detection limit, precision, and accuracy of the balance are not always evident in the literature, and the terminology used in the publicity brochures is not always descriptive or accurate. Some degree of standardization would be most helpful to the prospective buyer. In some cases, the sensitivity can be assumed to be about 1/100 of the minimum full-scale deflection mass range, since most charts are calibrated in 100 small divisions full scale, but digital readout systems can measure to a greater accuracy than this.

Sample capacities tend to be in the region of 50 μL for microbalance-based systems, giving a maximum sample mass of 40–100 mg, depending on the density of the sample (see Section II.A.1). Some equipment, such as that produced by Linseis, can take actual sample loads of up to 15 g. Based on the manufacturer's quotations, LPR values of 10^4 are typical of microbalances

(10 mg sample with sensitivity of 1 μg) and 5×10^5 for the larger balance systems (10 g sample with 20 μg sensitivity).

There are three main arrangements of the furnace relative to the weighing arm of the balance, one horizontal and two vertical. Lateral loading is a horizontal arrangement, in which the balance arm extends horizontally into a furnace aligned horizontally. Of the two vertical arrangements, one has the balance above the furnace with the sample suspended from the balance arm (bottom loading), and the other has the balance below the furnace and the sample resting on a flat platform supported by a solid ceramic rod rising from the balance arm (top loading). Advantages claimed are as follows:

Horizontal Mode

• Increased balance sensitivity because of long balance arm
• Permits rapid gas purge rates, up to 1 L/min, since gas flow stream perturbs balance arm less than for a vertical arrangement
• The influence of the Knudsen effect and convection errors can be ignored, i.e., eliminate "chimney" effects from the furnace
• Simplifies evolved gas analysis
• Reduces the condensation of evolved materials on the hang-down wires

Vertical Mode

• Enables much higher temperatures to be achieved, since the change in length of the sample holder does not affect the balance arm.

According to the survey of Escoubes et al. (134), most TG equipment (18 models) is based on the bottom-loading principle, although significant numbers of top-loading systems are evident (11 models). Only three models of lateral loading TG systems are apparent.

Some form of DTG capability is almost always an accepted part of a TG apparatus. When this is an electronic unit, the sensitivity is usually expressed as rate of mass change in milligrams/minute full-scale deflection or milligrams/minute/centimeter chart. All TG apparatus connected to microprocessor facilities will have derivative capability.

The ability to have controlled flowing atmospheres is usually available, with gas flow of up to 150 mL/min being typical. Some instruments offer threaded gas valve connections, which is obviously a superior arrangement. Work under vacuum down to 1 torr is common, and many thermobalances can operate down to 10^{-3} torr. One of the recommendations of ICTA (235) is that the sample atmosphere should be identified by pressure, composition, and purity, so some atmosphere monitoring device needs to be incorporated into the TG system. Not many instruments are fitted with gas flow meters, and this will be an additional requirement to monitor gas flow rate. Similarly, not many instruments are fitted with vacuum gauges. Only one balance is reported to be able to work at high pressure; the Thermomat S made by Netzsch has a hermetically sealed weighing compartment (323) and can operate up to 190 bar (134).

The most frequently encountered maximum furnace temperature is 1000°C, which can be achieved relatively cheaply with nichrome furnace windings. Higher temperatures up to 2400°C inevitably involve more exotic forms of heating, and hence these furnaces are rather more costly. Models made by Ulvac-Rico and Rigaku use infrared image furnaces for very high heating rates; the Rigaku model, for example, can reach 900°C from ambient temperature in 30 s and has a maximum operating temperature of 1350°C.

Although heating rates often increase in steps, say 1, 2, 5, 10, 30, 50, 100°C/min, microprocessor-controlled systems allow increments of 0.1°C/min, which can be useful in kinetic studies based on multiple heating rate methods (Section IX.B). Steps allow reproducibility from one experiment to another, whereas heating rates set by a dial may differ by small but significant values between runs. Certainly, the ability to cover a heating range of 1–100°C/min is useful.

Some instruments have been manufactured to overcome problems inherent in the designs described above. One area of concern is the various disturbances that can occur to the weight record, as outlined in Section V.A. These can be considerably reduced by use of a symmetrical thermobalance, in which both sample balance pan and counterbalance pan are suspended from either end of the balance arm into two identical furnaces. Both furnaces are programmed, so that buoyancy and convection effects are almost eliminated from the weight record. Seteram produces a range of these instruments, one of which operates from −196°C up to 1700°C (253). Rigaku manufactures a differential motion TG apparatus based on a top-loading arrangement in which the sample and counterbalance pans are supported on a system of beams and sub-beams so that the distance between the pans is small enough to be incorporated into the same furnace. Hence, again the pans are subjected to the same environment, which reduces errors in the weight record.

A magnetic suspension balance for use with highly corrosive atmospheres is manufactured by Netzsch based on the design of Gast (168). This is a single arm beam balance in which the pan is attached to a permanent magnet which is kept in suspension below an electromagnet attached to the hangdown wire. The distance between the two magnets is controlled electromagnetically. The sample chamber and the atmosphere containing the sample are thus isolated from the balance, and no attack on the mechanism can occur.

Although the range of TG equipment now available is extensive, there are still some instances when special requirements are necessary. These include working in very high vacuum or in corrosive or dangerous atmospheres, or

requiring high sensitivity and stability in the balance. The literature contains
many examples of these specialist TG apparatuses.

IV. EXPERIMENTAL VARIABLES

The shape of both TG and DTG curves is markedly dependent on various
experimental variables, such as (1) the form of the sample, e.g., its particle
size and morphology, (2) the mass of the sample, (3) the design of the
crucible, (4) the chart speed of the recorder, (5) the heating rate, and (6) the
atmosphere surrounding the sample. The effect of some of these variables
taken one at a time on the appearance of a TG curve is shown schematically
in Fig. 12. The combined effect of variation in heating rate and sample mass
on actual $T\alpha$ values for the decomposition of calcium carbonate has already
been mentioned (Table 2). It can be a very considerable effect, as further
indicated by TG and DTG curves for p,p' DDT (Fig. 13). Liptay (222) has
provided a compilation of such curves for many common inorganic and

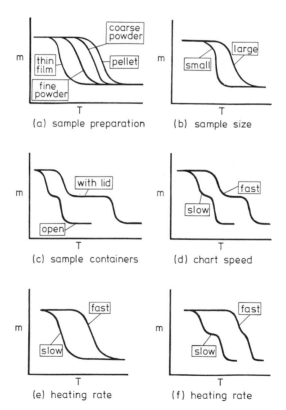

Fig. 12.   Schematic diagrams of the effect of experimental variables on the TG curve.

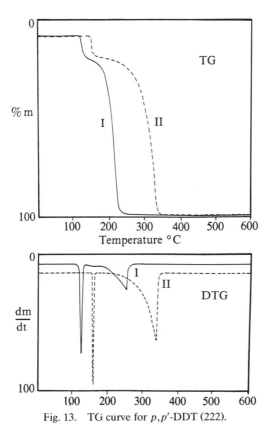

Fig. 13.   TG curve for p,p'-DDT (222).

organic materials. The effects of these procedural variables are discussed in turn in the sections that follow.

A. SAMPLE PREPARATION

The finer the particle size of a solid, the more reactive it is and hence, under isothermal conditions, the greater its rate of reaction, because of its increased surface area. It is to be expected that this increased reactivity will have a direct effect on the appearance of TG and DTG curves. Usually, the more finely divided the sample, the lower the temperature at the start and the completion of the reaction, as shown schematically in Fig. 12a. In practice, however, variation in the particle size may lead to two displaced, but more or less parallel, curves (Fig. 14) or to two curves markedly different in shape as well as displaced in temperature range (Fig. 15). Biotite (Fig. 15) is a mica and the flakes have very little adsorbed water. When ground up, the sample area is much increased and some structural damage occurs. Both

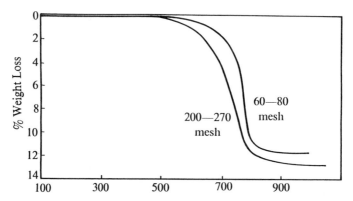

Fig. 14. TG curves for serpentine from Cardiff, Maryland. Samples of approximately 100 mg heated at 3°C/min in air. 180–250 μm (60–80 mesh); 53–74 μm (200–270 mesh). From B.N.N. Achar, G. W. Brindley, and J. H. Sharp, unpublished data.

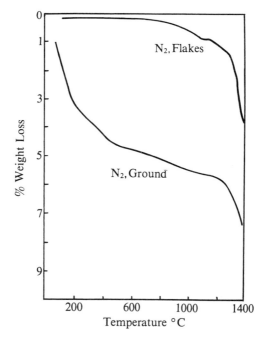

Fig. 15. TG curves for biotite from Bancroft, Ontario. Samples of approximately 400 mg heated at 1.1°C/min in flowing N_2. From A. D. White and J. H. Sharp, unpublished data.

phenomena lead to the presence of adsorbed water, and hence substantial low-temperature weight loss.

The use of pellets or deep beds or powder may lead to a change in mechanism from a rate-determining step based on the chemical reaction to one based on heat or mass transfer. There have been suggestions (189, 190) that heat and mass transfer may be rate determining even when thin layers are used under isothermal conditions, involving reactions previously supposed to be controlled by reaction at the interface between the reactant and the product.

One problem that must be avoided is the loss of part of the sample through ejection. This occurs most commonly when large crystals are heated or if liquid is formed, as, for example, in the decomposition of ammonium alum, which dehydrates and the anhydrous salt dissolves in the liberated water, which then boils away with much frothing.

Generally, it is desirable to use a relatively small sample and to avoid too great a temperature gradient through the sample. No set rules can be laid down, however, since each case should be considered on its merits, including the sensitivity of the balance, the expected change in mass, and the type of information required.

B. SAMPLE MASS

Temperature gradients are inevitable in a dynamic thermal method, because of the experimental conditions. There is a temperature gradient between the wall of the furnace and the center of the furnace tube, where the sample should be placed. At a constant heating rate this temperature gradient or lag is approximately constant. A second and more important temperature gradient is that within the sample, which varies according to the size of the sample and the enthalpy of the various reactions that occur within the sample. This temperature gradient is, of course, the essence of differential thermal analysis and often experimental variables are adjusted to maximize it. In TG, especially if kinetic parameters are to be determined, it is preferable to operate under conditions that maintain the temperature gradient at a minimum.

The magnitude of these temperature gradients was investigated by Newkirk (268), who measured the difference in temperature between thermocouples placed inside the crucible and those close to the furnace wall. Three runs were compared: one in which the empty crucible was heated, one in which the crucible contained 200 mg of calcium oxalate monohydrate, and the third in which it contained 600 mg of the same substance. The other experimental variables were unchanged. In the run with the empty crucible a temperature lag was soon observed and this remained approximately constant at about 12°C. In the presence of the sample, however, the temperature gradient changed with temperature and with the sample size. During endothermic dehydration the lag became 25°C with the 200 mg sample and over 30°C with the 600 mg sample; during the exothermic step (due to oxidation of carbon monoxide liberated while oxalate was converted to carbonate) the lag was reduced to a mere 2°C with the 600 mg sample; during the final endothermic step (due to the decomposition of calcium carbonate) the lag increased again to nearly 30°C with the 600 mg sample. Such variations in the temperature gradient affect the heat transfer characteristics between the furnace and the sample, and hence the temperature range in which changes in mass occur. However, the use of modern TG equipment based on a microbalance permits the use of 1 to 10 mg size samples, and thus temperature gradients within the sample are much less severe. In addition, the small thermal mass of the hang-down and sample means that thermal lag between sample and furnace is considerably reduced compared to the older, more massive TG systems.

Wendlandt (385) has suggested a means of reducing the thermal lag between the sample and a fixed measuring thermocouple located beneath the sample. This involves placing a second crucible and sample on the thermocouple plate, so that the thermal masses of each crucible and sample are essentially equal. Besides a more accurate record of the sample temperature, claimed to reduce the thermal lag to approximately 1 K as opposed to up to 10 K without the second crucible, better resolution of the TG curves for the dehydration of salt hydrates was achieved (385).

In general, for an endothermic reaction, the larger the sample weight, the higher is the temperature range at which the TG curve is observed (Table 2). The DTG curves for the decomposition of calcium carbonate shown in Fig. 16, however, clearly indicate that the onset temperature T_i is little affected by sample weight, but peak temperature T_p and final temperature T_f are displaced to higher temperatures in the case of the heavier sample. The

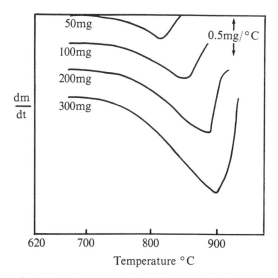

Fig. 16.   The effect of sample mass on the DTG curves of calcium carbonate (238).

magnitude of this effect is often underestimated and the data listed in Table 2 are typical of those for many decomposition reactions. Clearly, this effect must be borne in mind when comparing data obtained from one thermobalance with those from another.

Finally, the importance of the location of the measuring thermocouple in the thermobalance should be emphasized. In modern equipment the thermocouple is usually placed very close to the sample, but in older equipment it may be near the furnace wall and temperature lag may be considerable, perhaps as much as 40°C.

C. SAMPLE CONTAINERS

Both the geometry of the crucible and the material from which it is made can influence the shape of a TG curve. Crucibles are usually fabricated from either metal (often platinum) or ceramic (porcelain, alumina, or quartz) and their thermal conductivity varies accordingly. They may be conventionally shaped, i.e., deep with tapering sides, straight sided, in the form of a shallow dish, or even multiplate. The thinner the layer of sample, the smaller the buildup of gas within the sample bed and the greater the rate of evolved gases from the surface.

Another important factor is whether the sample container is open or closed with a lid. In the latter case, any evolved gases quickly displace the original atmosphere and may markedly affect the subsequent course of the decomposition. A possible consequence is a clearer plateau in a TG curve (Fig. 12c) when the sample is encapsulated, but less desirable effects may result and it is usual to leave sample containers open. Crucible lids are, however, frequently recommended for the thermal treatment of analytical precipitates to avoid any loss of sample through decrepitation or foaming.

Care must be taken to avoid the possibility of reaction between the sample and its container, as can occur when nitrates or sulfates are heated and in the case of polymers containing halogens, especially fluorine, phosphorus, or sulfur. Less obviously, platinum crucibles have been observed to act as a catalyst, causing change in the composition of the atmosphere; e.g., Pt crucibles promote conversion of SO_2 to SO_3. Hence, if sulfides are heated in air or oxygen, the weight gain due to sulfate formation is always greater if Pt crucibles are used, especially relative to ceramic crucibles. The reaction sequence is

$$MS + 1.5O_2 \longrightarrow MO + SO_2 \xrightarrow{O_2/Pt} MO + SO_3 \longrightarrow MSO_4$$

D. CHART SPEED OF THE RECORDER

With strip chart recorders it is possible to select a chart speed to suit the reaction sequences being studied. If the reactions are well separated, then it is usual to choose a chart speed that is a compromise between the most suitable length of chart for storage convenience and the temperature resolution required. If reactions are not well separated the chart speed can be increased so that the length of chart between reactions is increased. This is illustrated in Fig. 12*d*. It would be easier to draw tangents to the weight loss curves to estimate weight loss or extrapolated onset or offset temperatures at the faster chart speed than at the slower chart speed.

The chart speed can also be important in measuring induction time, that is, the time between insertion of the sample into the furnace and the onset of reaction. Here the chart speed is chosen to maximize the accuracy of the estimation, so that fast chart speeds are chosen for short induction times and slow chart speeds are chosen for long induction times.

E.   HEATING RATE

In Section II it was established on theoretical grounds that the temperature range over which a TG curve is observed is dependent on the heating rate. Indeed, it is found experimentally that the procedural decomposition temperature, the DTG peak temperature, the final temperature, and the temperature range all increase markedly as the heating rate is increased. The use of fast heating rates may lead simply to a displacement of the TG curve (Fig. 12*e*) or to a loss in resolution (Fig. 12*f*) if two or more steps of consecutive reactions merge together. The magnitude of the effect is indicated by the data listed in Table 2 for the decomposition of calcium carbonate.

Rapid heating rates can be used safely when the sample size is very small, but the combination of a fast heating rate and a large sample often leads to a TG curve significantly different from that obtained with a smaller sample at a slower heating rate. This observation is well illustrated by the TG curves (222) for *p,p'* DDT (Fig. 13) and acenaphthylene (Fig. 17). In the latter case the course of the reaction has changed because of the change in procedural variables. Another example illustrating a change in mechanism with heating rate for the oxidation of sulfides is given in Section VIII.A.6.

Sometimes to resolve two consecutive reactions it may be desirable to maintain a constant temperature for a period during a dynamic thermal experiment. This is particularly the case in analytical determinations, wherein the overlapping reactions may be due to two or more different components of the mixture. Thus, alunitic clays contain various proportions of the minerals alunite, kaolinite, and quartz, all of which give thermal effects in the range 500–600°C. To separate the loss of water from alunite and kaolinite completely from the loss of sulfur trioxide from alunite, Pekenc and Sharp (291) used TG with isothermal intervals. The method of analysis involved heating to constant weight at 300°C, followed by heating at 1°C/min up to 480°C and again maintaining until constant weight was achieved. Finally, the sample was heated to 1000°C. The heating at 300°C drives off any physically adsorbed

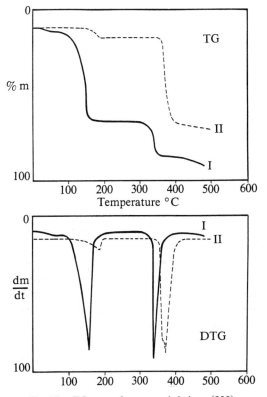

Fig. 17.   TG curve for acenaphthylene (222).

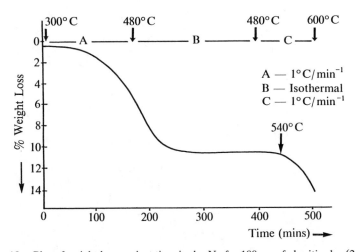

Fig. 18.   Plot of weight loss against time in dry N_2 for 100 mg of alunitic clay (291).

water; that at 500°C drives off the structural water associated with alunite and kaolinite; the final heating at 1000°C drives off the SO_3 formed from the sulfate ions of alunite. The method is reliable only if the weight losses are truly separated; no water must be retained beyond the isothermal heating at 480°C and no SO_3 lost until after that isothermal period. To establish this a sample was heated at 480°C to constant weight, which was achieved after 3.5 h; subsequently, the sample was heated at 1°C/min. The results are shown as a plot of weight loss against time in Fig. 18. It can be seen that loss of SO_3 did not commence until 540°C.

F. ATMOSPHERE

One other experimental variable that is of great significance is the effect of atmosphere on the appearance of TG and DTG curves. Many workers, especially in early studies, have operated their thermobalances in static air, but nowadays the use of controlled atmospheres is almost universally regarded as desirable and for many applications obligatory.

Two effects may be distinguished. The first is the suppression until a high temperature is reached of a decomposition by the presence of a volatile product in the atmosphere, e.g., the decomposition of calcium carbonate in various partial pressures of carbon dioxide, as shown in Fig. 19(238). This substantial effect is, in fact, closely related to that due to sample weight, since the increased amount of volatile product produced by heavier samples modifies the composition of the atmosphere close to the surface and within

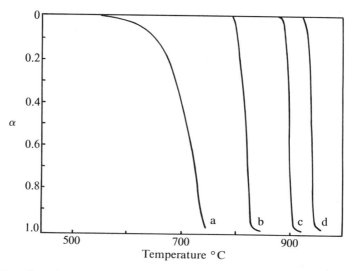

Fig. 19.   The effect of atmosphere on the decomposition of calcium carbonate. All four curves relate to 100 mg samples heated at 1.1°C/min. The atmospheres were (a) CO_2-free N_2, (b) 10% CO_2-N_2, (c) 50% CO_2-N_2, (d) 100% CO_2 (238).

the bed of the sample. When the atmosphere within the crucible is predominately that of the gaseous product, the term "self-generated" atmosphere is frequently used. Newkirk (270) has reviewed the effects on TG experiments of working in self-generated atmospheres. Paulik and Paulik (286) have used specially designed labyrinth crucibles containing baffles to deliberately inhibit gaseous diffusion away from the sample. After the commencement of reaction, the gaseous product within the crucible reaches atmospheric pressure and remains constant until the end of the process, which led to the technique being named "quasi-isobaric."

The second effect arises when the course of the reaction changes because the sample reacts with the atmosphere. Thus, when a polymer is heated in air it undergoes an exothermic oxidation (Fig. 20), but on heating in an inert atmosphere it undergoes an endothermic degradation, usually at a higher temperature. The course of the reaction may also be changed when micas are heated in air rather than in nitrogen, and the partial pressure of carbon dioxide has a surprising effect on the decomposition of dolomite, $CaMg(CO_3)_2$; both of these situations are discussed in Section VI.

The first effect, due to the presence of a partial pressure of air or evolved gas is a manifestation of the second law of thermodynamics. Indeed, thermodynamic data can be used to calculate theoretical decomposition temperatures, which can be compared with procedural decomposition temperatures obtained from thermogravimetry. The calculation may be illustrated for the decomposition of calcium carbonate (238):

Since

$$\Delta G^\circ = \Delta H^\circ - T\Delta S^\circ = RT \ln P_{CO_2}$$

$$= 40\,250 - 34.4T = 4.6T \log P_{CO_2}$$

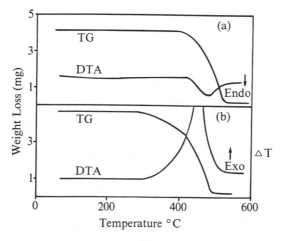

Fig. 20. Simultaneous TG–DTA of 10 mg samples of polyethylene heated at 20°C/min in (a) N_2, (b) air. From J. G. Dunn and I. D. Sills, unpublished data.

TABLE 8

Comparison of Thermodynamic and Procedural Decomposition Temperatures for
the Decomposition of Calcium Carbonate at Various Partial Pressures of CO_2 (238)

Atmosphere ($atm\ CO_2$)	Decomposition Temperature (°C)		
	Thermodynamic	Onset	$T_{0.1}$
0.1	759	780	798
0.3	821	830	850
0.5	852	865	887
0.7	873	885	905
1.0	897	910	930

which gives

$$T = 1170\ \text{K or } 897°\text{C at } P_{CO_2} = 1\ \text{atm}$$

and

$$T = 1032\ \text{K or } 759°\text{C at } P_{CO_2} = 0.1\ \text{atm}$$

Hence, the reduction in the partial pressure of CO_2 from 1 to 0.1 atm leads to a decrease of 138°C in the thermodynamic decomposition temperature. The procedural decomposition temperatures, while higher than the thermodynamic decomposition temperatures, can be seen to be affected by a similar order of magnitude (Fig 19; Table 8).

The procedural decomposition temperature is higher than the thermodynamic decomposition temperature because it is also influenced by problems of heat transfer and by kinetic factors. Indeed, it can be shown from thermodynamic reasoning that many reactions that do not proceed until relatively high temperatures should take place at much lower temperatures, sometimes even at room temperature, if it were not that the diffusion of ions through the reacting solid is so slow that the reaction proceeds to a negligible extent.

To illustrate this point, consider the reaction between calcium carbonate and silicon dioxide:

$$CaCO_3 + SiO_2 = CaSiO_3 + CO_2$$

Thermodynamic data are available for both the reactions:

$$CaCO_3 = CaO + CO_2$$
$$CaO + SiO_2 = CaSiO_3$$

and by combining these data we have for the formation of calcium metasilicate,

$$\Delta G° = 18\,950 - 34.3T$$

When $\Delta G° = 0$, $T = 552$ K or 279°C. Therefore, the presence of silica lowers the theoretical decomposition temperature in the presence of 1 atm of CO_2 from 897 to 279°C. It is emphasized again that this is a theoretical decomposition temperature, since kinetic factors determine that the reaction is too sluggish to be observed at 279°C; kinetic studies are almost invariably made under nonequilibrium conditions.

One final point should be made with respect to the effect of atmosphere: It is necessary to consider the possible interaction of any evolved gases with the apparatus. In addition to possible corrosive chemical reactions, there is the possibility of condensation of volatile products on cooler parts of the apparatus. These products may become volatile again as the temperature is increased. It is difficult to lay down general rules to avoid these circumstances, since each case needs to be considered on its individual merits. Sometimes modification to the design of the apparatus is required, but frequently the use of a flowing inert atmosphere is sufficient.

G. PRESENTATION OF TG DATA

The influence of the experimental variables on the shape of TG and DTG curves may be so great that care has to be taken when comparing the results obtained on one thermobalance using one set of experimental conditions with those obtained on a second thermobalance using a different set of experimental conditions. It is essential, therefore, that authors provide full details of the experimental conditions under which TG and DTG curves have been obtained when data are presented. The Standardisation Committee of ICTA has recommended that each TG curve should be accompanied by the following information (235):

1. Identification of all substances (sample, reference, diluent) by a definitive name, an empirical formula, or equivalent compositional data

2. A statement of the source of all substances, details of their histories, pre-treatments, and chemical purities, so far as these are known

3. Measurement of the average rate of linear temperature change over the temperature range involving the phenomena of interest

4. Identification of the sample atmosphere by pressure, composition, and purity; whether the atmosphere is static, self-generated, or dynamic through or over the sample. Where applicable the ambient atmospheric pressure and humidity should be specified. If the pres-

sure is other than atmospheric, full details of the method of control should be given.

5. A statement of the dimensions, geometry, and materials of the sample holder; the method of loading the sample where applicable

6. Identification of the abscissa scale in terms of time or of temperature at a specified location. Time or temperature should be plotted to increase from left to right.

7. A statement of the methods used to identify intermediates or final products

8. Faithful reproduction of all original records

9. Wherever possible, each thermal effect should be identified and supplementary supporting evidence stated.

10. Identification of the thermobalance, including the location of the temperature-measuring thermocouple.

11. A statement of the sample weight and weight scale for the ordinate. Weight loss should be plotted as a downward trend and deviations from this practice should be clearly marked. Additional scales (e.g., fractional decomposition, molecular composition) may be used for the ordinate where desired.

12. If derivative thermogravimetry is employed, the method of obtaining the derivative should be indicated and the units of the ordinate specified.

V.   SOURCES OF ERROR

A.   WEIGHT RECORD

Inaccuracies in the weight readings obtained from TG equipment can arise from several sources, as indicated in reviews by Sharp and Mortimer (336) and more recently by Massen and Poulis (246). The obvious first step for accurate mass readings is to calibrate the instrument. The most common technique, because of its simplicity and accuracy, is that of adjusting the balance against standardized masses. These are often provided as part of the accessories when a commercial thermobalance is purchased, but can be obtained from organizations such as ICTA and the National Physics Laboratory. Schwoebel (325, p. 89) has described a simple procedure for the construction of a set of standardized masses. An alternative procedure uses two buoyancy bulbs of the same mass but different volumes for calibration, and provides a simple test for the linearity, precision, and accuracy for any balance at pressures exceeding 1–2 kPa (98, p. 26).

The mechanism of the microbalance is susceptible to changes in temperature, and if long experiments are required then the microbalance environment should be thermostated. Any temperature fluctuation will result in drift in the zero point of the balance and hence decrease the precision of measurement. For high-accuracy work the microbalance should be housed in a thermostated enclosure (rather than relying on laboratory temperature control). Sharp and Mortimer (336) describe an enclosed microbalance controlled to ± 0.5 K, which over a 12 h period gave a reproducibility of approximately 0.5 mg at room temperature.

Spurious results can be obtained if electrostatic or magnetic forces are generated in the thermobalance and interact with the system. Magnetic effects can be overcome by winding the furnace in a noninductive manner, and electrical interference can be eliminated by grounding.

Loss of precision results from vibrational transmission from the environment to the thermobalance and can easily be recognized on a chart recorder as a persistent pen fluctuation. Schwoebel (325) has recommended that a seismic accelerometer be used to find the most suitable sites for microbalance installation, although if no choice of site is available then the balance may be rigidly mounted on a massive bracket attached to a main wall (336), or some form of spring or dampening mechanism may be inserted between the balance and balance platform (246).

Most of the other sources of error in the weight record derive from the forces that act on the balance as a result of the atmospheric environment, including buoyancy effects, convection in the furnace atmosphere, and various other molecular gas movements.

Errors to the weight record will arise if the established environment changes in any way during the course of the experiment. Hence, changes in the pressure of the gas in a static pressure experiment, or in the gas flow rate in a dynamic gas environment, will result in changes in the weight record that are not related to changes in mass of the sample. Hence, it is necessary to ensure that the environment remains consistent to obtain sensible results.

One of the major errors in thermogravimetry results from buoyancy effects, although the force arising can be calculated for an ideal gas environment from the equation

$$m = \frac{PM\,\Delta V}{RT}$$

where m is the mass of the displaced gas, P is the pressure in torr, M is the molecular weight of the gas, T is the temperature of the gas in K, and R is the gas constant of value 62 364 cm^3 torr/mol K. ΔV is the difference in volume between the sample holder, hang-down, and beam on either side of the balance pivot. In the older style thermobalances, ΔV was quite large and hence significant effects of the order of several milligrams increase in mass were apparent over a temperature range of ambient to 1000°C. In modern thermobalances based on microbalances, ΔV tends to be much smaller and hence the buoyancy correction tends also to be small. The only way to reduce the problem significantly is to use a symmetrical thermobalance in which ΔV is zero. The buoyancy value can be determined experimentally using a crucible containing an inert material such as calcined alumina as a substitute sample, and then following the mass change under experimental conditions identical to those used for the samples under investigation. The resulting curve may be used to obtain a corrected plot or to correct individual readings. Even then it should be remembered that during actual experiments the buoyancy of the sample itself will change, especially if there is a significant increase or decrease in sample mass through chemical reaction. These buoyancy effects have been well illustrated by reference to the decomposition of $CaCO_3$ (279).

The theoretical calculation using the equation above can be used to calculate the buoyancy effect, and significant deviations between observed and calculated results may give indications of other errors that will need investigation. The seriousness of the buoyancy problem will depend on the particular application. For routine quality control work involving the determination of major components in the sample, and using a microbalance, it is advantageous to ignore the buoyancy effect and display the weight change as a direct percentage of the sample weight, thus eliminating the need for calculation. The direct readout can be achieved with a suitable recorder or a digital printer.

Convection currents arise in TG equipment at pressures in excess of about 10 kPa, and cause fluctuations in the weight record as well as spurious weight changes. Czanderna and Wolsky (98, p. 37) cite an example of a fluctuation of ± 5 mg at 66.7 kPa in nitrogen rising to ± 15 mg at 101.3 kPa. The balance pan was at $-196°C$ and the balance beam at $27°C$.

Thermomolecular flow (TMF) (or Knudsen force) is caused if the sample temperature deviates from the temperature at which the balance is held at pressures of less than 1 kPa. The magnitude of the force is dependent on the gas, pressure, and temperature gradient of the system (98, p. 38). Again, this error can be best reduced by use of a symmetrical TG system. Haglund and Luks (186) examined mass changes due to TMF in a TG apparatus and found that at temperatures below 400°C and at low flow rates the necessary corrections amounted to less than 0.1 mg, but above 400°C and with gas flow rates above 150 mL/min the errors were up to 0.5 mg.

A novel effect has been reported during the dehydration of the mineral whewellite ($CaC_2O_4 \cdot 2H_2O$) (390, 279). At an initial pressure of 10^{-5} torr and a heating rate of 8°C/min, a weight gain was observed, followed by a decrease as water was lost from the system. The authors concluded that the weight gain was almost entirely due to the reimpact of water vapor molecules on the sample after escape from the sample, with about a 0.5% contribution from the recoil as the gas molecules left the sample. The effect was not evident at slow heating rates of 2°C/min. Absorption and desorption of gases onto the balance parts can occur, even on counterweights used in thermogravimetric experiments (311).

These few comments illustrate the need for care during TG experiments. It is useful for the operator to carry out blank runs and buoyancy checks under conditions identical to those used in a set of experiments, and also to calibrate the instrument at operating temperature to determine the magnitude of any errors. It is also useful to calculate the molecular weight (279) or purity of a well-characterized salt as a check on accuracy. These checks should be carried out for any different experimental conditions used, since the effects operating and their magnitude will alter with factors such as pressure, temperature, and gas flow.

B.   TEMPERATURE RECORD

Despite persistent efforts by an ICTA subcommittee and individual workers, accurate temperature calibration and even interlaboratory comparative temperature measurement of TG equipment remains problematic. The problem has been lucidly stated by Garn et al. (163):

> In most thermobalances the temperature sensor is not in contact with either the sample or the sample holder; the temperature of the sample is inferred from the measured temperature of sensor. Unless the temperature gradient within the furnace is known from experiment to be small, the operator has a considerable uncertainty in his estimate of the sample temperature. The actual error will vary with the distance and direction of the sensor with respect to the sample and with the geometry of the furnace-sample-sensor assembly. Particularly if the sensor is above or below the sample, the temperature difference between the sample and sensor will vary both with temperature and with heating rate. Still further, the heat transport through vacuum is different from that through air or nitrogen or from that through argon or carbon dioxide; the sample-sensor temperature difference will therefore be dependent upon the atmosphere and its pressure.

Early attempts at calibration depended on the chemical decomposition or dehydration of materials, but were abandoned because of the dependence of such reactions on procedural variables. Large variations, in excess of 25°C, were found between collaborators in round robin experiments.

More recently the ICTA committee has investigated the use of Curie point transitions in metal samples. In this method, a ferromagnetic material is placed in the sample crucible, and a magnet is located above or below the sample holder so that the magnetic flux is aligned with the gravitational field. The sample is then heated, and at a particular temperature the ferromagnetic material becomes paramagnetic, which results in a significant apparent weight change in the sample (276). Three temperatures have been defined on the heating cycle for this apparent weight change: the extrapolated onset temperature, the mid-temperature, and the extrapolated offset temperature of the ferromagnetic to paramagnetic conversion. Five metals have been examined, and are available from the USA National Bureau of Standards as ICTA Certified Reference Materials for Thermogravimetry GM 761. The materials chosen cover the temperature range 200–800°C. The values determined experimentally in a round robin investigation have been fully reported

TABLE 9
Measured Values for Temperature T_e Using ICTA Certified Reference Materials
for Thermogravimetry GM 761

	Range (°C)	Mean Value T_1 (°C)	Standard Deviation
Permanorm 3	242–263	253.3	5.3
Nickel	343–360	351.4	4.8
Mumetal	363–392	377.4	6.3
Permanorm 5	435–463	451.1	6.7
Trafoperm	728–767	749.5	10.9

SOURCE: Garn et al. (163).

by Garn et al. (163). Generally, precision was reasonable for intralaboratory comparisons, with a mean standard deviation over all participants of 3.6 ± 1.1°C. However, a considerable range of results was experienced for interlaboratory comparisons (see Table 9). The range of results increased as the temperature of the ferromagnetic temperature increased, and the standard deviation of the measured value also increased. The variation between laboratories was attributed to the different instruments used, and in particular to the different geometric relationships between the sample and the temperature measuring point. The use of the materials has been criticized because of the large differences that occur between values obtained by TG and other methods (130, 236). Garn (164) has pointed out that the published values are simply mean values derived from TG experiments.

Other methods for temperature calibration have been proposed. Wiedemann and Bayer (391) used a molecular beam of gases directed onto the sample holder and measured the apparent gain in weight. Gases used were CO_2, N_2, O_2, air, Ar, and ammonia, and weight changes were measured for constant pressures between 1×10^{-5} and 5×10^{-5} torr at different isothermal temperatures. According to the Maxwell theory, the measured weight changes are proportional to the square root of the absolute temperature. This condition was found to be satisfied providing the distance between the gas inlet and the sample plate was long enough (100 mm) to allow the gas molecules to attain the mean velocity corresponding to the temperature in the isothermal region. The method is limited to the vacuum region less than 10^{-3} torr. Standard deviations of the temperature of the order of ±5° and ±1° were obtained at 952 and 17°C, respectively (279).

McGhie et al. (236) have suggested a novel technique that permits the melting points of metals to be used. A weight is suspended within the sample crucible of the thermobalance by a thin wire of the temperature calibration material. The system is heated, and at the melting point of the wire the weight drops either onto the sample crucible, causing a momentary disturbance to the weight record, or through a hole in the base of the sample crucible, giving a large, rapid weight change. The latter procedure cannot be used if the temperature-measuring thermocouple is directly under the sample crucible. The authors claim superior precision for this method over others. The technique has been used to provide new estimates for the transition temperatures of the ICTA Standards (38).

A satisfactory calibration method is unlikely to emerge for universal temperature standardization because of instrument differences. The magnetic standards have value as the most proven method to date for intralaboratory calibration. This means that any work undertaken by TG that requires a high order of accuracy in the temperature measurement, such as in kinetic studies involving the calculation of rate constants and pre-exponential factors, is likely to be subject to significant error. For these studies it is much more appropriate to use simultaneous TG-DTA, where the thermocouple is in contact with the sample crucible and the instrument can be calibrated using ICTA standard for DTA.

VI. THE SCOPE OF THE METHOD

Any chemical or physical process that involves gain or loss of mass can be followed by thermogravimetry, and so the technique is suitable for the study of a wide range of problems across a broad spectrum of scientific interest. The most important reactions capable of being studied by TG are listed in Table 10. These general areas of application are of interest to all thermanalysts, and hence are discussed prior to the more detailed discussion of individual classes of materials in Section VIII.

A. THERMAL DECOMPOSITION

The temperature of a reaction can be defined in thermodynamic terms as the temperature at which the free energy of that reaction becomes equal to zero. This implies that under equilibrium conditions there is an unambiguous constant value for the decomposition temperature. In practice TG deter-

TABLE 10
Phenomena Capable of Study by Thermogravimetry

	Weight Gain	Weight Loss
Adsorption and absorption	x	
Desorption		x
Dehydration or desolvation		x
Vapourisation		x
Sublimation		x
Decomposition		x
Oxidation	x	x
Reduction		x
Solid–gas reactions	x	x

mines a temperature range over which a thermal decomposition reaction proceeds and this range is entirely dependent on the experimental conditions employed, such as heating rate, sample weight, atmosphere, and the design configuration of the instrument used.

The practical methods for assessing the thermal decomposition characteristics of a material have been discussed in Section II and illustrated in Figs. 3 and 5. The extrapolated onset and offset temperatures give reproducible results for the identification of the temperature at which reaction starts and finishes, and the temperature at which a certain fraction of weight loss has occurred (T_α) is very useful for comparative studies.

At this stage it is useful to consider the use of TG as an aid in assessing the thermal stability of materials. The technique is usually used on a comparative basis to provide a rapid method of ranking a group of materials. For this purpose T_α values are recommended, the value for α being chosen on the basis of inspection of the initial parts of the curves obtained from the different materials under identical experimental conditions. It must be emphasized that the values determined are not absolute values and cannot be used to indicate a temperature at which the material may be stored for a period of time or a reaction intermediate prepared.

The decomposition temperature determined under dynamic conditions gives no information about the effect of holding the material at lower temperatures for prolonged periods of time. Information of this kind can be obtained by heating the material to a specific isothermal temperature and following the weight changes as a function of time. Usually several temperatures are examined in separate experiments, thus generating a family of curves (see Section VI.C). This method is usually more successful for distinguishing between similar materials than the dynamic method.

An extension of the isothermal method is a test devised to measure the oxidative stability of materials, which involves heating the material in an inert atmosphere to a given temperature, establishing a stable baseline, and changing the atmosphere to oxygen or an oxidizing atmosphere. The time taken to produce the first major weight loss, called the oxidative induction period, is used as a measure of the oxidative stability (refer to Section VIII.B.2). An analogous DTA test has become established to measure the oxidative stability of polyethylene where an isothermal temperature of $200 \pm 1°C$ is used (77).

The technique of TG has advantages over other dynamic techniques in evaluating thermal stability in that quantitative results are given directly and instrumental factors such as baseline drift are minimized. The quantitative nature of TG can be further exploited to enable stoichiometric equations for the decomposition mechanism to be postulated. The mechanism may then be confirmed by analysis of the solid and gaseous decomposition products (see below). In some cases the decomposition reaction can be used for analytical determinations, e.g., the quality control analysis for $CaCrO_4$ (87).

B. OTHER CHEMICAL REACTIONS

In this section solid state reactions involving the evolution of a gas, interaction between the sample and the applied atmosphere, and the special case of TG as preparative method are discussed. A thermobalance can be used to follow reactions of the type

$$A(s) + B(s) \longrightarrow C(s) + D(g)$$

as, for example, in the curing of rubbers by elimination or condensation reactions. For this type of process TG can provide a directly quantitative and rapid means of assessing the extent of reaction (21). Examples involving the interaction of a sulfide and sulfate and the reaction of barium carbonate with oxides, e.g., Al_2O_3, are given in Section VIII.A.

A major use of thermogravimetry is to study the reaction between a sample and a reactive gas that is introduced into the sample chamber. The most widely used atmospheres have been those of air and oxygen to study the oxidation of metals, but modern apparatus permits the introduction of a wide range of corrosive gases or even vapors, both under high or low pressures. In this way the thermobalance can be used as a minireactor, for example, in catalyst studies. It is experimentally important in studies of this type to ensure that the reactive gas has easy access to the sample and this is considerably facilitated where the sample is spread in a thin layer in a shallow dish.

Frequently, the thermobalance is used to determine the temperature at which solid state preparative methods should be carried out. An account of the procedures required is given in the analytical section (Section VI.D). In cases where the preparative conditions are very critical, i.e., where there are two overlapping reactions, it is advantageous to carry out the preparation directly on the thermobalance. This technique is also valuable for providing samples of accurately known degrees of decomposition, for example, for preparing X-ray diffraction specimens when trying to determine a reaction stoichiometry.

C. PHYSICAL CHANGES

TG can be used to provide a quantitative measure of the rate of sublimation or volatilization under a wide range of experimental conditions and can be used to obtain heats of sublimation or volatilization, as detailed in Section IX for materials that do not decompose in the temperature range being investigated. When a sample is undergoing simultaneous decomposition and volatilization, the TG curve will reflect the sum of these two processes and it will not be possible to follow the decomposition reaction. This can be effected by carrying out simultaneous quantitative analysis of the evolved gases. It is important to ensure that the volatization product is completely removed from the hot zone, so that it does not undergo further reaction at a later stage in the heating program. The rate of sublimation or volatilization will be increased when small samples are spread over a large surface area, particularly in flowing gases or at reduced pressures. The use of a capillary type crucible will markedly reduce the rate of loss.

TG can be a useful technique to check on spurious results in DTA or DSC. The measurement of the enthalpy of melting of volatile materials requires the sample to be hermetically sealed into a sample pan, especially if repeat determinations are to be made. Thermogravimetry provides the ability to check whether significant volatilization occurs in the vicinity of the melting point using an open sample crucible, and to investigate the integrity of the seal in an enclosed pan. The use of TG to measure changes in the magnetic properties of materials has already been discussed in Section V.B.

D. ASSESSMENT OF ANALYTICAL PROCEDURES

1. Gravimetric Analysis

The precipitates that are of analytical importance are insoluble inorganic substances, such as halides and sulfates; complexes formed between organic ligands and metal cations; and a few ion association compounds, such as potassium tetraphenyl borate $K[(C_6H_5)_4R]$. These precipitates can, after isolation by filtration and washing, either be dried to constant weight in an oven or be ignited to constant weight in a muffle furnace. Thus, it is important to know the precise thermal conditions required to obtain the desired stoichiometric product. Duval has investigated some 1300 precipitates of analytical interest (126–128), and despite criticisms about his techniques, his work remains the most comprehensive.

When a precipitate forms a well-defined stoichiometric compound after removal of water it can be weighed in this form. In the case of metal complexes, advantage is taken of the high molecular weight of the complex, which reduces weighing errors in the calculation of the metal content. The constant-weight region in the TG curve after the evolution of water gives an indication of the drying temperature range (temperature A to B in Fig. 21). When no well-defined product is formed after drying at temperatures up to 200°C, the precipitate may need to be ignited. This can result in the decomposition of the precipitate or the removal of contaminants, such as occluded water or coprecipitated substances, or the volatilization of entrained solutes contained in the wash solution. Hence, the ignition of a precipitate causes many and varied weight losses to occur. The high-temperature plateau in the TG curve (between points C and D in Fig. 21) gives an indication of the ignition temperature.

If suitable plateaus in the TG curve have been located, samples can be heated in the thermobalance to specific temperatures within the plateau

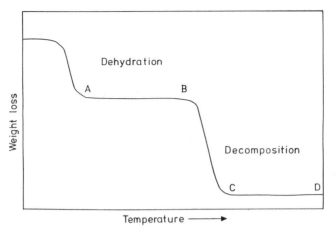

Fig. 21. TG curve for an analytical precipitate showing drying and decomposition ranges.

temperature range and then cooled back to ambient temperatures. This allows any weight gains, indicating perhaps uptake of water or carbon dioxide from the atmosphere, to be monitored. Significant weight gains may render the proposed weighing form invalid, and another weighing form on another plateau must be sought. The appearance of a plateau in a TG curve does not necessarily imply that a stoichiometric product has been formed, and the composition and, if necessary, the structure of the heated product should be checked by chemical and/or instrumental analysis.

The appearance of a plateau in the TG curve does not imply that the product is isothermally stable anywhere within the temperature limits of the plateau (341). Hence, as a final test, samples should be heated to specific temperatures within the plateau region and the weight change monitored as a function of time under conditions that approximate to those used in the analytical procedure. If drying or decomposition is carried out in static air in the analytical procedure then the TG experiments should likewise be carried out in static air. If for several different precipitates after heating at constant temperature, the precipitate analyzes to the required formula and structure, then it is reasonable to assume that the conditions are suitable for an analytical procedure. The final choice of temperature range can be selected to give the desired product within a reasonable time period, usually not more than an hour.

2. Thermal Stability of Analytical Standard Compounds

The preparation of a standard solution remains one of the most important steps in an analytical procedure. It is therefore essential to know the conditions required to produce a standard substance of precise composition.

In most cases this involves drying the material without decomposition, and TG can be used to assess the optimum temperature ranges to accomplish this, in a similar manner to the investigation of analytical precipitates. Again, Duval is often quoted (126–128).

3. Assessment of Volatile Metal Complexes for Analysis by GC

One analytical technique depends on the quantitative formation in solution of a metal complex followed by extraction into organic solvent and then injection into a GC. The complex thus needs to volatilize without decomposition at the operating temperature of the GC. TG has been used to assess the thermal stability and volatization characteristics of metal complexes, for example, for phenylmercury(II) compounds (24) and β-diketone complexes of neodymium, gadolinium, and erbium (338).

VII.   DERIVATIVE THERMOGRAVIMETRY

The definition of DTG and a discussion of the basic features of this technique have been provided in Section I.B. Although the DTG curve contains no more information than the TG curve, the DTG curve presents this information in a form that is more visually accessible (39). The DTG curve also allows the ready determination of the temperature at which the rate of mass loss is a maximum (T_p), and this provides additional characterization to the temperatures shown in Figs. 3 and 5. All temperatures respond to changes in experimental conditions, however, and T_p is no more definitive of a material than the other temperatures (269). The area under the DTG peak is directly proportional to the mass change.

A.   APPLICATIONS

1.   Separation of Overlapping Reactions

Reactions that occur within the same temperature region give TG curves that appear to the eye to consist of one continuous weight change. DTG curves, however, are discontinuous lines, and hence subtle mass changes are emphasized.

Several different sets of reactions can be distinguished (Fig. 22):

1. A single reaction that occurs over a small temperature range gives rise to a single well-defined peak (Fig. 22a).
2. Two reactions that are partially overlapping result in two peaks in which the minima does not return to the baseline (Fig. 22b), i.e.,

$$\frac{dm}{dt} > 0$$

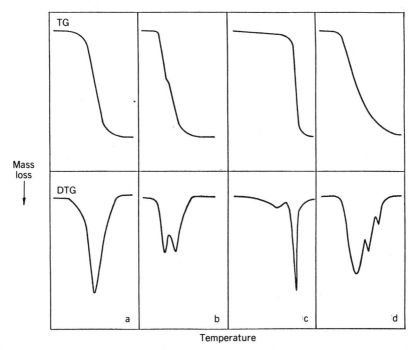

Fig. 22. DTG curves for various reaction types: (*a*) a single-step reaction, (*b*) two overlapping reactions, (*c*) two overlapping reactions that occur at different rates, (*d*) identification of minor reactions occurring close to a major one.

3. Two reactions take place, the first of which occurs slowly and over a wide temperature range and is followed by a fast reaction. This gives a small broad initial DTG peak followed by a well-defined one (Fig. 22*c*).

4. Minor reactions that occur during or near a major reaction can frequently be identified by the appearance of a shoulder or separate small peak in the DTG curve (Fig. 22*d*).

5. Slow reactions that occur over wide temperature ranges, in which other reactions happen, produce gradient changes in the DTG curve.

Examples 22*a* and *b* can be readily distinguished by TG, since distinct breaks occur between the mass changes. However, examples 22*c* and *d* are very difficult to distinguish by TG.

A good example of the use of DTG is provided by the oxidation of pentlandite (FeNi)$_9$S$_8$, which is one of the major nickel-bearing minerals. From the TG curve, one mass gain and three mass losses can be easily distinguished. The DTG curve, however, shows additionally the presence of

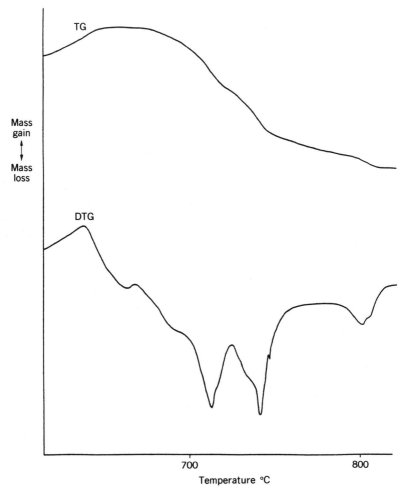

Fig. 23. TG and DTG curves for the oxidation of Nepean pentlandite from the Kalgoorlie region, Western Austalia. 10 mg sample heated at 10°C/min in a flowing air atmosphere (117).

shoulders and changes in gradient which indicate that other reactions are occurring in concert with the major ones (Fig. 23) (117).

2. Fingerprinting Materials

Because the subtleties of the TG curve are visually emphasized in DTG curves, the latter are frequently recorded as part of the characteristic information collected on a new, unknown, or standard material. Figure 24 shows some DTG curves of soil samples obtained from different sources, where it is evident that it is easier to compare DTG curves than TG curves.

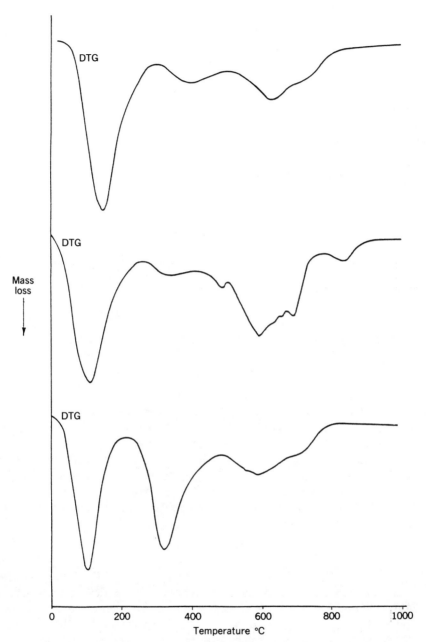

Fig. 24.   DTG curves for some soil samples illustrating the ease of comparison relative to TG curves. 30 mg samples heated at 30°C/min in air. From E. L. Charsley, unpublished data.

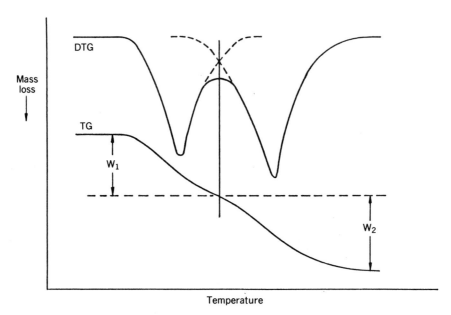

Fig. 25.   The use of the minimum in the DTG curve to separate overlapping weight losses.

3.   Calculation of Mass Changes in Overlapping Reactions

When overlapping reactions occur, it is sometimes difficult to locate on the TG curve an unambiguous point where one reaction ends and the other commences, thus making it difficult to calculate accurately the mass changes that arise from each of the reactions. It has been proposed that the minima in the DTG curve can be used as a means of selecting an objective (although still arbitrary) point on the TG curve (283). This approach is illustrated in Fig. 25 and has been shown to be the most reliable of three different methods used to calculate the oil and polymer contents in an oil extended EPDM rubber (357).

4.   Quantitative Analysis by Peak Height Measurement

As an alternative to calculating the weight changes as described in Section VII.A.3, DTG peak areas or even peak heights can be measured for quantitative purposes. The advantage of DTG in this application is that when no weight change is taking place, the baseline returns to the zero position and so only a simple baseline construction is required. Natural rubber (NR) and butadiene rubber (BR) have been analyzed in a rubber blend by this method (48) (see Section VIII.B.2), and also the degree of conversion in high alumina cement as an alternative to the DTA technique (75, 17) (see Section VIII.A.7).

5. Determination of Kinetic Data

The height of the DTG peak at any temperature gives the rate of mass change at that temperature. These values can be used to obtain kinetic information, since equations can be written of the form (cf. Eq. 4)

$$-\frac{dm}{dt} = Ae^{(-E/RT)}f(m) \tag{8}$$

This technique is less subjective than that of manually drawing tangents to the curve (see Section IX.B).

B. COMPARISON OF DTA AND DTG PEAK TEMPERATURES

It is evident from an inspection of simultaneous DTA–DTG curves that there is a close similarity, both in appearance and in the temperatures at which they occur, between corresponding DTA and DTG peaks. From theoretical consideration, however, an absolute coincidence is not expected, for the following reasons:

1. In the DTG curve, the temperature at which the maximum rate of reaction occurs is given by the DTG peak maximum. In the DTA curve, however, it has been shown (307) that the maximum rate of reaction occurs between the low-temperature inflection point and the peak temperature, but does not coincide with either. Thus, from these considerations, all other factors (sample mass, heating rate, etc.) being the same, it would be expected that

$$T_p(\text{DTA}) > T_p(\text{DTG})$$

2. In the DTG curve, the completion of reaction is signified by a return to the baseline, i.e., $dm/dt = 0$, and so the temperature at which the reaction ceases can be obtained from the extrapolated final temperature. In the DTA curve, however, it has been suggested that the completion of reaction occurs between the maximum and baseline on the downward slope of the DTA peak. Hence, from these considerations it would be expected that

$$T_f(\text{DTA}) > T_f(\text{DTG})$$

For an exothermic reaction, the DTA peak beyond the completion of reaction is a cooling phenomenon, and is a function of various instrumental factors, such as head configuration and heat capacity, and sample properties, such as heat capacity and mass.

VIII.   APPLICATIONS

The versatility of TG to study a wide range of physical and chemical processes, as discussed in Section V, translates to a wide range of applications from fundamental research work to routine quality control. An inspection of *Thermal Analysis Abstracts* (335) gives some indication of the relative importance of TG relative to other TA techniques. The number of references for 1977–1983 for TG, DTA, and TG-DTA are given graphically in Fig. 26. Citations involving TG have risen steadily since 1977; in fact, TG is the most often cited technique of all TA methods. Some increase in activity for simultaneous TG-DTA is also apparent.

It is also possible to infer from *Thermal Analysis Abstracts* something about the relative importance of a topic, because it is possible to cross-reference TG against a particular field of study. The most significant field of application is in inorganic chemistry, where a considerable amount of work has been done on the characterization and thermal decomposition reaction mechanisms and their associated kinetics for a variety of simple and complex inorganic compounds. The other major use of TG is in mineral characterization and analysis. Significant numbers of papers also appear each year in areas such as metallurgy, catalysis, plastics and rubbers, and coals and cokes.

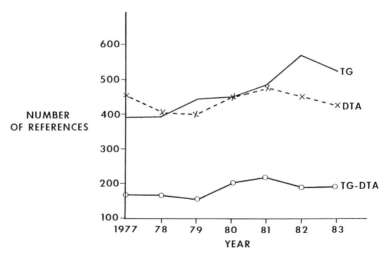

Fig. 26.   Citations in *Thermal Analysis Abstracts* for the techniques of TG, DTA, and TG-DTA between 1977 and 1983. From J. G. Dunn, unpublished data.

The remainder of this section is devoted to a discussion of these applications, illustrated by selected examples from the literature.

A. INORGANIC MATERIALS

1. Metals

The two major areas of application of TG in metallurgy are corrosion studies, particularly the formation of oxide layers, and extractive metallurgy. In the latter, a major concern has been the formation of metals, especially iron, by reduction of the metal oxides, although the thermobalance has also been used for small-scale feasibility studies on processes that allow metals to be segregated from low-grade ores and waste materials. From the weight gains (corrosion) or weight losses (reduction, volatilization), kinetic data can be calculated and information gained on the rate controlling step(s) and mechanism of reaction. Although the validity of absolute kinetic values obtained by TG methods is open to question, nevertheless the use of the thermobalance can provide useful relative information where the effectiveness of a series of, for example, reductants is being compared. The ease of reaction of a series of related materials with a single reactant can also be tested and compared.

Apparatus is often designed or modified for specific metallurgical investigations, for example, work in a highly corrosive atmosphere such as sodium vapor (254).

a. CORROSION STUDIES

TG has frequently been used to investigate the high-temperature behavior of metals and alloys in various corrosive atmospheres. Although the most widely studied systems involve oxygen, either alone or in concert with other constituents such as water vapor and halogens, other important atmospheres include sulfur or sulfur oxides, carbon oxides, and halogens. The main purpose of such studies has been to elucidate mechanisms of oxide formation or other corrosion products at the gas–metal interface. The usual procedure is to heat the metal or alloy rapidly in the desired atmosphere to a predetermined isothermal temperature and then measure the change in weight as a function of time. Several temperatures are usually examined in separate experiments and the specific weight gain, i.e., the weight gain per unit surface area, is fitted if possible to one of the four general equations (parabolic, linear, logarithmic, and cubic) that describe the behavior of the majority of reactions encountered. Sometimes the kinetic behavior fits more than one of the equations over different temperature ranges, which implies a change in mechanism for the growth of the surface coating.

Examples of simple oxidation studies include the oxidation of nickel wire (165), iron–chromium alloys (337), niobium–chromium–nickel alloys (103), zirconium alloys at temperatures between 600 and 900°C (217), and the oxidation of copper in CO_2 at 800–1000°C (367). The last study involved the use of a magnetic winch which allowed the insertion and removal of a specimen to and from the hot zone while under vacuum or carbon dioxide. Jehn (203) described a suspension balance that permitted the oxidation of iridium, rhenium, molybdenum, and tungsten at 1200–2300°C and pressures between 10^{-4} and 1 mbar.

The capability of versatile atmosphere control in the thermobalance can be used to advantage in corrosion studies. Finely divided metal powders, which undergo surface oxidation, can be pretreated by reduction in a suitable atmosphere. After cooling, the reductant can be removed by evacuation or purging and the oxidizing atmosphere introduced. Such a procedure was utilized in a study of the oxidation characteristics of some iron, copper, and nickel powders (185).

The effects of more complex atmospheres can also be readily investigated. The presence of water vapor in the oxidizing medium was shown to cause little change in the oxidation rate of iron–chromium alloys (377), but the rate of reaction between water vapor and aluminum was found to be faster than that between oxygen and aluminum (37). When halogens are introduced into the gas stream, the rate of oxidation usually increases (Fig. 27). The formation of a stable oxide layer prior to the introduction of the halogen usually gives protection, and no appreciable change in the rate may be observed. However, if the oxide layer is disrupted by thermal cycling in the thermobalance, then the rate increases again. This effect was observed in the oxidation

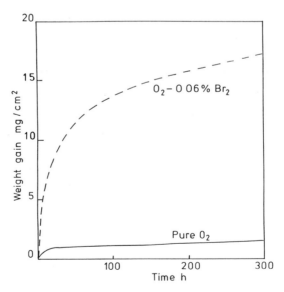

Fig. 27.   The effect of a 0.06% Br_2 addition to O_2 on the rate of oxidation of an austenitic steel at 750°C (5).

of an austenitic steel in the presence of bromine vapor (5). The effect of NaCl vapor on the corrosion behavior of Ni/Al alloys was studied in an air atmosphere (345) and the presence of NaCl was found to promote the growth of the Al_2O_3 layer.

Similar studies have been carried out in sulfur-containing atmospheres, for example, the reaction between titanium and sulfur vapor (124), nickel–chromium (398) and zirconium-based nickel alloys (355) in H_2–H_2S environments, and preoxidized cobalt–chromium alloy in H_2–O_2–H_2S vapor (199). The reaction of nickel with SO_2 and O_2/SO_2 mixtures was examined between 500 and 900°C (184), and a microgravimetric apparatus incorporating a liquid seal was used to follow the corrosion of nickel–chromium alloys between 600 and 1000°C in SO_2/O_2 mixtures (373). The addition of urethane to an SO_2 atmosphere was shown to increase the rate of corrosion for nickel–chromium and nickel–aluminium alloys (239). The presence of molten salts on the surfaces of metals often accelerate high-temperature corrosion, as indicated by several studies in the presence of SO_2 (239) and with oxygen (292), and for the corrosion of nickel by O_2–SO_2 mixtures at gas pressures ranging from 700 to 0.5 torr (216).

Metals are frequently "worked" at high temperatures during industrial finishing processes, and TG can be used to simulate such conditions. The oxidation of aluminium foil has been followed under conditions that approximate a high-temperature industrial annealing treatment (293). TG was used to follow the uptake of nitrogen in the nitriding of titanium–niobium alloy, and then used to study the oxidation kinetics (306). Thermal cycling was also carried out to sinter the oxide layer. Both pretreatments were found to reduce the rate of subsequent corrosion.

Besides investigating the relationship between variations in atmosphere and reaction rates, the gas composition can be kept constant and the metal composition varied. In this case comparisons can be made of the efficiency of various alloying agents in reducing, oxidizing, or other corrosive atmospheres. The addition of 0.1% quantities of lanthanide metals reduced the oxidation rate of copper by nearly one-half (218), but the addition of molydbenum to nickel–chromium alloy in excess of 3% increased the rate, at least in static air (292).

Thermogravimetry has been used as a means of checking the effectiveness of polymer coatings used to modify metal surfaces. The metal sheets were exposed to monomer vapor, and then subjected to a microwave plasma treatment to produce a thin, cross-linked polymer coating. The TG data provided quantitative information on the quantity of polymer present on the metal surface (322).

b. EXTRACTIVE METALLURGY

A major section of work of thus nature has been devoted to the reduction of metal oxides by various gaseous reductants. Studies on the reduction of iron oxides have been particularly evident. Since the reduction mechanisms of iron-bearing materials are quite well known, the main purpose of TG investigations is to determine the rate-controlling steps. On the other hand, the work on nonferrous materials has been concerned with the sequence and mechanism of reduction, since these are not well established. The preparation of metal catalysts by reduction has also received some attention, particularly the way in which the experimental conditions affect the surface area and structure of the final product (see Section VIII.A.5).

Shimokawabe et al. (340) examined the influence of preparation history on the reduction of α-Fe_2O_3 with hydrogen. The α-Fe_2O_3 samples were prepared by thermal decomposition of seven iron salts in a stream of oxygen, air, and nitrogen at temperatures of 500–1200°C. Generally, the higher temperatures and higher oxygen content atmospheres decreased the surface area and increased the size of oxide particles, and hence produced decreased reactivity to reduction. A simple interface for a TG–MS system was described by Szendrei and Van Berge (358), which allowed the gases evolved during the reduction to hematite with carbon to be determined. The TG–DTG curves closely followed the EGA profiles for CO_2 and CO and indicated the multistage nature of the reduction process. The presence of sodium carbonate in the system greatly enhanced the rate of reduction of Fe(II) → Fe(0) with a corresponding increase in the CO : CO_2 ratio, an effect that has been observed with other alkali compounds. The direct reduction of some iron ores using H_2, CO, and mixtures of H_2 and CO in the temperature range 200–870°C has been reported (197). The rate of reduction using H_2 was greater than with CO, and CO was less efficient as a reducing agent at temperatures below 870°C because of carbon deposition. Similar reductive studies have been carried out on cobalt ferrite (296) using hydrogen, and manganese oxide by graphite (305). Basu and Sale (22) have used thermogravimetry to prepare partially reduced copper–tungsten oxide mixtures that could be used as precursors for the direct production of Cu–W powders of unusual morphologies.

The usefulness of EGA for following reductive processes has also been exploited in a study of the reduction of nickel oxide by hydrogen in the presence and absence of magnetic fields (158). TG work, especially using metallic samples, will give rise to spurious results in magnetic fields, as already discussed (Section V.B). Hence, the reduction of nickel oxide was carried out in the absence of a magnetic field and both DTG and EGA H_2O profiles were recorded. Excellent correspondence between the two analytical methods was evident. The EGA technique was then used to follow the reduction of NiO, Co_3O_4, and Fe_2O_3 in the presence of magnetic fields (157).

Related to studies on metal production is the search for cheaper methods of processing low-grade ores that are not commercially viable by existing processes. As an illustration, the hydrochlorination of metal oxides to produce volatile metal halides, which are then reduced on carbon particles present in the solid mix, is discussed in some detail in the remainder of this section. The reactions may be summarized as follows:

$$2MCl + SiO_2 + H_2O \longrightarrow 2HCl\uparrow + M_2SiO_3 \tag{9}$$

$$2HCl + RO \longrightarrow RCl_2 + H_2O \tag{10}$$

$$C \quad + H_2O \longrightarrow CO + H_2 \tag{11}$$

$$H_2 \quad + RCl_2 \longrightarrow 2HCl\uparrow + R \tag{12}$$

The overall reaction is, therefore,

$$2MCl + RO + SiO_2 + C + H_2O \longrightarrow M_2SiO_3 + R + CO + 2HCl$$

where M and R are different metals, such as Na and Ni, respectively. Since the reactions all involve the production of volatile components, the rates, yields, and reaction temperatures can be observed by TG experiments.

The most usual low-cost source of HCl (Eq. 9) is either NaCl or $CaCl_2$. The high-temperature hydrolysis of NaCl has been investigated by measuring evolved HCl at various isothermal temperatures as a function of time (187) and producing HCl from $CaCl_2 \cdot 2H_2O$ by TG (244). In both cases, at the temperatures examined, HCl was evolved very rapidly. For the calcium salt, the temperature of 720°C produced a near stoichiometric yield based on the equation

$$CaCl_2 \cdot 2H_2O_{(s)} \longrightarrow CaO_{(s)} + H_2O_{(g)} + 2HCl_{(g)}$$

The presence of silica in the system results in the formation of sodium or calcium silicate, which makes the reaction thermodynamically more favorable at reaction temperatures:

$$CaCl_2 \cdot 2H_2O_{(s)} + SiO_{2(s)} \longrightarrow CaSiO_{3(s)} + H_2O_{(g)} + 2HCl_{(g)}$$

The HCl liberated during the hydrolysis reaction attacks the metal compound with the formation of volatile metal halides (Eq. 10). With copper (as in the Torco process) this occurs at about 550°C (245):

$$Cu_2O + 2HCl \longrightarrow \tfrac{2}{3}(CuCl)_3 + H_2O$$

The reduction of Cu(II) $\longrightarrow$ Cu(I) occurs prior to formation of the trimeric cuprous halide (55). NiO begins chlorination at temperatures as low as 370°C (259), but $NiCl_2$ is not volatilized from the ore until above 700°C (Fig. 28). A considerable amount of iron chloride is also volatilized with the

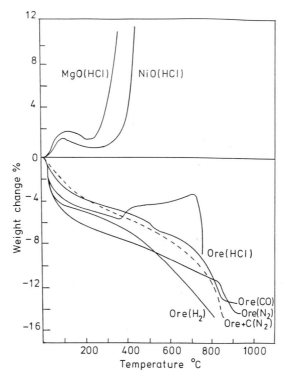

Fig. 28.   TG curves of the chloridization of NiO, MgO, and a garnierite ore and the reactivity of the ore under inert and reducing conditions (259).

nickel salt. Copper, nickel, and iron halides are all volatilized in good yields, but with chromium only 18% chlorination had occurred at 720°C. These yields can be estimated from the weight loss curves. The chlorination step in the chromium reaction has been stated to be diffusion controlled and is rate determining (244).

The final reactions involve the reduction by hydrogen of the metal chloride according to Eqs. 11 and 12. The final metal deposit on the carbon consists of pure copper from copper ores, but an alloy of nickel and iron from garnierite ores (259). With chromite ores, iron was volatilized almost completely onto the carbon at 700°C, but only 22% of the chromium was volatilized (244).

2.   Simple Inorganic Compounds

Thermogravimetry can be applied to the study of any inorganic reaction involving change in mass, such as oxidation, reduction, dehydration, and decomposition. Frequently two or more of these reactions occur simultaneously, although DTG peaks can sometimes be resolved by modifying the experimental conditions (see Section IV).

When an oxide of an element that displays several stable oxidation states is heated, redox processes are to be expected. Gross changes, e.g.,

$$\text{"FeO"} \longrightarrow Fe_3O_4 \longrightarrow Fe_2O_3$$

or

$$MnO_2 \longrightarrow Mn_2O_3 \longrightarrow Mn_3O_4 \longrightarrow MnO$$

are easily detected and by operating in a controlled atmosphere, detailed changes can be studied that allow the detection of the stoichiometry of the various oxides. Thus, FeO and MnO_2 are always nonstoichiometric and changes in mass vary in magnitude and temperature range according to the partial pressure of oxygen in the ambient atmosphere.

DTA curves of manganese dioxide heated in different atmospheres (Fig. 29) show clearly the dependence of the decomposition temperature on the partial pressure of oxygen. Different workers (107, 219, 313) agreed on the interpretation of the major peaks in the temperature ranges 450–600 and 750–1100°C, but disagreed on the interpretation of the endotherm at about 1200°C. Whereas Dollimore (107) suggested that this endotherm was due to the decomposition of Mn_3O_4 to MnO, Rode (313) interpreted it in terms of a phase transformation from one polymorphic form of Mn_3O_4 to another. The combination of high-temperature thermogravimetry and X-ray diffraction

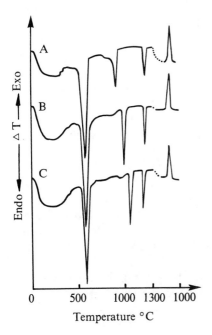

Fig. 29. DTA curves of γ-manganese dioxide in (a) N_2, (b) air, (c) O_2. Heating rate 1.1°C/min (364).

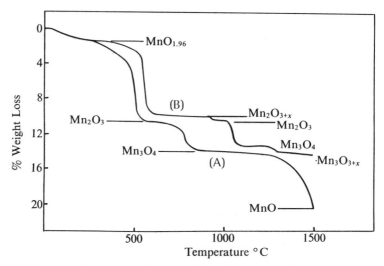

Fig. 30.   TG curves of γ-maganese dioxide in (a) N_2, (b) O_2. Heating rate 1.1°C/min (364).

(364) established conclusively that this conclusion is correct and that MnO is formed only above 1300°C in nitrogen and not at all below 1500°C in oxygen, as shown in Fig. 30. From the magnitudes of the mass changes observed, the stoichiometry of the various oxides can be determined.

The dehydration of simple inorganic hydrates may seem to be a straight-forward process, but often complications arise. Frequently, the dehydration takes place in steps, as in the case of $CuSO_4 \cdot 5H_2O$ (Fig. 31), where three stages can be distinguished involving the loss of two, two, and one molecules of water, respectively.

When some hydrates are heated they dissolve in their water of crystallization. Thus, $Mg(NO_3)_2 \cdot 6H_2O$ undergoes an aqueous fusion in the region of 80°C. Due to the formation of a film of the dehydrate on the surface of the melt, evolution of the water vapor is hindered and takes place in an uneven manner, giving rise to irregularities which may be observed on a simultaneous DTA–TG, as shown in Fig. 32.

When the effect of different partial pressures of water vapor on the dehydration process is examined, an unexpected effect is sometimes observed. This was first reported by Smith and Topley (346) who noted that as the partial pressure of water vapor was increased, the rate of dehydration decreased to a minimum and then increased to a maximum before decreasing again. A more recent observation of this effect is shown in Fig. 33 for the dehydration of calcium oxalate monohydrate (108). The explanation of this phenomenon is in terms of the crystallinity of the reaction product, which is amorphous at low pressures but crystalline beyond the pressure at which the rate increases.

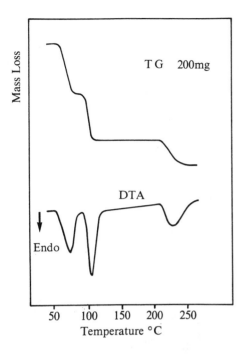

Fig. 31. TG–DTA curves for the dehydration of $CuSO_4$ $5H_2O$ in a 4-mg sample: heating rate, 10°C/min; N_2, 25 mL/min. From J. G. Dunn and I. D. Sills, unpublished data.

The distinction between dehydration and dehdroxylation is not always made, but is quite clear, since the former involves the loss of adsorbed water or water of crystallization, whereas the latter involves the loss of structural hydroxyl groups and must lead to the formation of a new compound rather than to the anyhdrous form of the reactant. It is not always easy to distinguish these processes, as in the case of clay minerals and cement hydrates, and indeed they may sometimes overlap, as in montmorillonite, but in simple inorganic compounds they are usually clearly distinguished. Many dehydroxylation processes, such as that of $Mg(OH)_2$, occur within a relatively narrow temperature range, but some lead to the formation of stable intermediates and hence to two-stage processes, e.g.,

$$2Al(OH)_3 \xrightarrow{-2H_2O} 2AlO(OH) \xrightarrow{-H_2O} Al_2O_3$$

The decomposition of carbonates and other oxysalts frequently follows a similar mechanism to that of dehydroxylation processes and can again lead to the formation of reaction intermediates. Although the final product is usually a metal oxide, exceptions do occur. Thus, various forms of TG curves have been reported (106, 229) for the decomposition of metal oxalates, as shown in Fig. 34. Although the final product of the decomposition of all oxalates in air is the oxide (Eq. 13), several oxalates decompose to form the metal in an

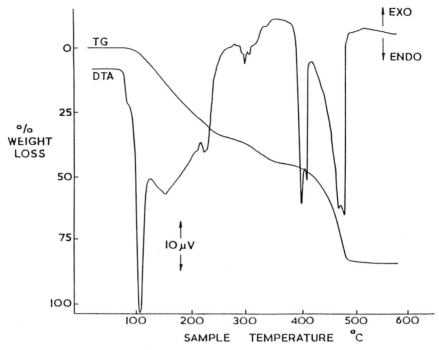

Fig. 32.   Simultaneous TG–DTA curve for magnesium nitrate hexahydrate, 9.5 mg heated at 20°C/min in air. From E. L. Charsley, unpublished data.

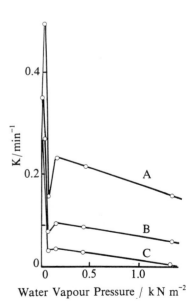

Fig. 33.   Variation of reaction rate constant with water vapor pressure: A, 120°C; B, 115°C; C, 110°C (108).

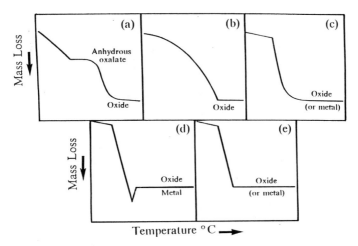

Fig. 34.   TG curves for the decomposition of various oxalates. Heating rate 2°C/min (229).

Oxalate	In Air	In N_2
Copper	d	e*
Zinc	a	a
Cadmium	e	e*
Aluminium	b	b
tin	e	e
lead	d	e*
Thorium	a	a
Antimony	d	e
Bismuth	d	e
Chromium	b	b
Manganese	a	a
Iron(II)	c	c
Iron(III)	c	—
Cobalt	a	a*
Nickel	a	a*

Compounds marked * give the metal as
the end product.

atmosphere of nitrogen (Eq. 14):

$$MC_2O_4 \longrightarrow MO + CO + CO_2 \tag{13}$$
$$MC_2O_4 \longrightarrow M + 2CO_2 \tag{14}$$

Whether the metal is formed depends on the value of the Gibbs free energy
for the reaction

$$MO + CO \rightleftharpoons M + CO_2$$

at the decomposition temperature and partial pressure of oxygen. These reactions are especially important to analytical chemists because of the frequent use of oxalic acid as a precipitating agent.

The temperature of decomposition of the oxysalt markedly influences its subsequent reactivity. At relatively low temperatures the oxide is formed with a high surface area and may have catalytic properties. Prolonged heating, especially at elevated temperatures, leads to sintering and an unreactive product of low surface area.

Many mixed metal oxides, some of which are industrially important ceramics, are prepared by the reaction between a carbonate and an oxide, e.g.,

$$BaCO_3 + TiO_2 \longrightarrow BaTiO_3 + CO_2$$

and

$$SrCO_3 + 6Fe_2O_3 \longrightarrow SrFe_{12}O_{19} + CO_2$$

These and similar reactions have been reviewed recently by MacKenzie (230). Thermogravimetry provides a convenient method for following the course of the reaction, but complementary methods must be used to establish the phase or phases formed. When barium carbonate is heated alone it does not decompose until above 1000°C, although it gives two endothermic peaks in the DTA curve due to polymorphic transformations at 810 and 975°C, which can lead the unwary into believing that decomposition has begun. In the presence of an oxide with which it reacts, e.g., Fe_2O_3, Al_2O_3, TiO_2, and ZrO_2, loss of carbon dioxide occurs at much lower temperatures. When $1:1$ molar mixtures are heated, some of these systems form the expected products ($BaFe_2O_4$, $BaAl_2O_4$, $BaZrO_3$), but others form an unexpected product (Ba_2TiO_4) or a mixture of products (several Sr-containing systems). Great care, therefore, has to be taken in carrying out kinetic studies; thus, a $1:6$ molar mixture of $BaCO_3:Fe_2O_3$ first forms $BaFe_2O_4$, which reacts further, but more slowly, to form $BaFe_{12}O_{19}$, which is the basis of ceramic permanent magnets. This is well illustrated by the isothermal results shown in Fig. 35 in which the conversion of the barium monoferrite formed as a reaction intermediate into barium hexaferrite was followed by both X-ray diffraction and saturation magnetization (64).

Even greater care must be exercised in the interpretation of the TG curves of the related Sr-containing systems. The effect of iron(III) oxide on the decomposition of strontium carbonate is shown in Fig. 36 (359). Although the presence of iron oxide dramatically lowers the decomposition temperature of the strontium carbonate, some of the latter decomposes directly to strontium oxide. This reacts with the strontium ferrite phase to form another ferrite richer in strontium. The system is further complicated because the particular strontium ferrites that are found depend upon the oxygen content of the

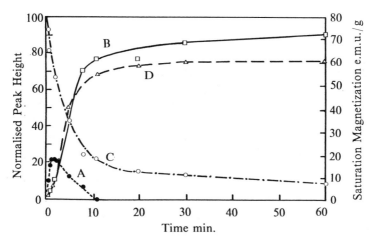

Fig. 35. X-ray phase analyses and magnetization data for $1:6$ $BaCO_3/Fe_2O_3$ mixture heated in air at 1000°C: (A) $BaFe_2O_4$ content (by XRD), (B) $BaFe_{12}O_{10}$ content (by XRD), (C) Fe_2O_3 content (by XRD), (D) Saturation magnetization of the reaction mixture (229).

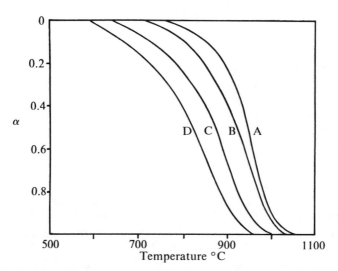

Fig. 36. Fraction reacted α versus temperature in N_2 atmosphere for (A) $SrCO_3$, (B) $6:1$ molar mixture of $SrCO_3:Fe_2O_3$, (C) $1:1$ molar mixture of $SrCO_3:Fe_2O_3$, (D) $1:6$ molar mixture of $SrCO_3:Fe_2O_3$ (359).

atmosphere (30) and differ from the monoferrite formed in the barium system (359). In nitrogen, the first product is either $Sr_7Fe_{10}O_{22}$ (225, 261) or $Sr_4Fe_6O_{13}$ (213) and the side product is $Sr_2Fe_2O_5$; but in air oxidation, some iron(IV) occurs with the formation of $SrFeO_{3-x}$ (182). These complexities illustrate the futility of trying to obtain meaningful kinetic parameters merely from a study of the rate of loss of carbon dioxide.

The observed decomposition temperatures depend not only on thermodynamic factors and procedural variables (as discussed in Sections I and IV), but also on the geometrical contacts between the particles. Hence, as the amount of Fe_2O_3 in the mixture is increased, the lower is the temperature at which loss of CO_2 is observed. A similar observation is made when Fe_2O_3 is replaced by Al_2O_3 (359). For this reason alumina should not be used as a diluent in DTA for samples containing carbonates.

The examples cited, of necessity, cover only a limited range of simple inorganic compounds. They have been selected to show that the uncritical interpretation of TG data alone can lead to erroneous conclusions, and that the technique is most useful when used in conjunction with other experimental methods. This conclusion applies similarly to many further examples omitted for lack of space. Many of these are discussed in the books cited in Section I, in specialist texts, such as those by Garner (166) and Solymosi (348), and in the reviews by MacKenzie (229, 230).

3. Coordination Compounds

The range of reactions exhibited by simple inorganic compounds, namely oxidation, reduction, dehydration, and decomposition, can also be observed and studied by TG in coordination compounds. The studies are often complicated, however, by the ligand, which can also undergo different reactions.

a. Thermal Decomposition

Complexes in which the metal–ligand bond is relatively weak tend to decompose by expulsion of one or more ligands, and quite well-defined weight losses are evident in the TG curve. For reversible reactions, this type of decomposition is very dependent on procedural variables (Secton IV.F), particularly atmosphere control, and small changes in experimental conditions can cause significant variations in the decomposition mechanism and the temperature at which decomposition occurs. This is illustrated in Table 11 and Fig. 37, for $[Ni(\beta\text{-picoline})_4](SCN)_2$ (242, 208, 351), where the conditions used produce mechanisms ranging from a single well-defined step in which all four ligands are evolved, to three weight loses. The decomposition temperature also varies between 30°C in vacuum and 185°C in an essentially self-generated atmosphere of the ligand. It has also been suggested that for a series of complexes, a change in experimental conditions may not only alter the decomposition temperatures, but may also alter the

TABLE 11
Effects of Experimental Conditions on the Decomposition of Ni $(\beta$-picoline)$_4$(SCN)$_2$

TG Curve (Reference)	Sample Weight (mg)	Heating Rate (°C/min)	Crucible Type	Atmosphere	Stiochiometry of Decomposition	Decomposition Temperature (°C)
(242)	50	5	—	Air	$-3, -1$	150
1A (208)	100	3	Pt	N$_2$	$-2, -1, -1$	125
1B (351)	500	6	Pt with lid	Ligand	$-1, -2, -1$	185
2 (351)	500	(5)	Labyrinth	Ligand	$-1, -2, -1$	185
3 (351)	2.59	0.5	Pt	Air	$-2, -2$	110
4 (351)	10.6	2.5	Pt	Vacuum	-4	30

Note. Ligand atmosphere refers to a large sample under essentially static atmosphere conditions. TG curve 2 was obtained under quasi-isothermal and quasi-isobaric conditions.

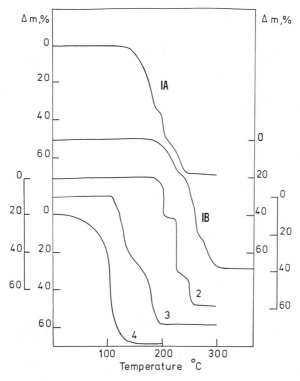

Fig. 37. Variation in the decomposition mechanism of [Ni(B-picoline)$_4$] SCN$_2$ with atmosphere (for key see Table 11) (351).

relative order of the temperatures, since not all complexes respond in the same kinetic or thermodynamic manner to the changed circumstances (274).

After the expulsion of the weakly bound ligands, the molecule can undergo further reaction. With $[Ni(py)_4](SCN)_2$, for example, after a two-stage decomposition, each of which evolved 2 mol of pyridine, the thiocyanante decomposed in air to form at various temperatures NiS, NiO, dicyan, and sulfur dioxide (381).

If the metal ion can exist in more than one oxidation state, then reduction of the cation may occur during decomposition. Co(III) (381) and Pt(IV) (155) coordination compounds exhibit this behavior. In the decomposition of $K_2[Pt(CN)_4Br_2]$ in air, cyanogen was evolved in at least two steps (155):

$$2K_2[Pt(CN)_4Br_2] \xrightarrow{360°C} K_4[Pt_2(CN)_6Br_4] + (CN)_2$$

$$500°C \downarrow$$

$$2Pt + 4KBr + 3(CN)_2$$

In this sequence the cation was reduced from Pt(IV) to Pt(0), with a Pt(III) intermediate stage tentatively proposed. The KBr volatilized above 700°C and caused a further weight loss in the TG curve (Fig. 38).

The central cation can also be oxidized during decomposition, which is again demonstrated by a cyano compound. Finely divided mixtures of the

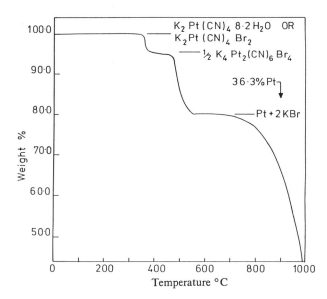

Fig. 38.   TG curve in air at 10°C/min of $K_2Pt(CN)_4Br_2$, 12.02 mg (155).

stoichiometric oxides of Eu(III) and Fe(III) were formed on the decomposition in air of $NH_4[EuFe(II)(CN)_6] \cdot 4H_2O$ (154). Further heating produced the ferrite $Eu_2Fe_2O_6$, which is of interest as a car emission control catalyst.

Some complexes contain both oxidizing and reducing species, and decompose even in inert atmospheres at relatively low temperatures. For example, nitrate as a counter ion may oxidize organic ligands associated with the central cation, forming a wide range of products. From a pyrolysis–GC–MS study in helium of the decomposition of Ni $[aniline_2(NO_3)_2]$, no fewer than 43 decomposition products were identified (226).

Complexes containing multidentate ligands may also decompose by loss of ligands in a stepwise manner (381). The decomposition of $[Co(en)_3]X_3(X = Cl, Br)$ has been studied by TG, DTA, thermomagnetometry, reflectance and infrared spectroscopy, mass spectrometry, and pyrolytic techniques (93). The initial decomposition step in nitrogen was the loss of one mole of the bidentate ligand:

$$Co(en)_3X_3 \longrightarrow \textit{trans-}[Co(en)_2X_2]X + en\uparrow$$

Then $\textit{trans-}[Co(en)_2X_2]X \longrightarrow CoX_2 + (NH_4)_2CoX_4 + \text{organic products}$

$$(NH_4)_2CoX_4 \longrightarrow CoX_2 + 2NH_4X\uparrow$$

$$CoX_2 + 2H^+ \longrightarrow Co + 2HX\uparrow$$

In vacuum the end products were the same but no reaction equivalent to the second stage was evident.

When the metal–ligand bond is strong, fragmentation of the ligand itself can occur in preference to metal–ligand cleavage, and the TG curve shows a gradual weight loss beyond the decomposition temperature. This kind of decomposition is encouraged by an oxidizing atmosphere when the ligand degrades more readily by oxidative attack than by pyrolytic decomposition. The decomposition temperature is lower in air or oxygen than in an inert atmosphere.

b. THERMAL STABILITY

The decomposition temperatures obtained from the TG curve are frequently used as a means of assessing the thermal stability of a complex and to arrange a series of complexes in order of thermal stability. Some of the problems associated with obtaining reproducible measurements have already been outlined in the earlier part of this section, and in Sections IV and VI, but have not always been appreciated. Hence, it is not unusual to find values for the procedural decomposition temperatures that vary by as much as 100°C for the same 8-hydroxyquinoline complex and also produce differing relative stability orders for the same series of related complexes (46, 126, 74,

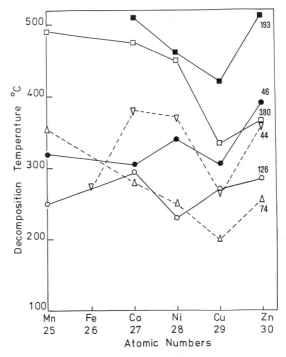

Fig. 39.   Procedural decomposition temperatures for some 8-hydroxyquinoline complexes determined by TG. (*a*) 8-hydroxy quinoline complex $M(OX)_2$ (46, 126, 74, 380), (*b*) bis(8-hydroxy-5-
quinolyl)methane polymers (193), (*c*) 8-mercaptoquinoline complexes $M(B)_2$ (44).

380, 193, 44) (Fig. 39). It is possible to obtain reproducible decomposition
temperatures under rigorously controlled conditions, but it should be appreciated that these values have no absolute significance.

Besides the simple reporting of thermal stability data, attempts have been
made to correlate the thermal stability of a series of related complexes to
some molecular property of the complex. For such results to be valid, the
complexes should ideally have the same empirical formula, be isostructural,
and decompose by the same mechanism. Although the first two criteria are
usually obeyed, it is not often that the decomposition products are analyzed
to check that the same mechanistic decomposition pathway has been followed.

Correlations that have been sought between thermal stability and molecular property include the influence of cation size on the thermal stability of
metal oxalates (106), the size of the outer sphere cations on nitrate complexes (43), electronegativity of the central cation on 8-hydroxyquinoline
complexes (193), ionization energy of the central cation (10), and equilibrium
stability constants of pyridine complexes (221). In the last case the thermal
stabilities of $M(py)_2(SCN)_2$ complexes (M(II) = Ni, Co, Mn, Zn, Cd) were
reported to increase with increasing equilibrium stability constant. However,

with some chelate ligands, such as salicylaldoxime and dimethlglyoxime, the DTG peak temperatures decreased with increase in stability constant. It was suggested that the pyridine complexes decomposed by ligand expulsion, but the chelate complexes decomposed by ligand degradation. Hence, the DTG value represented the thermal stability of the ligand rather than the strength of the metal–ligand bond.

As might be expected the thermal stability of a complex increases as the ligand is changed from nonchelating to chelating agent. Thus, $(Co(NH_3)_6)Cl_2$ is reported to decompose in nitrogen at 85°C, compared with a value of 250°C for $[Ni(en)_3]Cl_2$ (170); and the $Co(py)_2Cl_2$ complex decomposes at 102°C, compared with 268°C for $[Co(bipy)_2]Cl_2$ (278).

In complexes of formula $M(NH_3)_6X_2$ ($X = Cl^-, Br^-, I^-$), the usual stability order is $Cl < Br < I$ (381).

c. PREPARATIVE APPLICATION

Either the thermobalance or information gained from TG studies can be used to prepare complexes by thermal dissociation. One of the more popular applications involves ejection of one or more coordinated ligands followed by insertion of a counter ion:

$$[Co(NH_3)_5H_2O]X_3 \xrightarrow{\text{heat}} [Co(NH_3)_5X]X_2 + H_2O\uparrow$$

X can be a wide range of anions, including CN^-, SCN^-, Br^-, and Cl^- (381). Similar products can be formed by ejection of a bidentate chelating agent. The range of products from a single precursor can be greatly extended by heating the complex in the presence of an ammonium or alkali metal halide, for example, with $Cr(en)_3Br_3$ (381):

$$NH_4F + [Cr(en)_3]Br_3 \xrightarrow{70-120°C} cis[Cr(en)_2F_2]F$$
$$NH_4Cl + [Cr(en)_3]Br_3 \xrightarrow{25-165°C} cis[Cr(en)_2Cl_2]Cl$$

Reactions between complexes of formula $[Co(NH_3)_4X_n]$ ($n = 1, 2$) and NH_4X and HX have also been examined (71). The ammonium halide, besides being the source of the required halide ion, has also been assumed to act as a catalyst in such reactions. However, in the reactions of $[Cr(en)_3]Cl_3$ with NH_4X ($X = Cl, Br, I$) it was deduced from the TG curves that only the chloride salt had any catalytic effect (195). Heating a compound such as $Mn(phen)_2Cl_2$ can produce $Mn(phen)Cl_2$ (25), thus changing the coordination number from six to four.

Some dimeric Cr(III) compounds of general formula $Cr_2Cl_6L_4$ (L = pyridine, 3-picoline-, 4-picoline; L_2 = bipyridine) have been prepared by thermal dissociation of the octahedral monomer (57).

$$[Cr(py)_3Cl_3] \xrightarrow{200°C} [Cr_2(py)_4Cl_6] + 2py\uparrow$$

Many other examples in which thermal dissociation produces stable intermediates can be found in the preceding sections.

4. Analytical Precipitates

The uses of TG in the assessment of such factors as drying temperatures
and weighing form of analytical precipitates have already been indicated in
Section VI.D. Some caution is required in the acceptance of temperatures
recorded in the early literature associated with the assessment of analytical
precipitates, and the particular case of oxime complexes has been discussed
in Section VIII.A.3.b. Generally, reported temperature values for the drying
of precipitates tend to be high, partially because quite large samples (about
300 mg) were required and excess water was not removed prior to the TG
experiment. Figure 40a shows the TG curve obtained for the drying of
magnesium oxinate (118).

$$Mg(C_9H_6NO)_2 \cdot 2H_2O \longrightarrow \quad Mg(C_9H_6NO)_2$$
$$\downarrow$$
$$MgO + \text{decomposition products}$$

The suggested drying temperature range for the complex is 206–300°C (128).
However, the interpretation of the beginning of the plateau region as the
minimum drying temperature is clearly incorrect, since the temperature
interval 122–201°C is a reflection of the experimental conditions used. In

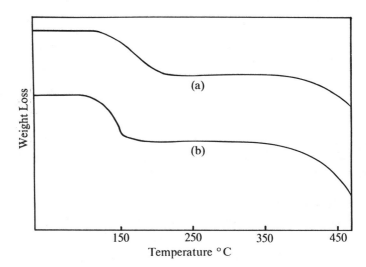

Fig. 40.   TG curve of magnesium oxinate: (a) 300 mg sample, heating rate 10°C/min; (b) 10 mg
sample, heating rate 10°C/min (118).

practice, a drying temperature of 150–160°C for 1 h is adequate, a compromise in terms of reasonable drying time with little risk of decomposition. For comparison, a TG curve of a smaller sample of the same complex is shown in Fig. 40*b*. Here the drying region lies between 110 and 150°C.

Some examples of this application have been given for the estimation of thallium(III) by precipitation as barium bisoxalato-thallate(III) (317), manganese(II) by precipitation with benzoyl-*m*-nitroacetanilide (243), and copper by precipitation with 4-amino-5-mercapto-3-methyl-4, 1, 2-triazole (152).

One of the more difficult tasks of the analytical chemist is to asses the best temperature range to ignite the precipitate. The ignition characteristics are significantly dependent on the precipitating conditions, and several precipitates should be examined using slightly differing conditions. The classical precipitating technique of slow addition of the precipitating agent to a hot, well-stirred solution containing the analyte, or by the technique of precipitation from homogeneous solution, tends to yield reasonably well-crystallized products that require a relatively low ignition temperature. A less well-crystallized precipitate will tend to be more strongly contaminated by adsorbed and occluded impurities, and may require higher ignition temperatures to produce a weighing form of acceptable purity. Studies of this nature have been carried out on alumina precipitated by different methods, but conflicting results were obtained. One study showed that constant weights were achieved in the range 280 to 1031°C (122), and another between 650 and

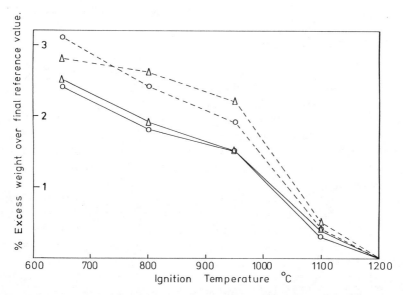

Fig. 41. Effect of atmosphere exposure on the weight of ignited aluminium oxide precipitates (258): o , urea-basic succinate method; △, urea method; —, precipitate transferred directly from furnace to desiccator; - - -, precipitate placed in desiccator after 3 min cooling in atmosphere.

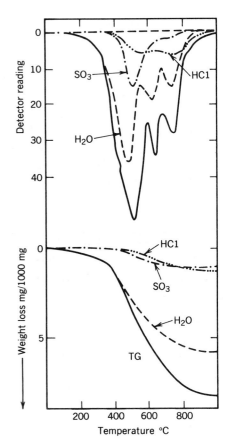

Fig. 42. TG–EGA traces for BaSO$_4$ demon-
strating the evolution of adsorbed salts (284).

950°C (137). A minimum ignition temperature of 1050°C was recommended
(133). In another investigation heating was carried out at constant temperature at 50°C intervals above 600°C, and weight losses were still occurring at
the 900°C limit of the thermobalance (258). Based on the isothermal studies,
a minimum ignition temperature of 1200°C was recommended. The hygro-
scopic nature of the alumina ignited at the lower temperatures was also
demonstrated (Fig. 41).

   Further information on the nature of the contaminants can be obtained
from a TG-EGA investigation, as demonstrated for BaSO$_4$ precipitated by
various methods (284). The TG–DTG–EGA traces for BaSO$_4$ precipitated
from various concentrations of hydrochloric aid in the presence of salts such
as NH$_4$Cl and (NH$_4$)$_2$SO$_4$ are shown in Fig. 42. Absorbed contaminants are
released at differing temperatures depending on the precipitating conditions.

   The appearance of a plateau in the TG curve has sometimes been taken to
indicate a suitable weighing form, but this is not necessarily valid since
nonstoichiometric species can be formed. Nonvolatile contaminants can also
be present.

5. Catalysts

Thermoanalytical techniques have been used successfully to simulate industrial processes involving catalyts, particularly for heterogeneous systems in which solid–gas reactions occur. TG can be used to study most aspects of the catalytic cycle, from catalyst preparation to poisoning and regeneration, and also to compare the effects of variables such as catalyst composition, operational temperatures, and gas flow rates on performance. Since most TG work is small scale and experiments can be carried out rapidly, much useful information can be gathered cheaply and quickly prior to large scale tests. This work is further facilitated by use of TG–EGA systems to measure quantitatively the product yield. XRD, infrared spectroscopy and chemical analysis are frequently used to characterize the catalyst material at various stages in the cycle. Substantial reviews of the use of thermal analysis in studying catalytic processes have been published by Dollimore (109) and Bond and Gelsthorpe (45), and a shorter summary by Habersberger (181).

a. CATALYST PREPARATION

The aim in catalyst production is to prepare a material that has a high surface area of active sites. Hence, it is necessary to investigate those factors that affect surface area and to optimize them. Many metal catalysts, for example, are prepared by impregnating a support material with a solution of a salt of the metal, and then drying, calcining, and reducing the support–salt mixture. Since each of these steps involves a weight loss, TG can be used as a technique to follow these processes (Fig. 43). Conditions of heating rate, gas flow rate, and the temperature range associated with each step can be varied

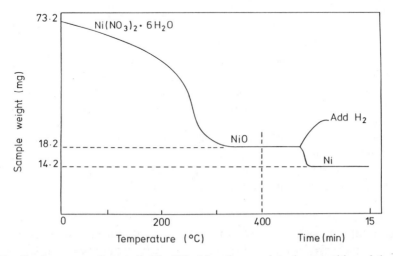

Fig. 43. Catalyst preparation studied by TG. After the complete decomposition of the metal salt in air at about 300°C, the temperature is stabilized at 400°C and the atmosphere changed to H_2, causing reduction to Ni (20).

and the effect on the surface area of the metal determined by standard methods. Experimental studies have shown that, generally, low rates of reduction produce a high surface area of active sites. Thus, in the preparation of a Ni/Al_2O_3 catalyst, the optimum surface area of nickel was achieved by direct reduction of a $Ni(NO_3)_2/Al_2O_3$ preparation at a slow heating rate (20). The temperature of decomposition of the parent compound also affected the surface area, with the compounds decomposing at a lower temperature yielding a higher surface area. This has been demonstrated in the production of Co_3O_4 (294) and $\alpha\text{-}Fe_2O_3$ (339) catalysts from various salts. The reactivity of the catalyst is usually demonstrated by an inspection of the TG curves, since the more active catalysts promote reaction at lower temperatures. Similar procedures have been described for the preparation of catalytic electrodes (23) by studying the decomposition of inorganic salts such as tungstates and vanadates. The authors also describe a symmetrical thermobalance for the determination of surface area and pore size.

Cimino et al. (85, 86) have reported on the use of TG and photoelectron spectroscopy to study the interaction between a rhenium catalyst and several supports, and a similar study appears for a $NiO\text{-}Al_2O_3$ system (347). In the last paper various $NiO\text{-}Al_2O_3$ mixtures were calcined at 1000°C for 3 h and then reduced by H_2. The TG–DTG curves indicated that as the Al_2O_3 ratio increased, higher temperatures were required to cause reduction of NiO, indicating an increasing interaction between the NiO and Al_2O_3 lattice. The preparation of some iron-based Fischer–Tropsch catalysts was carried out by thermal cycling in sequential oxidizing and reducing conditions using Mossbauer spectroscopy and XRD to characterize the phases formed at each stage (370).

b. GASEOUS ADSORPTION / DESORPTION

The first step in a heterogeneous catalytic reaction is the adsorption of gaseous molecules onto the surface of the active catalytic sites. Since the activity of the catalyst is dependent on the number of available adsorption sites, which in turn is related to the surface area of the catalyst material, it is of some importance to be able to measure the latter. Volumetric and gravimetric techniques are frequently used for this purpose, although TG has the major advantage of continuously monitoring the weight gain that results from gas adsorption, and this gives information on the rate of gas uptake as well as the attainment of equilibrium conditions. From the gain in weight at a specific temperature it is possible to calculate the specific surface area of the catalyst material, providing the temperature used is that which gives monolayer coverage and that the number of molecules of gas adsorbed per unit of catalyst is known.

The total area of the catalyst material, i.e., catalyst plus support medium, is usually measured by the physical adsorption of an inert gas typically at subambient temperatures using a vacuum microbalance. A useful section on

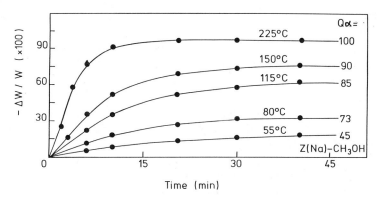

Fig. 44. Vacuum desorption isotherms for methanol from a zeolite catalyst (318).

thermobalances used for the measurement of total surface areas is given by
Keattch and Dollimore (214).

To measure the surface area of the active catalyst only, gases that are
chemisorbed by the catalyst are used; hydrogen and carbon monoxide are
frequently used to calculate from weight gain data the specific surface area of
metal catalysts, usually at ambient temperatures. The determination of the
surface area of palladium is a well-known example (72). A further advantage
of TG is that the surface of the catalyst can be cleaned prior to adsorption by
a readily reproducible temperature degassing technique. Typically, the metal,
or supported metal catalysts, may be heated to 100°C in vacuo prior to
introduction of the gaseous adsorbate.

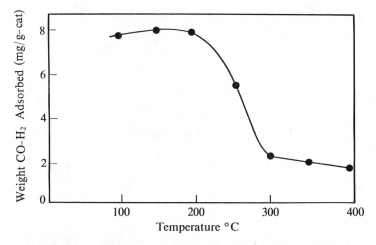

Fig. 45. Monitoring the adsorption of vapor onto a catalyst by TG at several set temperatures.
Above about 200°C the molecules gain enough thermal energy to diffuse away from the catalyst
surface, but reach a limiting diffusion rate above 300°C (136).

Desorption isotherms can also be monitored. The vacuum desorption behavior of methanol adsorbed onto a zeolite catalyst between 47 and 227°C (318) is shown in Fig. 44. Approximately 40% of the methanol remained adsorbed, which was explained by pore filling.

It is sometimes useful to monitor the adsorption behavior of the catalyst in the temperature range at which reaction occurs. To do this, the reaction gas mixture can be passed over the catalyst at specific temperatures and the weight changes followed until equilibrium is achieved. Such a curve for a H_2/CO mixture passed over a Ni/Al_2O_3 catalyst (136) is shown in Fig. 45. A rapid weight loss can result if the surface area of the metal is reduced by sintering, or if the decomposition temperatures of catalyst material is exceeded. In the example cited, it is suggested that the weight loss is caused by diffusion of product molecules from the catalyst surface. The hydrogen and oxygen adsorbing properties of some tin–platinum reforming catalysts at 23 and 450°C have been investigated by TG (263).

c. Catalyst Efficiency

The effectiveness of a range of catalysts can be compared using TG if the reaction involves a weight change, because the more effective the catalyst the lower will be the temperature at which the weight loss commences. In the oxidation of hydrocarbons such as methane and ethane with V_2O_5, there is a weight loss as oxygen is consumed and V_2O_5 is reduced (266). Small amounts of precious metals present in the V_2O_5 promote the reaction, Pd being the most effective addition. The effect of various metal oxides on the oxidative decomposition of polybutadiene indicated that the catalytic effect was independent of conductor type, and appeared to be associated with those oxides for which the forbidden zone was less than 1.9 eV (96). Inui et al. (201) studied the synergistic effects of composite catalysts on the direct hydrogenation of carbon. They concluded that a three-component system of either Ni or Co as the catalyst substrate combined with a small amount of La_2O_3 and a platium group metal exerted a synergistic effect within the range 300–800°C. The effect of metal oxides on the decomposition of $KBrO_3$ was studied, and the kinetic parameters of the catalyzed and uncatalyzed reactions calculated by the methods of Freeman–Carroll, Coats–Redfern, and Horowitz–Metger (209). A series of CoMo oxide catalysts supported on Al_2O_3 were investigated for their hydrogen reducing capabilities, in the composition range $r = Co/(Co + Mo) = 0.00$ to 1.00 (153). Inhibition of the reducing capability was observed in catalysts of composition range $r = 0.25–0.75$.

d. Catalyst Stability

The structure of the active form of the catalyst may be unstable beyond a certain temperature range, thereby limiting its usefulness. If a change occurs that involves a gaseous emission, then this can be detected by TG.

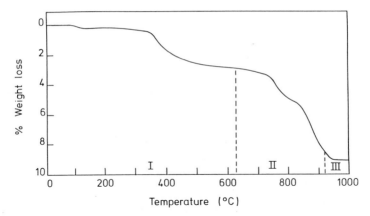

Fig. 46.   The use of TG to evaluate the effective operating range of a catalyst (375). TG curve of La$_{0.8}$K$_{0.2}$MnO$_3$ at 20°C/min, 10% H$_2$ in N$_2$: region I, active perovskite structure; region II, partial decomposition to MnO and La$_2$O$_3$; region III, only MnO and La$_2$O$_3$ detectable.

The TG curve of a potential NO$_x$ reducing catalyst of formula La$_{0.8}$K$_{0.2}$MnO$_3$ heated in nitrogen (375) is shown in Fig. 46. Beyond 620°C the active perovskite structure is unstable, with eventual decomposition to La$_2$O$_3$ and MnO. The loss of oxygen at 350°C was shown to be reversible, an important point when considering catalyst regeneration. Phase changes in the catalyst cannot, of course, be detected by TG, but DTA is often a suitable technique to monitor such transformations.

e.   CATALYST POISONING AND REGENERATION

The occupation of active sites on the catalyst by strongly adsorbed foreign molecules, or deposited solid, necessarily reduces the catalytic activity. The adsorption or deposition process can be observed by TG. Reactions that involve organic molecules frequently deposit coke on the catalyst surface, and deposition can be followed as a function of temperature or time, as in a study of a silica–alumina catalyst used in petroleum processing (196). The variation in the weight of coke deposited as a function of time at certain temperatures is shown in Fig. 47. The deposition of carbon is also an important consideration in the methanation of CO. The kinetics of this process as a function of P_{CO}, P_{H_2}, H$_2$/CO, H$_2$S, H$_2$O, and temperature were determined for aluminum-supported nickel and nickel bimetallics (161). The results showed that relatively high partial pressures of CO and low partial pressures of H$_2$ are necessary to produce massive carbon deposition. The catalyst support was found to have little influence on carbon deposition over nickel catalysts used for steam reforming (137).

An example of a product causing poisoning is given by Inui et al. (202). During the oxidation of propylene to acrolein, using a cuprous oxide catalyst,

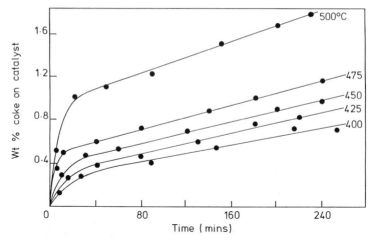

Fig. 47.   The poisoning of a catalyst by coke deposition as a function of temperature (196).

the acrolein formed initially was adsorbed strongly on to the catalyst and retarded subsequent reaction.

When a catalyst is deactivated by adsorption/deposition processes, TG can be used in some cases to examine the regenerative possibilities. For antimony(III) and iron(III) oxides deactivated by water adsorption, the weight loss between 150 and 250°C corresponding to dehydroxylation correlates well with the activity of the regenerated catalyst (321). TG–DTA was used to study the loss of catalytic activity of La_2O_3 through carbonation and hydration, and the results were used to determine the temperature conditions required to regenerate the active La_2O_3 (32).

6.   Minerals

From their inception thermal methods have been extensively used in mineralogy. When a mineralogical specimen is taken from the field to the laboratory it is first characterized to determine whether it is monomineralic and to determine its chemical composition and purity. Several experimental methods are used in conjuction to characterize the sample and these usually include X-ray diffraction, optical and/or electron microscopy, and chemical analysis, whether by traditional solution methods or by instrumental methods such as X-ray fluoresecence and atomic absorption spectroscopy. TG, DTA, and perhaps evolved gas analysis and even thermosonimetry may be placed along with IR spectroscopy in a second group of experimental techniques that are frequently used to supply additional information about the sample. The combination of TG and DTA, either as a simultaneous technique or in separate experiments, can provide information about the impurities present in a mineral specimen. For example, the presence of quartz associated with a clay mineral is readily detected by DTA through its reversible polymorphic transformation at 573°C, while the weight loss observed for the dehydroxylation reaction in the TG curve can be compared with that calculated for the ideal composition of the clay mineral to provide an estimate of the amount of clay present.

Mineralogical analyses still usually quote values for $H_2O -$ and $H_2O +$, i.e., the weight loss that occurs on heating in an oven at about 110°C overnight and the further weight loss that occurs on heating to a high temperature, often unspecified but usually not in excess of 1000°C. Although this procedure provides some information about the presence of adsorbed water, structural hydroxyl groups, carbonate ions, etc., it should be interpreted with great caution. In a smectite it is very difficult to separate the weight losses due to the loss of interlayer water and structural hydroxyl groups. In a mineral containing fluoride, species such as SiF_4 or alkali fluorides may be lost through volatilization. In an iron-containing mineral, oxidation will affect the magnitude of the observed weight loss (see below). If a small amount of carbonate is present this may be responsible for an appreciable proportion of the observed weight loss supposed to be from the structural hydroxyl groups of the principal constituent.

A better method of determining the hydroxyl content of a mineral is by means of the Penfield tube (180), although a TG curve at a slow heating rate to allow separation of the various reactions will provide more information. The qualitative and quantitative determination of clay minerals in soils has been reviewed by Mackenzie and Caillere (232). More specifically, TG studies of ammonium zeolites X and Y have enabled Rees (308) to quantify the relationship between the number of water molecules and the number of ammonium ions in the unit cell. MacKenzie (228) has shown that hydroxyl groups are not always lost according to the simple equation $2OH^- \longrightarrow O^{2-} + H_2O$, but that loss of hydrogen (dehydrogenation) may occur. Poppl et al. (295) quote the TG data shown in Table 12 to indicate that the observed weight loss from kaolinite may vary according to the amount of dehydrogenation as opposed to dehydroxylation that occurs and according to the nature of the metakaolinite, which Pampuch (281) has shown is not necessarily an anhydrous phase.

TABLE 12
TG Data of Kaolinite

Atmosphere	600°C	900°C	1100°C
N_2	12.10	13.31	13.31
Air	10.97	11.09	11.09
CO_2	9.31	9.48	9.48
10^{-5} torr	13.64	13.91	13.96

SOURCE: Poppl et al. (295).

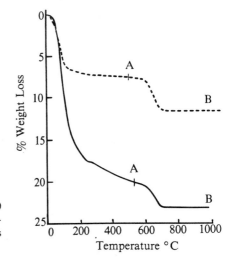

Fig. 48.   TG curves for Rb-montmorillonite (- - -)
and Mg-montmorillonite (—). The point corresponding to complete removal of sorbed water is
indicated by *A* (232).

The TG curve for montmorillonite varies with the exchangeable cation and
it has been established that sometimes, as in the curve for the Mg-exchanged
montmorllonite shown in Fig. 48, dehydroxylation commences before dehydration is complete (231, 232).

When iron is present, the change in oxidation states can provide a
mechanism for dehydrogenation (3, 145, 51, 388); thus,

$$4Fe^{2+} + 4OH^- + O_2 \longrightarrow 4Fe^{3+} + 4O^{2-} + 2H_2O$$

Although water is liberated, only a small weight loss is observed as the
oxygen is supplied by the atmosphere. In the absence of hydroxyl groups,
oxygenation can occur; thus,

$$4Fe^{2+} + O_2 \longrightarrow 4Fe^{3+} + 2O^{2-}$$

This process involves a gain in weight and the addition of oxide ions to the
surface of the mineral, similar in mechanism to the oxidation of a metal. The
effect of these reactions can be clearly seen in TG curves of amphiboles and
micas carried out in controlled atmospheres; a typical example is that of
biotite shown in Fig. 49. The loss of mass in nitrogen was greater than in air,
because dehydroxylation had occurred in the former case, but most of the
hydroxyl groups were converted into oxide ions by dehydrogenation in the
latter case. Around 950°C in air a small gain in mass was observed, indicating
that oxygenation also occurred. Above 1000°C, and especially on holding the
sample under isothermal conditions at 1450°C in nitrogen, further loss in
mass occurred due to volatilization of species such as alkalis and SiF_4.

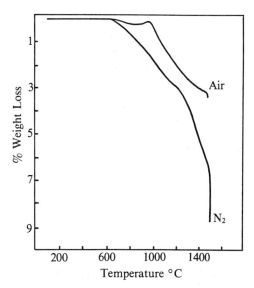

Fig. 49. TG curves for lithum biotite from Trelavour Down, Cornwall, UK. Approximately 400 mg sample heated at 1.1°C/min in flowing N_2 and static air. From A. D. White and J. H. Sharp, unpublished data.

It should be remembered that many silicates and other minerals contain both hydroxyl groups and variable valence cations, at least in small amounts, so that quantitative calculations based on TG or weight loss data should allow for oxidation reactions. This is frequently difficult and it is still common practice to assume that a mineral has a perfect anionic framework. Although it may be reasonable to base the stoichiometry of an amphibole and a mica on 24 oxygen atoms, it is clearly unreasonable to assume that they contain 22 O^{2-} and 2 OH^- ions, and 20 O^{2-} and 4 OH^- ions, respectively. Careful TG studies carried out in controlled atmosphere can provide alternative assumptions based on a sounder foundation.

Oxidation can also be observed in minerals that do not contain hydroxyl groups. $FeCO_3$, siderite, provides an example. In an oxidizing atmosphere, siderite decomposes according to the following equation (156):

$$2FeCO_3 + \tfrac{1}{2}O_2 \longrightarrow Fe_2O_3 + 2CO_2$$

The mechanism of dehydroxylation of hydroxides and hydroxyl-containing silicates has received considerable attention, especially since Brindley (53) and Ball and Taylor (14) independently proposed the idea of an inhomogeneous mechanism, involving the formation of donor and acceptor regions and counter migration of ions, to explain the observed topotaxy. TG studies alone have little to contribute to the determination of the reaction mechanism, but have a part to play as an ancillary technique in support of X-ray diffraction and electron microscopy. It is established (51) that the trioctahedral minerals, based on divalent cations such as Mg^{2+} and Fe^{2+} in octahedral sites, recrystallize almost immediately after dehydroxylation and frequently display good topotactic relations to the parent mineral; whereas the dioctahedral minerals, based on trivalent cations such as Al^{3+} and Fe^{3+}, form almost anhydrous, poorly crystalline, metastable reaction intermediates, such as δ- and γ-Al_2O_3, γ-Fe_2O_3, metakaolinite, and pyrophyllite dehydroxylate. On the other hand, the kinetic mechanism, usually studied by isothermal weight loss methods, seems to involve the movement of a reaction interface for both magnesium and aluminium hydroxides, but rather more complex mechanisms for both magnesium and aluminium layer silicates. Green (178) has reviewed the decomposition of magnesium hydroxide, and important papers on the mechanism of this and related dehydroxylation processes have been published by Niepce and coworkers (271, 272).

The mechanism of decomposition of dolomite, $CaMg(CO_3)_2$ is also of interest, since a satisfactory theory should explain the apparently anomalous dependence of the decomposition on the partial pressure of carbon dioxide. Many authors have pointed out that the first stage of the decomposition is apparently unaffected by increase in the partial pressure of carbon dioxide, but the second stage is, as expected, displaced to higher temperatures as the partial pressure of carbon dioxide is increased. This is clearly shown in the DTA curves (see Fig. 50), taken from Stone (354). Careful examination of the DTA curves indicates that the first peak actually occurs at a higher

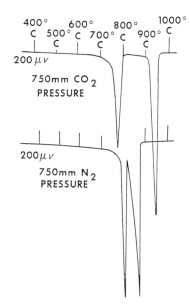

Fig. 50. DTA curves for decomposition of dolomite in atmosphere of N_2 and CO_2 (354).

TABLE 13
The Effect of Atmosphere on the DTG and DTA Peak Temperatures of Dolomite

Atmosphere	DTG 1°C/min		DTA 10°C/min	
	1st Peak	2nd Peak	1st Peak	2nd Peak
N_2	745	805	802	882
N_2–10% CO_2	720	830	780	885
N_2–20% CO_2	695	830	768	892
N_2–30% CO_2	705	848	772	902
100% CO_2	730	925	773	943

SOURCE: McIntosh (237).

temperature in nitrogen than in carbon dioxide. This anomaly has been confirmed by McIntosh et al. (237), who determined DTA and DTG curves in carefully controlled atmospheres. The results shown in Table 13 indicate that, whereas the second peak is displaced progressively to higher temperatures, the first peak is first displaced to lower temperatures and then to higher temperatures as the partial pressure of CO_2 is increased. The anomalous behavior of the first peak is thought to be due to kinetic factors rather than thermodynamic factors (237). Berg (31) has shown that at very high pressures of CO_2, the DTA curve of dolomite shows a peak indicating the breakdown into $MgCO_3$ and $CaCO_3$ prior to decomposition. Since the magnesite has been stabilized to temperatures in excess of its thermodynamic decomposition temperature by incorporation into the crystal structure of dolomite, it decomposes rapidly on formation, whether it is present in an atmosphere of nitrogen, 10% CO_2, or 100% CO_2. The kinetics of the reaction depend on the nucleation of magnesium oxide and the rate of subsequent growth of the nuclei, which is a process very dependent on the atmosphere. For reasons not fully understood, the rate of reaction is more rapid in low partial pressures of CO_2 than in the absence of CO_2. The reaction is also dependent on the partial pressure of water vapor (237), which is similarly likely to affect the kinetics of nucleation and growth.

TG and other thermal methods have also been applied to other minerals such as borates, phosphates, and sulfates. There is insufficient space to review all of these thoroughly, but this section is completed by a discussion of the application of TG to oxidative studies of sulphide. The reactions can be complicated, because of the formation of various intermediate species, some of which may be nonstoichiometric. Sulfides are strongly corrosive and will vigorously attack sample crucibles, including platinum, especially at elevated temperatures. This is less of a problem in oxidizing conditions than in an inert atmosphere, since most sulfides oxidize by 600°C. The oxidation of sulfides provides a good example of the need to control all the procedural variables, since relatively small changes in any of them can significantly influence the extent of reaction and even in some cases the mechanism of reaction.

The initial reaction of sulfides under oxidizing conditions is often the formation of the sulfate, accompanied by a gain in weight. Both FeS (215) and NiS (116) exhibit this behavior. The sulfation reaction is enhanced by the use of platinum sample crucibles, since the platinum catalyses the conversion of SO_2 to SO_3. For NiS the reaction sequence is reported to be (116)

$$NiS \qquad + 2O_2 \longrightarrow NiSO_4 \qquad < 490°C$$
$$NiS \qquad + \tfrac{3}{2}O_2 \longrightarrow NiO + SO_2 \qquad > 490°C$$
$$NiO + SO_2 + \tfrac{1}{2}O_2 \longrightarrow NiSO_4$$

The thermal stability of metal sulfates differs considerably; whereas $FeSO_4$ begins to decompose around 630°C, $NiSO_4$ is stable up to 800°C. Oxidation of Fe(II) to Fe(III) also occurs and the reactions are (215)

$$6FeSO_4 + \tfrac{3}{2}O_2 \xrightarrow{575-625°} [Fe_2(SO_4)_3]_2 \cdot Fe_2O_3$$
$$[Fe_2(SO_4)_3]_2 \cdot Fe_2O_3 \xrightarrow{625-675°} 3Fe_2O_3 + 6SO_3$$

Alternatively, the simple sulfide species can undergo rapid oxidation, accompanied by weight loss, around 550°C, according to the equations

$$4FeS + 7O_2 \longrightarrow Fe_2O_3 + 4SO_2$$
$$3FeS + 5O_2 \longrightarrow Fe_3O_4 + 3SO_2$$
$$2NiS + 3O_2 \longrightarrow 2NiO + 2SO_2$$

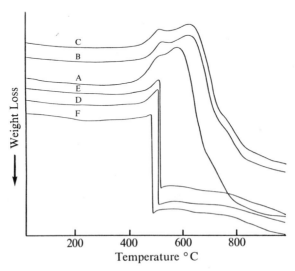

Fig. 51.   TG curves of nickel sulfide concentrate in oxygen. Heating rate (°C/min): *A*, 1; *B*, 5; *C*, 10; *D*, 15; *E*, 30; *F*, 50 (119).

These reactions do not proceed to completion even when held under isothermal conditions for long periods. Unreacted sulfide is still present probably because of the formation of protective coatings of oxide and sulfate (117, 200). At higher temperatures disruption of the oxide film or decomposition of the sulfate allows further oxidation to occur.

In a study of the oxidation is some nickel sulfide concentrates (119, 120) consisting of pentlandite, violarite, nickel ferrous pyrrhotite, pyrite, talc, and antigarite, it was demonstrated that small variations in heating rate and sample mass could lead to significant changes in the TG curves. At heating rates of between 1 and 10°C/min or less, and with sample sizes of 15 mg, a typical stepwise oxidation process occurred in which the mineral phases present were sequentially converted to sulfates (weight gain) or to oxides (weight loss) (see Fig. 51A–C). In oxygen, the major weight loss region commenced at about 600°C at a heating rate of 10°C/min. At heating rates of 15, 30, and 50°C/min, a striking change in the reaction profile was observed (see Fig. 51D–F). The formation of sulfate was still evident, although decreasingly so as the heating rate increased. The weight gain region was followed by a very rapid weight loss, which tended to occur at increasingly lower temperatures as the heating rate increased, and up to 170°C below the onset temperatures observed for the profiles obtained at heating rates of 10°C/min or below. A small weight gain beyond this large weight loss indicated formation of nickel sulfate. These reaction profiles are

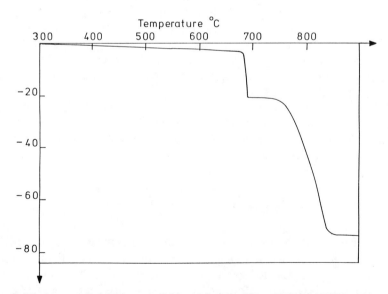

Fig. 52. Solid state reaction between stoichiometric quantities of NiS and $NiSO_4$ observed by heating the mixture in a thermobalance under N_2. Heating rate 10°C/min, sample mass 10 mg (116). Note the weight loss beyond 800°C due to the decomposition of $NiSO_4$, indicating that the solid–solid reaction was incomplete.

typical of ignition reactions where high heating rates produce a fast self-sustaining reaction that is highly exothermic (121).

The transition from stepwise to ignition reaction was influenced also by the sample mass, larger masses requiring lower heating rates to produce the ignition reaction. Thus, at a heating rate of $10°C/min$ in oxygen, a 25-mg sample showed an ignition reaction commencing at 470°C, whereas with a 5-mg sample, no ignition reaction was observed at a heating rate of $20°C/min$ (119).

Solid state reactions, such as that between $NiSO_4$ and NiS, can also occur, as illustrated in Fig. 52. Rapid weight loss occurs at 680°C according to the equation

$$3NiS + 2NiSO_4 \longrightarrow Ni_3S_2 + 2NiO + 3SO_2$$

as established by mixing the constituents and heating under nitrogen, although again the reaction does not proceed to completion (116).

7. Cements

Although there is no book equivalent to that written by Ramachandran (304) on DTA of cements, TG and DTG have been extensively applied to the study of the chemistry of cements. TG can be used to determine the purity of the raw materials used in the production of portland cement. The limestone or chalk can be monitored to ensure that a consistent batch is supplied to the rotary kiln. Gypsum or hemihydrate is always added to the cement clinker to retard the hydration of tricalcium aluminate. Hadrich and Kaisersberger (183) have reported a TG method for the quantitative determination of the various phases in the system $CaSO_4 \cdot nH_2O$, involving the use of both dry and humid atmospheres, and a method for the determination of these phases in cement has also been published (344).

The principal phases present in ordinary portland cement (OPC) are Ca_3SiO_5 (or C_3S), β-$Ca_2SiO_4(C_2S)$, $Ca_3Al_2O_6$ (C_3A), and $Ca_4Al_2Fe_2O_{10}$ (C_4AF), although all, especially C_4AF, vary in composition because of solid solution. The hydration of the calcium silicates follows the equations

$$2C_3S + 6H \longrightarrow C_3S_2H_3 + 3CH$$
$$2C_2S + 4H \longrightarrow C_3S_2H_3 + CH$$

where H represents H_2O. Since the calcium silicate hydrate is poorly crystalline and of uncertain composition, the formation of calcium hydroxide provides a method for following the kinetics of the hydration processes. Thermal methods, based on either TG or DTA, are here in direct competition with quantitative X-ray diffraction (QXRD) and chemical analysis after extraction with organic solvents. Midgley (257) has applied all four methods to a sample of OPC hydrated for 28 days at 18°C after mixing with a

TABLE 14
$Ca(OH)_2$ Concentrations (wt %) Present in a Sample
of Portland Cement Paste (257)

Method	wt %	Standard Deviation
DTA	15.8	0.2
TG	15.3	0.05
QXRD	12.0	0.4
Glycol extraction	19.9	0.3
Alcohol–glycerol extraction	12.7	—

water : cement ratio of 0.36. Although the two thermal methods gave similar results, the other two methods gave very different results (Table 14).

The reasons for these differences are as follows. QXRD determines only the crystalline calcium hydroxide and hence gave the lowest estimate of calcium hydroxide, since amorphous calcium hydroxide may also be present in the cement paste. Repeated extractions with ethylene glycol gave results varying from 9.9 to 19.9%, which indicates that the calcium silicate hydrate phase was attacked by the extracting solution, hence ultimately yielding a high value for the calcium hydroxide content. The two thermal methods gave reasonably good agreement between the extremes of the alternative methods. Midgley conclude that the thermal analysis techniques gave the most reliable results and it seems likely that TG will be used increasingly to follow the kinetics of hydration of portland cement and pure calcium silicates.

Pozzolanic cements, based on volcanic ash, have been used since Roman times and high-quality cements incorporating volcanic ash are produced nowadays, for example, in Mexico. In other countries, such as the United Kingdom, where electricity is produced from coal, there is considerable interest in the production of extended cements, such as that based on a blend of ordinary portland cement and pulverized fuel ash (PFA). All of these materials involve a reaction between the calcium hydroxide, liberated during the hydration of OPC, with the highly reactive silica-rich phase in the PFA to produce still more calcium silicate hydrate. Once again thermal methods provide a convenient method for following this reaction by monitoring the formation and subsequent disappearance of calcium hydroxide. Buttler and his coworkers (61–63) recommend a method in which the calcium hydroxide is first converted into calcium carbonate using a CO_2 atmosphere and is then determined by thermogravimetry.

DTG curves have been used extensively by Collepardi et al. (91, 92) to follow the hydration of tricalcium aluminate and of tetracalcium aluminoferrite in the presence of lime and gypsum. As the reaction proceeds the amount of gypsum decreases, whereas the amount of ettringite, $Ca_3Al_2O_6 \cdot 3CaSO_4 \cdot 32H_2O$, first increases and then decreases as it is converted into monosulfate, $Ca_3Al_2O_6 \cdot CaSO_4 \cdot 12H_2O$. These phases, and lime, can read-

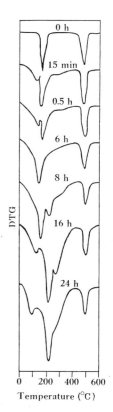

Fig. 53. DTG curves of hydration products from a 5:1:1 mixture of Ca$_3$Al$_2$O$_6$:Ca(OH)$_2$:gypsum (91).

ily be identified in the DTG curves shown in Fig. 53. Thus, the DTG curve at zero time shows the presence of gypsum (160°C) and calcium hydroxide (500°C), whereas that after hydration for 15 min also indicates the presence of a small amount of ettringite (130°C). After hydration for 6 h, the amount of ettringite observed has increased, while the amount of gypsum has decreased until it has all been consumed. Afte 8 h a peak at 210°C due to monosulfate was observed as well as peaks due to ettringite and calcium hydroxide. On further hyration this peak grew in intensity, while that due to ettringite decreased.

TG and DTG have also been applied to the study of phases formed in nonportland cements, e.g., C$_3$AH$_6$, C$_{12}$A$_7$ (1) and MgNH$_4$PO$_4 \cdot 6$H$_2$O (2). The most important application, however, is without doubt that to the conversion of high alumina cement (HAC). HAC is based on CaAl$_2$O$_4$, which is a different calcium aluminate from Ca$_3$Al$_2$O$_6$ found in OPC. On hydration at temperatures below 20°C, CaAl$_2$O$_4$ of CA forms CAH$_{10}$ and alumina gel. At slightly higher temperatures C$_2$AH$_8$ may also be formed. CAH$_{10}$ and C$_2$AH$_8$ are metastable, hexagonal hydrates of low density. On heating, or more gradually at low temperatures, these hexagonal hydrates convert to the stable cubic hydrate, C$_3$AH$_6$, with formation of gibbsite, Al(OH)$_3$, or AH$_3$

according to the equations

$$3CAH_{10} \longrightarrow C_3AH_6 + 2AH_3 + 18H$$
$$3C_2AH_8 \longrightarrow 2C_3AH_6 + AH_3 + 9H$$

The conversion is accompanied by a considerable change in volume that leads to increased porosity of the converted cement. This in turn can lead to substantial loss in strength and increased permeability, epecially if a relatively high water : cement ratio was used initially. Further details of the conversion process and its consequences can be found elsewhere (169, 255, 256, 282, 361).

In 1973–74 three accidents occurred in the United Kingdom in buildings constructed with HAC concrete beams. There is still controversy as to the reasons for the accidents. They may be due partly to poor engineering design and partly to the conversion reactions discussed above. This led to a national program of testing of HAC concrete involving the use of thermal methods. Recommendations for the testing of HAC concrete samples were published by the Thermal Methods Group (393). Most of the testing was carried out with DTA, but some use of DTG was made and it has been suggested that the DTG method is to be preferred (75, 17).

A typical DTG curve is shown in Fig. 54. Since C_3AH_6 reacts during natural exposure to atmospheric CO_2, the amounts of CAH_{10} and AH_3 are used to determine the degree of conversion:

$$DC = \frac{\text{Amount of } AH_3}{\text{Amount of } AH_3 + \text{Amount of } CAH_{10}}$$

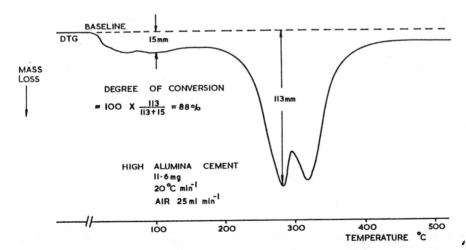

Fig. 54. The estimation of the degree of conversion of high alumina cement by DTG: 11.6 mg cement heated at 20° C min in air (25 mL/min). From E. L. Charsley, unpublished data.

Since the peak due to CAH_{10} overlaps with that due to alumina gel, and the peak due to AH_3 with that from C_3AH_6, peak heights from the DTG (or DTA) peaks were used to estimate the amount of these phases present. Using the example shown in Fig. 54 the percentage conversion can be written as

$$\% \, DC = 100\left(\frac{113}{113 + 15}\right) = 88\%$$

The degree of conversion is, therefore, an empirical scale, which can be used to estimate the likelihood of further conversion that might weaken the concrete or make it more permeable and, therefore, more vulnerable to chemical attack.

B. ORGANIC MATERIALS

1. Simple Organic Compounds

TG has been used mainly to study those organic compounds that decompose prior to volatilization and to investigate reactions between an organic compound and other substances. The techniques of DTA and DSC, since they can detect phase transitions and melting and boiling points, tend to yield alternative information to that from TG.

As an illustration of a decomposition reaction, many organic carboxylic acids dehydrate and decarboxylate on heating. The compound benzene-1,2,3-tricarboxylic acid dihydrate undergoes loss of water at 70°C followed by decarboxylation at 199°C (144).

When the organic material decomposes to a known stable product, or is completely volatilized, the weight loss information can be used for analytical purposes. It is also possible to determine the components in a mixture, even when the components interact, providing that the reactions are known or can be investigated, and that they occur at sufficiently differing temperatures for the weight losses to be determined with accuracy. This principle is well demonstrated by a series of papers on the thermal behavior of mixtures containing salicylic acid, sodium salicylate, sodium carbonate, and sodium hydrogen carbonate, substances that are involved in the commercial manufacture of sodium salicylate. By means of TG, the product yields and compositions of multicomponent mixtures can be monitored at various stages in the production process (299).

A novel approach using nonvolatile organic ligands as a matrix allows the thermal decomposition of volatile organic compounds to be observed (4). The nonvolatile matrix controls by diffusion the vaporization of the volatile compound, thus allowing an increased temperature to be reached prior to complete vaporization. Various chlorobenzenes decomposed with the evolution of HCl.

Vacuum TG has also been used to volatilize organic molecules adsorbed onto solid material, as in the case of polycyclic aromatic hydrocarbons (PAHs) adsorbed into soot particles (104). The evolved PAHs were analyzed by a coupled quadrupole mass spectrometer. The weight loss allowed a mass balance check to be made on the quantities of volatiles determined by QMS.

2. Plastics and Rubbers

Several review articles on polymers have appeared that contain references to TG applications (84, 224, 132, 174, 21) and a recent text is concerned wholly with thermal characterization of polymeric materials (369). The biennial reviews "Analysis of High Polymers" have already been mentioned (90, 260).

The major application of TG to plastics and rubbers is in the assessment of thermal stability, which has been critically reviewed by Still (353), and in the study of decomposition processes. Kinetic data for decomposition reactions are frequently determined. As an extension of kinetic studies, TG is sometimes applicable to the prediction of the thermal life of materials as a rapid alternative to established ASTM methods. The qualitative and quantitative analysis of rubbers by TG has received much attention.

a. THERMAL STABILITY

TG can be used to compare rapidly the thermal properties of a range of polymers for a particular application. After the loss of small molecules such as water, solvents, and monomers, the first significant weight loss of the sample is taken as the end of the thermal stability region. An extrapolated decomposition temperature (T_e) or the temperature at which a specific weight loss occurs may be determined as indicated in Section VI.A.1. This value is then used to rank the materials in relative order of thermal stability (Fig. 55). A comparison of the stability in nitrogen and air or oxygen gives an indication of the role of oxygen in the breakdown mechanism. Some plastics are remarkably unaffected by the change in atmosphere (for example, nylon), while others begin to decompose at much lower temperatures in an oxidizing atmosphere.

Thermal stability measurements have been made on a large variety of plastics and rubbers by this technique, and some useful examples are given for elastomers (179), phenol formaldehyde and epoxy resins (132), PVC (395), and some P-N-containing polymers (69).

The procedural decomposition temperature provides the maximum operational temperature at which the polymer can exist, and is of value when the material is being used as a sacrificial material, for example, in re-entry vehicles in space programs (149). However, this test gives no indication of the long-term behavior at lower temperatures. This information is more meaningfully provided by heating the material to specific temperatures and monitoring the weight change as a function of time. Figure 56 shows a family of

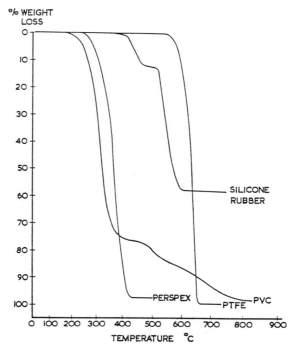

Fig. 55. Based on the TG procedural decomposition temperatures in air, the thermal stability of the polymers increased in the order PVC < perspex < silicone rubber < PTFE. From E. L. Charsley, unpublished data.

curves generated by this technique for viton, where it is evident that although this material may be usable up to 300°C, beyond that serious decomposition occurs in a relatively short time (19).

The oxidative stability of a material can be measured quantitatively by heating the material in an inert atsmosphere to a given temperature below the decomposition temperature, establishing a stable baseline, and then switching the atmosphere to oxygen. A small weight gain is first observed due to adsorption of oxygen. The time taken to produce the first weight loss after the admission of oxygen, called the induction period (t in Fig. 57), is a measure of the oxidative stability at that temperature (9, 194). The TG procedure has also been used as an analytical technique for the determination of low levels of antioxidants (0.001–0.06%) in polyethylene (9), where a linear relationship between the induction period and antioxidant concentration was obtained.

The effects of various additives on the thermal stability of plastics and rubbers can be compared by TG to untreated materials. An effective antioxidant should increase T_e (under the same experimental conditions), decrease the weight loss at a given temperature, or increase the time to the initiation of oxidative degradation. Typically, papers discuss the variation of thermal

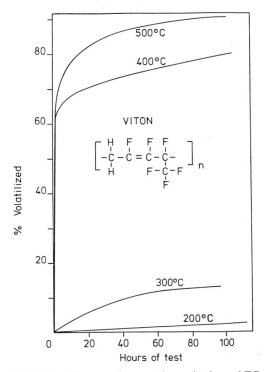

Fig. 56.   Thermal stability studies on polymers using an isothermal TG technique (19).

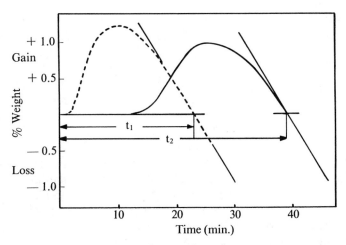

Fig. 57.   TG curves at 200° C of the oxidation of polyethylene in an oxygen atmosphere (9): ---, without antioxidant; —, with antioxidant.

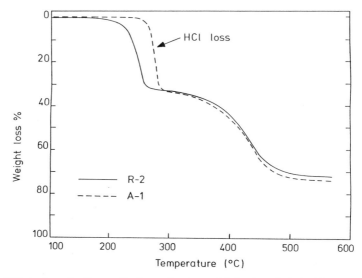

Fig. 58. Effect of production methods on the thermal behavior of vinyl chloride-acrylonitrile copolymer (205).

stability with the concentration of thermal stabilizers, such as organotin compounds in PVC (396), and with antioxidants in polypropylene (9) and in elastomers (69). The effect of inorganic salts on the stability of viscose rayon has also been noted (67), and metal chelates on the oxidation of polyolefins at high temperatures (26).

The thermal stability of a material can be affected by structural factors, which in turn may be dependent on production methods. An alternating vinyl chloride/acrylonitrile copolymer (A1) was 40°C more stable in nitrogen than a random structure (R2), and the decomposition temperature also increased with relative viscosity (205). The first weight loss (Fig. 58) corresponded to the elimination of HCl and was used to calculate the vinyl chloride content.

It is of some technological importance to be able to predict the long-term life expectancy of materials used, for example, as insulating enamels. This topic has been reviewed by Flynn (142). The usual test is ASTMD-2307, which takes as its criterion the temperature at which 20 000 h of usage would cause twisted pairs of the wires to fail. A minimum testing time of one year still produces 10–20°C scatter in the final results. TG methods have been developed as accelerated test procedures to replace the ASTM method. The TG methods are based on the assumption that the decomposition process that occurs at the decomposition temperature is the same as that occurring at the temperature normally experienced by the enamel. This can be expressed as

$$\text{Time to failure} = \frac{1}{kT}$$

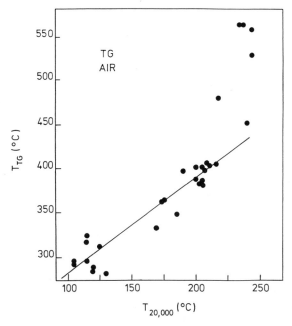

Fig. 59. Correlation between the T_5 value determined by TG and the $T_{20\,000}$ value determined by ASTMD-2307 for some magnet wire enamels (58).

where k is the velocity constant for the decomposition, and T is the temperature at which the enamel is held. This test procedure appears to be limited to decomposition processes that occur by a single well-defined mechanism, i.e., when the TG curve shows only one smooth weight loss.

Various TG parameters have been used in the calculation of thermal life. One method measured the temperature at which the first detectable weight loss was observed at various heating rates, and then derived the required temperature by extrapolation to zero heating rate (101). In a more fundamental treatment an equation was derived that allowed the $T_{20\,000}$ values to be calculated from TG curves obtained at various heating rates (368). Excellent agreement was obtained between $T_{20\,000}$ values and those calculated from TG curves for enamels that decomposed in a single step. Another approach used the temperature at which 5% of the material had decomposed (58). Again good correlations were obtained between $T_{20\,000}$ values and those derived from the TG results (Fig. 59).

In a critique Flynn (142, 143) has pointed out that failure of materials may result from processes that are only tenuously linked or even unrelated to weight-loss kinetics, and the use of TG data for lifetime predictions must be considered in relation to the application. If TG data are applicable, then studies should be made over a wide range of conditions, and especially at various heating rates, so that changes in mechanism can be established, and competing reactions and rate-controlling steps recognized.

b. Decomposition Reactions

The decomposition of plastics has been extensively studied by thermal techniques, for example, polybutadiene (49), PVC (396), polyephichlor-hydrin (262), polymer mixtures (138), and polyester polyurethanes (177) and poly-imides (397). Decomposition products can be isolated by trapping the volatiles from various temperature regions and then subjecting them to analytical techniques such as GC–MS (see, for example, Chatfield and Einhorn (80)), or they can be identified by coupled systems such as TG–MS. This information can be used to study mechanisms and rates of decomposition, and also yields practical information such as the toxic hazards likely to be encountered if polymeric materials are present during fires (129). McNeil (240) has introduced the technique of Thermal Volatilization Analysis (TVA), which monitors the pressure of volatiles evolved from a polymer subjected to a linear temperature program and contained in a continuously evacuated system. A cold trap condenses out less volatile materials, so that the Pirani gauge responds only to products volatile under vacuum at ambient temperature. Thus, the technique offers alternative information to that provided by DTG curves, which give the total rate of mass loss.

The addition of flame-retardant compounds to fibers used in clothing manufacture or to plastics used in building and furniture construction has been increasingly investigated (290). The major flame retardants used are aluminium oxide trihydrate and antimony oxide, the latter in conjunction with organic halogen compounds and organic phosphates. Unlike the addition of antioxidant or thermal stabilizers discussed above, the flame retardant may cause a decrease in the decomposition temperature. Its main functions seem to be to alter of the mode of decomposition, and particularly to reduce the rate of decomposition. Autoignition of the material may also be prevented. TG and DTG can be used to examine the effect of flame retardant addition on the rate of weight loss. Figure 60 illustrates the effect of a commercial flame retardant on the decomposition of cotton (34). The additive caused the initial weight loss to occur some 20°C lower than in the untreated cotton. However, the decomposition mechanism appeared to have altered, since much less weight was lost in this first step. The second weight loss occurred at a much slower rate than in the untreated cotton, reducing the rate of evolution of volatile flammables. The residue was also greater, indicating that the volume of volatile material had been reduced. Plots of rate of weight loss for polyester–cotton material treated with tetrakis (hydroxymethyl) phosphonium chloride urea-poly(vinyl bromide) anti-flame agent showed a similar trend (267). A TG test has been proposed as an alternative to the pit test for flammability studies (35).

c. Analysis of Rubbers

The composition of a rubber is complex, and may contain polymers, inert inorganic fillers, zinc oxide, reinforcing agents, such as carbon black and graphite, extending oils, cure residues, and protective agents. The analysis of

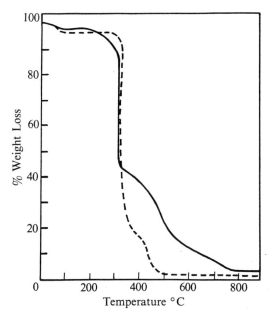

Fig. 60. TG curves of cotton in air (30 mL/min, 6°C/min)(34): ---, untreated cotton; —, cotton treated with a flame-retarding agent (34).

these components is demanding and much attention has been given to the quest for more automated, less time-consuming alternative procedures. TG provides such a method, at least for the analysis of bulk constituents (248–250, 357, 48, 289). The role of thermal methods in the analysis of vulcanizites has been discussed by Tidd (362) and an extensive review has been written by Brazier (50).

Figure 61 shows an idealized curve typical of an oil-extended ethylene–propylene–diene (EPDM) rubber. In nitrogen, the volatilization of the extending oil, together with other volatiles, such as stabilizers and unreacted curing aids, causes the first weight loss. The second weight loss is caused by polymer decompositions, which for noncarbonizing rubbers allows the polymer content to be calculated. If carbonates are used as fillers, a further weight loss may be evident at around 700°C. If the atmosphere is changed to oxygen, carbon black is oxidized, and inorganic fillers and zinc oxide remain as ash. Thus, the oil, polymer, carbon black, and ash can be determined in one TG run.

Some formulations contain both carbon black and graphite. TG can be used to calculate both values, by cooling the thermobalance back to 300°C before changing to an oxidizing atmosphere. On reheating, the carbon black combusts at lower temperatures than graphite.

DTG curves have been recorded (343) and stored as a means of finger-printing various rubber formulations to assist in the identification of polymers in blended rubbers, and much of the published data has been collected by

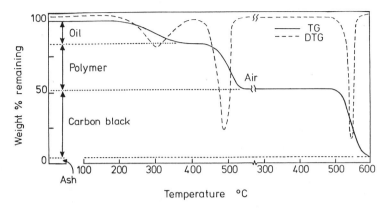

Fig. 61. Analysis of EPDM rubber for volatilies, polymer, carbon black, and ash. The first two components are determined by heating the rubber in N_2 and then oxidizing the carbon black in air. The residue is inorganic oxide ash.

Brazier (50) in his review. The DTG peak heights can be used for quantitative purposes. When a reproducible temperature and similar sample sizes are used, the peak heights of natural rubber (NR) and butadiene rubber (BR) are consistent to better than 1%. The DTG curve for the NR–BR blend and the ratio of peak heights of NR to BR are shown in Fig. 62a and b (48). Similar results were obtained for an NR–SBR blend, but in an NR–SBR–EPDM blend only the NR could be determined, because the DTG peaks for the other two components overlapped (48).

Clearly, the methods outlined above should not be applied indiscriminantly to rubbers, since many components could have weight losses that occur in the same temperature range. A thorough feasibility study using individual components is required prior to adoption as a routine analytical technique. Nevertheless, the rapid analysis time, which can be in the region of 30 min per sample for TG, does offer advantages that may outweigh a certain lack of accuracy.

TG and DTG curves have been obtained in oxygen, but the results are generally less reproducible relative to nitrogen, mainly because the exothermic oxidation of the polymer is so difficult to control. Depending on the sample packing, contact with the atmosphere and heating rate, consecutive runs on similar quantities of the same material may produce conditions that cause rates of reaction to differ considerably, which results in differing shapes, decomposition temperatures, and DTG peak temperatures.

DTG peak temperatures have been used to identify the extending oils used in EPDM rubbers and plasticizers in NBR materials (357). The type of carbon black used in rubber formulations can be identified by a number of time-consuming methods, and a TG method that is an extension of the analytical scheme discussed at the beginning of this Section has been proposed as a rapid alternative (247). One of the essential differences between

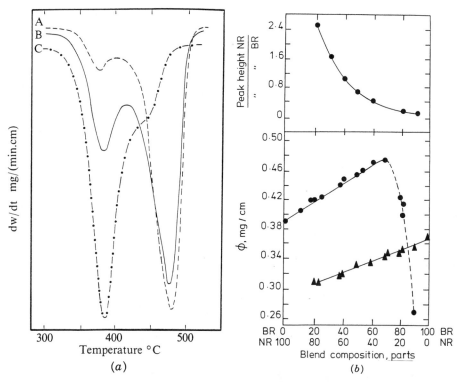

Fig. 62.   The use of DTG peak heights to determine NR and BR in a NR–BR blend. Uncured blends containing 50 phr HAF–LS. Heating rate 10° C/min, N_2 atmosphere. *A*, 10:90 NR: BR; *B*, 40:60 NR:BR; *C*, 90:10 NR:BR (48). (*a*) DTG curves; (*b*) ratio of peak heights of NR to BR as a function of composition.

carbon blacks is surface area, which is influenced by the preparative technique. A high correlation has been found between the surface area of carbon blacks and the temperature at which 15% weight loss, T_{15} (a purely arbitrary figure), occurs (247, 248, 289) (Fig. 63). This relationship was reported to be valid for both free and compounded carbon blacks, in a variety of elastomer formulations, although the T_{15} values are generally lower in compounded rubbers relative to free blacks. More recent work on a wider range of carbon blacks than previously studied suggests that the relationship is not linear (362), and one report considers that the scatter of results is so great even for free carbon blacks that the method is of little value (350). It has been shown that the T_{15} value is significantly affected by the method used to cure the rubber (248), as well as the maximum temperature achieved during the pyrolysis step, and the flow rate of air and the heating rate used during the oxidation step (78, 79), so care must be exercised in the determination. An alternative procedure using an isothermal oxidation method has been proposed (324). From their results they concluded that the relationship

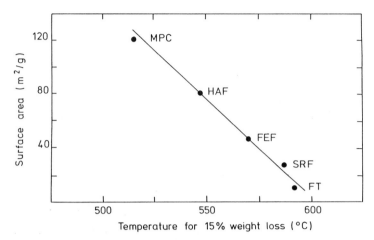

Fig. 63. Correlation between T_{15} and the surface area of carbon blacks used in rubber formulations (362).

between rate of oxidation and surface area held only for some carbon blacks significantly separated in surface area. Also, the rate can be quite different for carbon blacks of the same type but produced by different manufacturers. The technique, therefore, cannot be recommended as suitable for the identification of a carbon black in an unknown formulation, but can be used as a routine quality control check on batch rubbers. Identification of the individual carbon blacks in rubbers containing more than one carbon black has also been attempted (249).

3. Natural Products

Some attention has been given to the thermogravimetric behavior of natural materials. Since the interpretation of results obtained from such studies tends to be complex, it is usual to investigate the behavior of the isolated or synthetic pure bulk components prior to assessment of the natural product.

A book edited by Shafizadeh et al. (331) contains many examples of the application of thermal methods to wood. TG can be used to study the pyrolysis and oxidative degradation of wood, and the decomposition products can be analyzed by EGA. Much of the work is concerned with a fuller understanding of the mechanism and kinetics of decomposition and also the variation that occurs with change in proportion of the major constituents or additives. Tang (360) has reported the effect of various salts on the decomposition of woods, mainly to assess their flame-retardant properties. TG curves in vacuo for wood in an untreated state are compared in Fig. 64 with the same wood treated with a variety of inorganic compounds. An initial weight loss (not shown) is due to water volatilization, and above 220°C pyrolytic

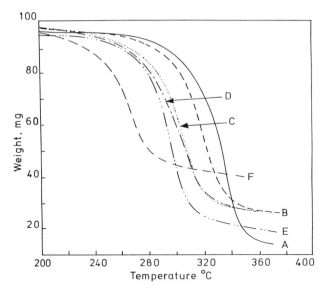

Fig. 64. Comparative TG curves in vacuum, for wood, and wood treated with various inorganic salts (360): *A*, untreated; *B*, 2% $Na_2B_4O_7 \cdot 10H_2O$; *C*, 2% NaCl; *D*, 2% $KHCO_3$; *E*, 2% $AlCl_3 \cdot 6H_2O$; *F*, 2% $NH_4H_2PO_4$.

degradation of the wood commences. The presence of the various salts decreases the temperature at which decomposition commences and also reduces the degree of volatilization. This behavior is similar to that shown by textiles treated with flame retardants (see Section VIII.B.2.a). Tang concluded that cellulose is the controlling influence in the decomposition of wood, whether untreated or treated, a view supported by other work (192), although an increase in the lignin content did tend to increase the rate of decomposition above 330°C. Calculations showed that the decomposition activation energy decreased from 67.0 kcal/mol at 3.7% lignin to 41.4 kcal/mol at 23% lignin.

In oxygen (Fig. 65) the decomposition may occur via a two-stage process, i.e., preignition and post ignition volatilization (360). Generally, the tendency of any material to ignite increases with heating rate and the availability of oxygen, and for diffusion-controlled reactions by a sample with high surface area.

The TG curves obtained from the pyrolysis of tobacco are in themselves of little value, since only the gross weight losses are evident. DTG coupled with an evolved gas detector (EGD) provided rather more information about temperature regions in which mass loss and gaseous evolution rates are at a maximum (Fig. 66) (59). By replacement of EGD with a GC some of the individual gases were identified. Use of a Carbosieve B column allowed the evolution of CO, CO_2, and H_2O to be monitored, and a Poropak Q column coupled with a FID permitted the analysis of hydrocarbons.

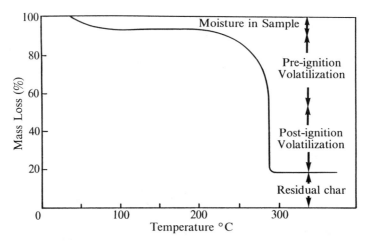

Fig. 65. Oxidation of untreated ponderosa pine. Heating rate 18° C/min, 200-mg sample; O₂ flow rate 90 mL/min (360).

An interesting use of TG and DTG is the assessment of crystallinity of cellulosic materials (65). When a series of cotton celluloses, having differing degrees of crystallinity, was subjected to TG and DTG, the TG decomposition temperature and the rate of weight loss varied with the amount of amorphous material present in the sample (Fig. 67). The TG decomposition temperature decreased with decrease in crystallinity, as did the temperature at which the maximum rate of weight loss occurred. The DTG curve also indicated that as the amorphous nature of the material increased, the

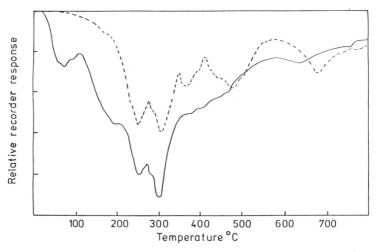

Fig. 66. DTG–EGA analysis of pyrolyzed tobacco (59): —, DTG; ---, EGA.

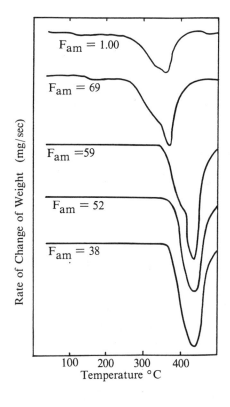

Fig. 67. DTG curves of cotton celluloses having different degrees of crystallinity. F_{am} is the fraction of amorphous material present in the sample (65).

decomposition tended to occur in at least two stages, since shoulders imposed on the low-temperature side of the main peak become evident. The crystallinity of the material was also determined by moisture regain measurements and XRD.

4. Coals, Cokes, and Carbons

Coal is still one of the major sources of energy, hence its thermal behavior is of considerable interest. Although the factors that are important tend to be associated with the thermal energy evolved during combustion, information that can be gained by DTA and DSC, TG has made important contributions to the analytical chemistry of coal and to gasification studies.

Cumming and McLaughlin (97) have provided a substantial account of the analytical applications of TG and DTG to coal, involving proximate analysis, burning profile tests, and a volatile release profile test.The proximate analysis of coals is carried out by a TG method similar to that used for the analysis of rubbers (see Section VIII.B.2.c). The coal sample is heated in nitrogen to 105°C and held until constant weight is achieved. This weight loss is proportional to the moisture content of the coal. The sample is then heated to 900°C and held at this temperature for 7 min. This second weight loss corresponds to the evolution of volatile matter. The furnace temperature is then lowered to 815°C and the atmosphere changed to air, which burns off the fixed carbon and leaves the ash as a residue. Other authors (quoted in Ref. 97) have published similar methods. Fourteen coals of widely differing composition were analyzed by the British Standard Specification 1016, Part 3 (1973), and by the TG method. The ash content was very similar by both methods. Seven of the coals gave low volatile matter and high fixed carbon by the BS method relative to the TG values, with the largest deviation being -1.8% volatile and $+2.0\%$ fixed carbon (absolute values) for the same sample. Five of the coals gave high volatiles and low fixed carbon, and two coals gave positive relative values for both volatile and fixed carbon. The major advantage of the TG method is its speed, with one analysis taking about 35–40 min to perform.

The burning profile test is a characterization method involving the rate of weight loss of a coal sample burning in air plotted against temperature. The original test involved a 300 mg sample, but the TG technique uses only 15 mg (97). The temperature at which the maximum rate of combustion occurs (the DTG peak temperature) is taken as a measure of combustability, with lower temperatures indicating more easily combustable coal. The test is dependent on a number of procedural variables, especially heating rate and sample mass, since changes in these parameters can change the reaction from a slow combustion process to an ignition reaction (121). In the latter case, the oxidation occurs very rapidly and over a narrow temperature range, so that only a single DTG pack may be evident. Some TG curves obtained under combustion as opposed to ignition conditions are given in Fig. 68.

The volatile release profile is similar to the combustion profile except that the former is carried out in N_2. Thus, the profiles consist of peaks due to the emission of water and volatiles as well as organic molecules from the pyrolytic degradation of the coal structure (97). The major use of the burning profile and volatile release profile is for intralaboratory comparison of coals from different sources and prediction of behavior in industrial coal-fired boiler furnaces.

Aylmer and Rowe (8) have adapted the proximate analysis method by including a magnetic step that permits the determination of pyrite. After the proximate analysis has been carried out, a magnetic field is switched on and any apparent increases in weight are attributed to paramagnetic species such as Fe_2O_3. The sample is then heated to 400°C and H_2 admitted, which reduces all iron species to Fe(0) and produces a large apparent weight gain caused by the ferromagnetic effect. After cooling to room temperature the field is switched off and the apparent weight loss is used to calculate the pyrite content. Any significant paramagnetic species present needs to be corrected for. The authors analyzed 30 coal samples by the ASTM and TG magnetic method for proximate analysis and pyrite and generally obtained good agreement.

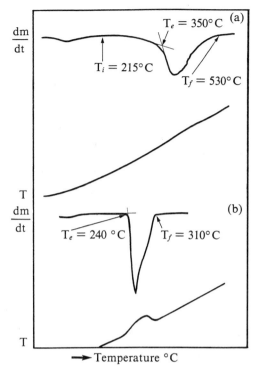

Fig. 68. DTG curves for a coal heated under different experimental conditions: (*a*) 10° C/min, 75 mL/min air, 11.1 mg; (*b*) 20° C/min, 27 mL/min O_2, 11.9 mg. From J. G. Dunn and I. D. Sills, unpublished data.

TG-DTA studies provide information on gasification of coals and chars, either under pyrolysis of oxidative conditions (125, 378). The derivative weight loss can be used to obtain kinetic information on gasification, and may be important when coals from a variety of sources are being assessed. The effects of impurities and additives can be examined, as well as changes in behavior after periods of storage (379). The deposition of carbon due to the cracking of methane has been found to be important in the reactivity of chars (212). The presence of inorganic materials that act at catalysts is also important. The reactivity of cokes and char can also be lowered by the addition of B_2O_3 (70), which appears to occupy surface sites and hence reduce the available surface for reaction. The direct hydrogenation of coal has been studied using a symmetrical thermobalance system capable of operation up to 1000°C and 5 MPa (171, 172).

TG has been used to investigate the thermal regeneration of spent activated carbons (356), by loading active carbons with 32 different single-component organic compounds. The TG curves could be classified into three groups by shape, which were related to the ease of volatization of the organic molecules: Group I organics were volatile, group II organics were easily decomposed, and group III organics gave high residuals even up to 800°C.

Thermal methods have been used to investigate the mineral content of coals (277). During the combustion of coals atmospheric conditions may range from oxidizing to reducing and these atmospheres may affect the behavior of the minerals present. Thee conditions can be simulated on a thermobalance. For example, the decomposition temperature of gypsum is lowered to 780°C in 12% $CO-N_2$, 10% H_2-N_2, and 18% $CO_2-2\%$ $CO-N_2$ mixtures compared to values of 1000°C in neutral or oxidizing atmospheres (277). In 12% $CO-N_2$, the decomposition appeared as a two-stage process. The TG–DTG results can also be used for the quantitative analysis of minerals when reactions are not overlapping.

5. Petroleum Products and Related Materials

The use of thermal methods of analysis to characterize petroleum products has been extensively reviewed up to 1978 by Noel and Cranton (275), and although the majority of references are concerned with DTA or DSC, some TG applications are apparent. In addition, TA applications to oil shales and sands have been reviewed by Rajeshwar (302), and for these materials TG is the dominant technique used.

Because the transformation of the organic matter (or kerogen) contained in petroleum source rocks, oil shales, and tar sands into hydrocarbons is primarily dependent on temperature, the extent of hydrocarbon formation and other volatile materials can be followed by TG and DTG, and the products can be analyzed by EGA. This information is valuable because it can be used to estimate the hydrocarbon generating capacity of the source as well as to classify the kerogen type (123, 303, 220, 320). In general, kerogen combusts in air or oxygen in two distinct thermal regions, the first producing mainly oxidative decomposition products of aliphatic materials, and the second producing oxidation decomposition products of aromatic compounds. The pyrolytic decomposition produces two main weight losses, the first corresponding to the evolution of water and CO_2 from the decomposition of oxygen-containing molecules, and the second from the releasse of hydrocarbons. The relative proportion of the pyrolysis weight losses indicate either an oxygen-rich (type III) or hydrogen-rich (type I) kerogen, with type II kerogen being intermediate between I and III.

Some information on the maturity of samples of the same type can also be inferred from TG data, since the least mature kerogen generates the greatest quantity of hydrocarbons and hence shows the greatest weight loss. Although it is advantageous to examine the source material directly, in some cases minerals present in the material decompose in the same temperature regions as the kerogen decomposes and introduce errors into the results. Hence, either demineralization or solvent extraction of the kerogen is necessary, and these are time-consuming steps. Jha (204) has reported that the thermal degradation of heavy oil core samples has been investigated in He, N_2, and air using TG and DSC. TG and DTG curves showed four weight loss regions occurring in four different temperature zones. In inert atmospheres the four temperatures zones produced reactions of volatization, thermolysis of the heavy oil, and volatilization, cracking, and coking. In air, oxidation and combustion occurred.

The decomposition kinetics of oil shale kerogen has also been widely investigated by TG, especially for Green River oil shales (see Rajeshwar and Dubow (301) and references therein). A simple TG method for estimating oil yields has been reported to give good correlations with Fischer assays for oil shale kerogens of the same type (315).

TG has been used to study the vaporization characteristics of fuels, crude oil fractions, greases, and aliphatic and aromatic organic compounds (reported in Noel and Cranton (275)). The technique has been used to study pyrolysis reactions of pitch derived from hydrocracked bitumen (326, 150) and provides a diagnostic test for the suitability of various pitches as matrices for carbon–carbon composites (73). DTG and TG have been used to identify adhesion agents in bitumen (111).

Most TG studies of oils and waxes have investigated their thermal stability, although some have involved characterization and analysis. Several tests are available for the estimation of the storage quality of vegetable oils. These tests attempt to simulate, under accelerated conditions, the normal storage environment of the oil. The tests usually depend on exposing the oil to contact with increased partial pressure of oxygen, with perhaps mild heating (up to 100°C), and the time taken for oxygen uptake to occur is correlated with the storage quality of the oil. In the TG method, oxygen is in contact with the oil under ambient atmospheric conditions, but the temperature is raised beyond 100°C. A TG curve for an oil (273) treated in this manner is shown in Fig. 69. The temperature at which the weight gain commenced, T_i, was found to be the best indicator of storage quality. It has been subsequently pointed out (188) that oils with widely differing storage properties had a T_i value range of only 20°C, and hence good experimental technique is essential for differentiation. Isothermal tests have also been used, (analogous to the oxidative stability test discussed in Section VIII.B.2), in which the oil is heated to a specific temperature under nitrogen and then the atmosphere is changed to oxygen. The time taken for the oil to oxidize is used as an indication of storage quality (188). Reasonable correlation was found between this method and the active oxygen method. A rapid method for the determination of the auto-ignition temperature of heat transfer oils has been published (68).

The thermal stability of chlorinated paraffins, used as secondary plasticizers for PVC, has been established by TG runs in nitrogen (349). The

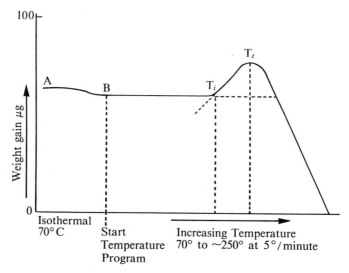

Fig. 69.   TG curve for an oil sample. Typical sample size of 1.5 mg (273). *A* to *B*, loss of any solvent; *B* to T_i, induction period; T_i, initiation temperature; T_t, temperature of maximum gain in sample weight.

decomposition of the waxes is similar to that of PVC, with dehydrochlorination occurring at just above 200°C, followed by pyrolysis of the residual material.

Some indication of the molecular weight of a wax and its molecular weight range can be gained from TG runs. The volatilization of waxes will occur at increasing temperature with increasing molecular weight (42), and the temperature range over which the volatization occurs gives some idea of the molecular weight distribution of the waxes (95). The variation that occurred in the TG curves for a binary mixture of paraffin and carnauba waxes permitted quantitative analysis to be carried out (95). The behavior of some dental waxes was also examined in this last study.

6.   Medical and Biological

The most obvious application of TG to medical science is in the identification and determination of inorganic and organic compounds in solid materials such as calculi and bones. A useful review has been published (28). The thermal properties of materials such as waxes, enamel, and dentin are of interest to the dental profession.

A.   INORGANIC COMPOUNDS

The compounds present in gallstones, kidney stones, and other forms of calculi are usually $CaC_2O_4 \cdot 2H_2O$, $CaC_2O_4 \cdot H_2O$, $Mg(NH_4)PO_4 \cdot 6H_2O$, uric acid, and secondary and tertiary calcium and magnesium phosphates. On

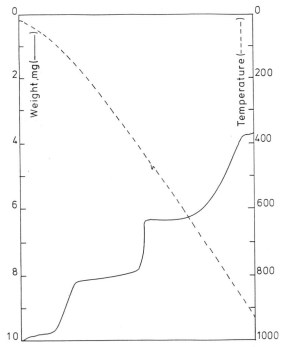

Fig. 70. TG record of a renal calculus containing 93% calcium oxalate. The weight losses correspond to Ca(COO)$_2$ 1.5H$_2$O $\xrightarrow{100°\,\text{C}}$ Ca(COO)$_2$ + 1.5H$_2$O, Ca(COO)$_2$ $\xrightarrow{400°\,\text{C}}$ CaCO$_3$ + CO, and CaCO$_3$ $\xrightarrow{550°\,\text{C}}$ CaO + CO$_2$. Sample weight, 10 mg; heating rate, 30° C/min; atmosphere, O$_2$ at 30 mL/min (314).

heating, these compounds lose water or decompose stoichiometrically, so that the weight loss can be used to determine the component in the original sample. Although interference between these compounds is possible, in the majority of cases the stone consists of either predominantly CaC$_2$O$_4$ · H$_2$O and CaC$_2$O$_4$ · H$_2$O, or a mixture of oxalates and Ca$_3$(PO$_4$)$_2$, which simplifies the analytical problem. The TG method allows differentiation of mono- and dihydrates of calcium and of urea and urea monohydrate; such differentiation is not readily accomplished by other methods. The TG approach is also rapid, with one analysis being carried out in under 300 min.

In one study, 502 characteristic curves were obtained for stones that consisted primarily of CaC$_2$O$_4$ (mixed mono- and dihydrate), NH$_4$MgPO$_4$ · 6H$_2$O, CaHPO$_4$ · 2H$_2$O, uric acid, and crystine (314) (Fig. 70). Quantitative analysis on stones that were essentially either calcium oxalate or calcium urate showed correlation coefficients of 0.947 and 0.97, respectively, with values obtained by chemical analysis. Successful analyses were possible on stones containing pure calcium oxalates and mixtures of calcium oxalates and phosphate, uric acid and ammonium urate, calcium oxalate and urate, and cystine. Difficulties were encountered in the analysis of stones for $NH_4MgPO_4 \cdot 6H_2O$ since some $Ca_3(PO_4)_2$ was inevitably also present and may also contain $CaHPO_4 \cdot 2H_2O$. The latter loses water in the same region as the magnesium salt and hence high results were obtained. The $Ca_3(PO_4)_2$ is not decomposed at least up to 900°C and so the residue is heavier than expected for $NH_4MgPO_4 \cdot 6H_2O$ alone.

Stones obtained from donors in Eastern Europe have been analyzed by simultaneous DTA–TG–DTG, the DTA providing additional means of identification (28, 29). Calcium oxalates again predominate (71% of 2295 stones examined) and calcium oxalate was also an important secondary component. Of these stones, 23.7% contained appreciable quantities of urea. TG methods appear to be generally applicable and perhaps should be more commonly used in the routine analysis of calculi, a view that is supported in a review of the topic by D'Ascenzo et al. (99). A range of hard and soft tissues, such as bone, enamel, and dentin as well as stones have been analyzed for inorganic and organic constituents by TG–DTA (6).

B.   ORGANIC COMPOUNDS

A DTA–TG–DTG study of proteins has been reported (33) and curves have been presented for cytochrome C, ribonuclease, fibrinogen and fibrine, and the collagen-containing materials gelatin, rat skin, and Achilles tendon. The collagen-containing materials exhibited differences in DTG curves with age, through the formation of covalent cross-links between adjacent peptide chains. A loss of water caused a DTG peak at 320°C; two further DTG peaks at 580 and 620°C were attributed to the breakdown of unaged and aged (more highly cross-linked) collagen, respectively. Plots of weight loss at these two temperatures against age of tissue from human Achilles tendon showed significant mathematical relationships, although no correlation coefficients were quoted. A similar study on rat skin (198) showed a high-temperature component that gave a DTG peak at 420°C, which was gradually replaced by a peak at 360°C as the skin aged. Again, some relationship between age and weight loss was evident but not strong. Although of interest, the scatter on these results seems to indicate that any dating method developed would not be of sufficient accuracy to be useful.

Thermal analysis techniques including TG/DTG have been used to study the effects of adsorbed body fluids on 40 heart valve poppets implanted for periods of between 22 and 79 months (36). Differences in DTG peak temperatures were found between new and used poppets, presumably through the alteration of polymer properties as a result of the adsorbed materials. The quantity of adsorbed liquid was calculated from the TG curve and found to agree well with values found by solvent extraction techniques.

7. Pharmaceuticals

Most of the applications of thermal techniques to pharmaceuticals are centered around identification, purity, polymorphism, and the compatibility of mixtures, properties that are best studied by DTA or DSC. Consequently, TG is used only to provide supplementary information, such as the water content of the drug, the temperature at which drying can be carried out without decomposing the required compound (47), or estimating the degree of solvation by organic molecules (176). It is difficult to use TG for purity check purpose, since many of the impurities are volatile or decompose in the same temperature ranges as the major constitutent, although in pharmaceutical preparations the inorganic filler content can sometimes be determined from the weight remaining at the end of a heating program. A combination of DTA and TG was found useful in the identification of some relatively simple analgesics (383) and DTA–TG–DTG were used to study the decomposition of some medical compounds based on complexes of Bi(III) (298). However, for more complex mixtures involving vitamin preparations, thermal methods were less useful, and assignment of reactions to DTA peaks and TG weight losses was difficult. Some problems were caused by the relatively low concentration of vitamin in an inert inorganic matrix (94). In a study of 31 drugs by DTA, TG, and DTG, 18 gave purity values that were in agreement with the calculated value (300).

IX. DETERMINATION OF PHYSICOCHEMICAL DATA

A. THERMODYNAMICS

The use of thermogravimetry for the determination of vapor pressures by the Knudsen effusion method has been discussed by Wiedemann (389). The technique involves measurement of the rate of loss of molecules of the evaporating material in an effusion cell through a calibrated orifice under reduced pressure. The measurements were made at 10^{-6} torr under isothermal conditions by simultaneously recording the weight loss, the sample temperature, and the pressure outside the effusion cell. The latter was mounted on a modified Mettler Thermoanalyser in place of the standard sample holder. A cross section of the system is shown in Fig. 71. The body A of the cell is made from aluminium. The copper–constantan thermocouple E measures the temperature of the sample F, and the precious metal thermocouple D is used to control the furnace temperature. The interchangeable orifice B is made of 0.01 mm-thick Ni–Cr foil with hole diameters ranging from 1 to 3 mm. Samples normally of about 500 mg are introduced at C and a balance sensitivity of 1 mg full-scale deflection is used. Both screw covers are sealed with Teflon rings and the cell has a working range from -100 to $+200°C$.

Four organic materials, p-chorophenyl-N',N'-dimethyl urea (a herbicide), p-phenacetin, anthracene, and benzoic acid, were studied below their melting points in the range 250–400 K (a Pyrex cell was used for the former since it appeared to decompose in the aluminium cell) and equilibrium vapor pres-

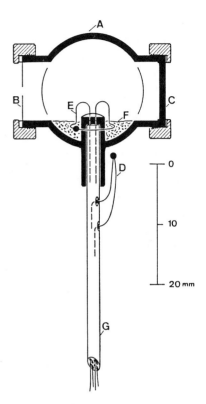

Fig. 71. Schematic diagram of a system used to measure vaporization rates based on a Kundsen cell (389). See text for explanation.

sures were calculated according to the ideal Knudsen equation

$$p = \frac{\Delta m}{\Delta t}\frac{1}{q}\sqrt{\frac{2\pi RT}{M}}$$

where p is the vapor pressure in dyn/cm^2, q is the orifice area in cm^2, $\Delta m/\Delta t$ is the rate of weight loss in g/s, T is the temperature in K, and M is the molecular weight of the specimen.

The data were found to fit the Clausius–Clapeyron relationship $\ln P = B + A_1/T$, enabling direct calculation of the heat of sublimation ΔH_s or, above the melting point, of the heat of vaporization ΔH_v. It is suggested that if measurements are made over a narrow temperature range either side of the melting point the heat of fusion can be determined by the difference between ΔH_s and ΔH_v and this is illustrated for benzoic acid.

For the method to be successful it is necessary that the temperature be kept constant to within 0.1°C and that the pressure and the zero point of the balance do not drift over the period of the experiment, normally hours. Ashcroft (7) has used a simpler method to determine enthalpies of sublimation by thermogravimetry using the Langmuir method based on the rate of escape of molecules from a surface under vacuum conditions. According to the Langmuir method, the rate of sublimation per unit area of substance is related to vapor pressure by the equation

$$\frac{dm}{dt} = \left(\frac{M}{2\pi RT} \right)^{1/2} P$$

Samples of 50–100 mg were examined at 5 or 6 constant temperatures covering a 20–30°C temperature range on a full-scale deflection of 1 mg holding for approximately 10 min at each temperature. By choosing temperatures for which the rate of weight loss was low and the overall loss was kept under 2%, good straight-line Clausius–Clapeyron plots $\log[m(T)^{1/2}]$ against $10^3/T$ were given with slopes corresponding to the heats of sublimation reproducible to approximately 5%. Reasonable agreement was obtained with literature values for a number of the compounds studied and it is interesting to note the value obtained for benzoic acid agreed with that obtained by Weidemann by the Knudsen method in a similar temperature range, to better than 3%. Knudsen effusion studies have been carried out on SnS, and the heat of sublimation and entropy of sublimation calculated to be 220.4 kJ/mol and 162.4 J/K mol^{-1}, respectively (392).

Till (363) determined the enthalpy and entropy for the decomposition of barium peroxide by measuring the decomposition temperature of barium peroxide in equilibrium with a known partial pressure of oxygen. The measurements were made between 670 and 843°C.

B. KINETICS

The standard method of determining kinetic parameters is by means of a series of isothermal experiments carried out at controlled partial pressures of any volatile reaction products. This procedure has three principal disadvantages. First, it is often difficult to ascertain precisely the time and temperature of the onet of the reaction, since a cool sample has to be inserted into a hot furnace and takes a finite time to reach the isothermal condition of the experiment. Second, the method is laborious, since it requires several experiments each at a different temperature. Third, each experiment requires a separate sample, which must react in a manner identical to that of the other samples in the set. If the purpose of the investigation is to determine the kinetic mechanism, then the isothermal method is recommended. Sometimes the first disadvantage can be overcome by heating the sample to the required isothermal temperature in a relatively high pressure of an evolved volatile to suppress the reaction, then reducing the pressure to allow the reaction to commence.

Frequently, however, the purpose of the study is to obtain kinetic parameters to provide a quantitative comparison of similar samples obtained under different preparative conditions, or to extend knowledge of a particular reaction to partial pressures not previously investigated. Moreover, the sample may be in short supply or difficult to reproduce exactly. In such circumstances, a single TG curve covers the entire temperature range without any missing regions and can yield kinetic data from a single sample.

The basic equations of the DTG and TG curves (Eqs. 5 and 6) were derived in Section II, with the implicit assumption that the temperature of the sample is uniform at any specific instant in time. Under dynamic thermal conditions it is inevitable that there is some temperature gradient within the sample, but this was shown in Section IV to be a minimum if either a small sample, a slow heating rate, or both are used. Many kinetic studies in the literature are of no value whatsoever because of appreciable temperature gradients within the reacting sample. The gradient is likely to be worse at low fractions reacted when the rate of reaction is rapid.

It can be seen from Eq. 6,

$$g(\alpha) = \int \frac{d\alpha}{f(\alpha)} = \frac{A}{\beta} \int e^{-E/RT} dT = kt \tag{6}$$

that it is necessary to know the function $f(\alpha)$ (and hence the related function $g(\alpha)$), to establish it, or to use a method of determining the kinetic parameters E and A, without knowing it. All three approache are employed (139, 115).

Only occasionally has the function been known with confidence, although it has frequently been assumed. It may be known when the isothermal method is used in conjunction with the dynamic thermal method to establish the kinetic mechanism under one set of isotherms and isobaric conditions and then to use TG to obtain kinetic parameters under other sets of isobaric conditions, as for the dehydroxylation of kaolinite (54).

Many attempts have been made to determine the form of $f(\alpha)$ from TG curves either by trial and error procedures or by means of a method, such as that due to Freeman and Carroll (147), that purports to determine it. These procedures are fraught with dangers and have undoubtedly led to wrong conclusions. It is not always valid, for example, to assume that the best least-squares analysis is given by the correct kinetic mechanism. Supporting evidence from another experimental method, especially a direct observation such as a microscopic technique, is desirable but rarely obtained. Many methods are restricted to the determination of an order of reaction, which is of limited significance when applied to solid state reactions (334, 319, 173, 54), which may be controlled by nucleation, diffusion, or heat transfer processes, many of which cannot be expressed in the form of the order of reaction equation. Some mechanisms that do apply to solid state reactions and their functions $g(\alpha)$ are listed in Table 1.

Methods of determining kinetic parameters without the necessity of knowing $f(\alpha)$ and $g(\alpha)$ have been developed (139, 148, 280) and applied mostly to polymer degradation reactions. They have been claimed to be superior to the isothermal procedure (280), but this claim has been disputed (227). The methods involve several TG curves obtained at different heating rates, so that they are as time-consuming as the isothermal procedure and several samples must be used, but the entire temperature range is scanned without any missing regions in each run. The methods are outlined under category 4 below.

C. CLASSIFICATION OF METHODS

The right side of Eq. 6 has no exact integral and the various approximations to this integration or methods of avoiding it have led to the proliferation of methods of kinetic analysis of a TG curve that have been proposed since 1960. Doyle stated in 1966 (115) that although at first it may appear that these are different, there has been more repetition than invention, and this statement is even more valid today. Various classifications have been proposed and that which follows is similar to, but not identical with, that used by Flynn and Wall (139). The methods may be classified into four categories:

1. Integral methods that involve calculation of the exponential integral
2. Differential methods that involve the rate of weight change
3. Methods that involve a reference temperature, usually T_m, the temperature at which the maximum rate occurs (the DTG peak temperature)
4. Methods that involve several TG curves obtained at various heating rates.

Many methods can be classified under more than one heading and classes 3 and 4 can be subdivided according to whether the method is integral or differential.

Cameron and Fortune (66) have subdivided integral methods into three groups:

1A. Methods using tabulated values of the exponential integral, e.g., Doyle (112), Zsako (399), and Satava and Skvara (319)
1B. Methods using a simple approximation for the exponential integral, e.g., van Krevelen et al. (371) and Doyle (113)
1C. Methods using a series expansion to approximate the exponential integral, e.g., Doyle (114) and Coats and Redfern (89).

These methods either require prior knowledge of the kinetic mechanism, i.e., the form of the functions $f(\alpha)$ and $g(\alpha)$, or attempt to determine it by

laborious trial and error procedures. Under favorable circumstances, however, the activation energy can be determined with considerable precision.

The differential method in its most general form (54, 333) is derived from
Eq. 5 by taking logarithms:

$$\ln\left[\frac{d\alpha/dT}{f(\alpha)}\right] = \ln\left(\frac{A}{\beta}\right) - \frac{E}{RT}$$

The left side of the equation is plotted against $1/T$ using the appropriate
form of $f(\alpha)$, and E is determined from the slope of the linear plot. The
disadvantage of the method is the need to determine $d\alpha/dt$, the tangents to
the TG curve, at a series of values of T. Experimental scatter, which
inevitably accompanies this procedure, often renders the application of the
differential method difficult. Johnson and Gallagher (206), however, have
used sophisticated computations to develop the differential method to such
precision that it has advantages over the integral methods.

The most used method of kinetic analysis of a TG curve is that due to
Freeman and Carroll (146). The method has been described (139) as a
difference–differential method, since it depends on taking the difference of
the differential terms, which doubly magnifies the experimental scatter and
makes accurate evaluation of kinetic parameters difficult. The method has
been of great historical importance in pioneering the application of TG to
the determination of kinetic parameters, but so many more recent methods
are superior that it should no longer be used (139, 333, 387).

Methods involving a reference temperature (category 3) are not recommended for the determination of kinetic data since they place too much
reliance on data obtained at one point.

The final group, which involves methods based on multiple heating rates,
has been developed since 1965 almost exclusively by workers involved with
polymer degradations. There are two subgroups according to whether the
method is differential or integral.

The differential method, developed by Friedman (148), is based on Eq. 7.
Values of $d\alpha/dT$ and their corresponding values of T are determined at a
series of values of α from each of several experiments at different heating
rates. As long as the kinetic mechanism does not change, fixing α also fixes
$f(\alpha)$. Therefore, plots of $\ln \beta(d\alpha/dT)$ against $1/T$ for each value of α have
slopes E/R.

The integral method was developed by several sets of workers (140, 280,
309) and is usually known as Ozawa's method. It can be shown (140, 280) that
plots of $\log \beta$ against $1/T$ for various values of α give a set of straight lines
with slopes E/R.

It would be convenient if one method could be chosen as superior to all
others. The choice is subjective and depends on the extent of knowledge that
is available about the kinetic mechanism (i.e., the form of $f(\alpha)$ and $g(\alpha)$)

and the type of kinetic data that is required, i.e., approximate or precise. Sestak (327) has pointed out that the most serious errors are introduced not by the choice of method, but by faulty temperature measurement (which distorts the value for E) and nonlinearity of heating rate (which distorts the form of $f(\alpha)$). Great care must be given to such experimental matters if satisfactory kinetic data are to be obtained.

X. THERMOGRAVIMETRY IN COMBINATION WITH OTHER THERMAL METHODS

A. TG–DTA

A comparison of the types of phenomena that can be studied by the two techniques is shown in Table 15. It can be seen that the two techniques are essentially complementary. TG is intrinsically quantitative but is limited to reactions that cause a change in weight, while DTA is more universally applicable detecting any reaction that occurs with a change in enthalpy but requiring careful calibration before quantitative measurements can be made. In principle, it has long been possible to detect small changes in mass as long as appropriate steps are taken to avoid problems caused by vibration and buoyancy, whereas amplification of DTA signals is relatively recent. In practice, however, simultaneous TG–DTA often requires a small sample and fast heating rate to obtain the best quality DTA curve, but a larger sample and slower heating rate to optimize the TG curve. These requirements can primarily be attributed to design limitations that have been overcome in the most up-to-date equipment, so that it should no longer be valid to argue that the techniques can be combined only to their mutual detriment.

TABLE 15
Phenomena Capable of Study by Simultaneous TG–DTA

	Change in Weight		Thermal Effect	
	Gain	Loss	Endothermic	Exothermic
Adsorption and absorption	x			x
Desorption		x	x	
Dehydration or desolvation		x	x	
Crystalline transition			x	
Melting			x	
Vaporization		x	x	
Sublimation		x	x	
Decomposition		x	x	x
Oxidation	x	x		x
Reduction		x		
Solid–gas reaction	x	x	x	x

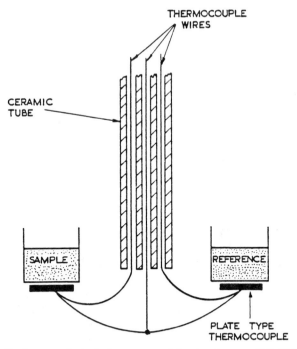

Fig. 72. Schematic diagram of a typical arrangement for simultaneous TG–DTA measurement.

In general, a simultaneous TG–DTA equipment is based on the addition of a DTA head assembly to a standard thermobalance. The differential circuit is connected to the measuring system, using 1/1000 in. wires or similar means. A typical modern TG–DTA assembly based on an electronic microbalance is shown schematically in Fig. 72. As can be seen the only additional feature required is a dc amplifier to amplify the DTA signal. It will also be necessary to use a multipen recorder or a multipoint recorder to monitor a minimum of three outputs.

An illustration of the advantage of a simultaneous technique is provided by the decomposition of CrO_3. The TG curve shows that this gives a complex three-stage weight loss to form Cr_2O_3 as the final product. The DTA curve shows a sharp endothermic reaction in the region of 200°C that is not accompanied by a change in weight. This could be attributed to either a solid–solid phase change or fusion reaction; visual observation supports the latter case. The DTA curve shows a complex pattern consisting of two exothermic reactions followed by an endothermic reaction. The simultaneous TG trace allows these DTA peaks to be unambiguously assigned to the three weight losses. Considerable care would be needed to correlate these reactions if the runs had been carried out on separate TG and DTA apparatus, and with more complex materials absolute correlation is not always attainable under these conditions.

A second example is illustrated by the fusion reaction of KCl, which was being assessed as a possible enthalpy standard for DTA. The TG curve in a flowing atmosphere shows a weight loss due to sublimation below the melting point, and heating through the melting point to obtain a stable baseline causes further weight loss. This type of information is difficult to obtain by indirect weighing after samples have been removed from the DTA apparatus, since an unknown amount of further sublimation and volatilization will have occurred during the cooling cycle and hence will give rise to a further unknown weight loss.

B. TG–EGA

TG is probably the most useful thermal analysis technique to interface with gas analysis equipment, since it gives a quantitative measure of the evolved volatiles. The relevance of mass spectrometry to thermal analysis has been reviewed (110). The two most useful techniques are those of simultaneous TG–mass spectrometry and simultaneous TG–gas chromatography–mass spectrometry. In the former case the volatiles are passed from the thermobalance chamber into a fast scan mass spectrometer operating either at atmospheric pressure by means of a suitable interface or at the operating pressure of the mass spectrometer. For complex decomposition processes, for example, polymer decomposition, it is advantageous to separate the components before analysis in the mass spectrometer, and a discrete sampling system is used to provide samples for the GC. Where one is interested in measuring a particular gas it may be preferable to use a specific gas analyzer calibrated for this purpose. Examples of this are the determination of water using a conductivity cell, or carbon monoxide analysis using infrared spectroscopy. The combination of TG and EGA is a powerful technique for studying materials that sublime or vaporize with simultaneous decomposition. The TG curve enables the total effect due to both processes to be measured, whereas with suitable traps the EGA will measure only the degree of decomposition, enabling an estimate to be made of the relative importance of sublimation or vaporization and decomposition. Gallagher (159) has described a number of applications of EGA to inorganic materials and concluded that the specificity and sensitivity of EGA techniques such as mass spectrometry considerably enhance the TA methods.

C. OTHER TECHNIQUES

Combinations of TG (or TG–DTA) with other techniques have been evident and widespread. The major aims of these combinations have been to improve the specificity of the system and to increase the data collected from a given sample, without the need to carry out separate experiments that

TABLE 16
Simultaneous (or Combined) Techniques Used in Association with TG

Technique	Reference
X-ray diffraction	18
Molecular beam	391
Calorimetry	135
Photoluminecence	52
Dilatometry	285
Electrical conductivity	82, 210
Emanation thermal analysis	13, 60, 131
Thermomagnetic analysis	342, 394
Photometric	223, 207
Thermobaragravimetric	15, 251, 374
Titrimetry	285, 287, 138
Mass spectrometry	395, 110
Gas chromatography	40, 83
Gas chromatography–mass spectrometry	71
Thermoluminescence	156, 342, 394

might involve variations in experimental conditions, and hence make correlations between the techniques difficult. Combined techniques have been reviewed by Paulik and Paulik (288). A tabulation of TG combined with other techniques is given in Table 16, derived from that review and more recent literature.

ACKNOWLEDGMENTS

We express our gratitude to Professor E. L. Charsley, Head of the Thermal Analysis Consultancy Service, Leeds Polytechnic, Leeds, UK, for many useful discussions and suggestions for material to be included in this work, and for provision of some of the diagrams.

REFERENCES

1. Abdelrazig, B. E. I., K. M. Parker, and J. H. Sharp, in B. Miller, Ed., *Thermal Analysis: Proc. 7th ICTA*, Vol. 1, Chichester, UK: Wiley-Heyden, 1982, p. 571.

2. Abdelrazig, B. E. I., J. H. Sharp, P. A. Siddy, and B. El Jazairi, *Proc. Br. Ceram. Soc.* **35**, 141 (1984); ibid, *Cem. Concr. Res.*, 18, 415 (1988).

3. Addison, C. C., W. E. Addison, G. H. Neal, and J. H. Sharp, *J. Chem. Soc.* 1468 (1962).

4. Adonyi, Z., and G. Korosi, in D. Dollimore, Ed., *Proc. 1st European Symp. Thermal Analysis*, London: Heyden, 1976, p. 200.

5. Antill, J. E., and K. A. Peakall, *Corros. Sci.* **16**, 435 (1976).

6. Aoki, H., T. Ban, M. Akao, and S. Iwai, in H. Chihara, Ed., *Thermal Analysis: Proc. 5th ICTA*, London: Heyden, 1977, p. 9.

7. Ashcroft, S. J., *Thermochim. Acta* **2**, 512 (1971).

8. Aylmer, D. M., and M. W. Rowe, *Thermochim. Acta* **78**, 81 (1984).

9. Bair, H. E., Thermal analysis of additives in polymers, in E. A. Turi, Ed., *Thermal Characterization of Polymeric Materials*, New York: Academic, 1981, Chapter 9.

10. Bakcsy, G., and A. J. Hegedus, *Thermochim. Acta* **10**, 399 (1974).

11. Balek, V., *J. Mater. Sci.* **5**, 166 (1970).

12. Balek, V., *Anal. Chem.* **42**, 16A (1970).

13. Balek, V., *Thermochim. Acta* **22**, 1 (1978).

14. Ball, M. C., and H. F. W. Taylor, *Mineral. Mag.* **32**, 754 (1961).

15. Bancroft, G. M., and H. D. Gesser, *J. Inorg. Nucl. Chem.* **27**, 1537 (1965).

16. Bargues, M., in C. Eyraud and M. Escoubes, Eds., *Prog. Vac. Microbalance Tech.*, London: Heyden, Vol. 3, 1975, p. 275.

17. Barnes, P. A., and J. H. Baxter, *Thermochim. Acta* **24**, 427 (1978).

18. Barret, P., in C. Eyraud and M. Escoubes, Eds., *Prog. Vac. Microbalance Tech.*, Vol. 3, London: Heyden, 1975, p. 205.

19. Barron, S., *J. Fire Flammability* **7**, 387 (1976).

20. Bartholemew, C. H., and R. J. Farrauto, *J. Catal.* **45**, 41 (1976).

21. Barton, J. M., W. A. Lee, and W. W. Wright, *J. Therm. Anal.* **13**, 85 (1978).

22. Basu, A. K., and F. R. Sale, *J. Mater. Sci.* **14**, 91 (1979).

23. Behret, H., H. Binder, and E. Robens, *Thermochim. Acta* **24**, 407 (1978).

24. Belcher, R., *Chromatographia* **9**, 201 (1976).

25. Bell, C. F., and R. E. Morcom, *J. Inorg. Nucl. Chem.* **36**, 3689 (1974).

26. Benbow, A. W., C. F. Cullis, and H. S. Laver, *Polymer* **19**, 824 (1978).

27. Bentlage, H., and W. Wilkens, in S. C. Bevan, S. J. Gregg, and W. D. Parkyns, Eds., *Prog. Vac. Microbalance Tech.*, Vol. 2, London: Heyden, 1973, p. 239.

28. Berenyi, M., and G. Liptay, *J. Therm. Anal.* **3**, 437 (1971).

29. Berenyi, M., *Hung. Sci. Instrum.* **38**, 101 (1976).

30. Beretka, J., and T. Brown, *Aust. J. Chem.* **24**, 237 (1971).

31. Berg, L. G., Simple salts, in R. C. Mackenzie, Ed., *Differential Thermal Analysis*, Vol. 1, London: Academic, 1970, pp. 343–361.

32. Bernal, S., F. J. Botana, R. Garcia, and J. M. Rodriguez-Izquierdo, *Thermochim. Acta* **66**, 139 (1983).

33. Bihari-Varga, M., *Hung. Sci. Instrum.* **23**, 23 (1972).

34. Bingham, M. A., and B. J. Hill, *J. Therm. Anal.* **7**, 347 (1975).

35. Bingham, M. A., and B. J. Hill, *J. Therm. Anal.* **9**, 71 (1976).

36. Biolsi, M. E., *Thermochim. Acta* **9**, 303 (1974).

37. Blackburn, P. E., and J. B. Hudson, *J. Electrochem. Soc.* **107**, 944 (1960).

38. Blair, R. L., and P. G. Fair, *Thermochim. Acta* **67**, 233 (1983).

39. Blandenet, G., J. Detheve, and J. Peron, *Chim. Anal.* **50**, 385 (1968).

40. Blandenet, G., *Chromatographia* London: **5**, 184 (1969).

41. Blazek, A., *Thermal Analysis*, London: Van Nostrand Reinhold, 1973.

42. Boelter, J., *Fette*, Siefen, Anst, 593 (1973).

43. Boguslawska, K., and A. Cyganski, *J. Therm. Anal.* **9**, 337 (1976).

44. Bohn, P. W., and R. S. Bottei, *J. Therm. Anal.* **13**, 231 (1978).

45. Bond, G. C., and M. R. Gelsthorpe, *Therm. Anal. Abstr.* **14**, 205 (1985).

46. Borrel, M., and R. Paris, *Anal. Chim. Acta* **4**, 267 (1950).

47. Brancone, L., and H. J. Ferrari, *Microchem. J.* **10**, 370 (1966).

48. Brazier, D. W., and G. H. Nickel, *Rubber Chem. Technol.* **48**, 661 (1975).

49. Brazier, D. W., and N. V. Schwartz, *J. Appl. Polym. Sci.* **22**, 113 (1978).

50. Brazier, D. W., *Rubber Chem. Technol.* **53**, 437 (1980).

51. Brett, N. H., K. J. D. MacKenzie, and J. H. Sharp, *Quart. Rev. Chem. Soc.* **24**, 185 (1970).

52. Breysse, M., L. Faure, C. Claudel, and J. Veron, in S. C. Bevan, S. J. Gregg, and N. D. Parkyns, Eds., *Prog. Vac. Microbalance Tech.*, London: Heyden, Vol. 2, 1973, p. 229.

53. Brindley, G. W., *Prog. Ceram. Sci.* **3**, 1 (1963).

54. Brindley, G. W., J. H. Sharp, and B. N. N. Achar, in J. P. Redfern, Ed., *Thermal Analysis: Proc. 1st ICTA*, London: Macmillan, 1965, p. 180; in *Proc Intl Clay Conf*, Jeruslalem, ed. L. Heller and A. Weiss, Eds., *Proc. Intl. Clay Conf.*, Vol. 1, 1966, p. 67.

55. Brittan, M., and R. Lubenberg, *Trans. Inst. Min. Met.* **80**, C156 (1971).

56. Brown, H. A., Jr., E. C. Penski, and J. P. Callahan, *Thermochim. Acta* **3**, 271 (1972).

57. Brown, D. H., and R. T. Richardson, *J. Inorg. Nucl. Chem.* **35**, 755 (1973).

58. Brown, G. P., D. T. Haarr, and M. Metlay, *IEEE Trans. Electron. Insul.* **EI-8**, 36 (1973).

59. Burton, H. R., Thermal decomposition and gas phase analysis of carbohydrates found in tobacco, in F. Shafizadeh, K. V. Karkanen, and D. A. Tillman, Eds., *Thermal Uses and Properties of Carbohydrates and Lignins*, New York: Academic, 1976, pp. 375–310.

60. Bussiere, P., B. Claudel, J. P. Renouf, Y. Trambouze, and M. Prettre, *J. Chim. Phys.* **58**, 668 (1961).

61. Buttler, F. G., and S. R. Morgan, in H. Wiedemann, Ed., *Thermal Analysis: Proc 6th ICTA*, Vol. 1, Basel: Birkhaeuser, 1980, p. 381.

62. Buttler, F. G., and S. R. Morgan, *Proc. 7th Intl. Cong. Chem. Cement*, Paris, Vol. 2, II-43 (1980).

63. Buttler, F. G., and E. J. Walker, in D. Dollimore, Ed., *Proc. 2nd European Symp. Thermal Analysis*, London: Heyden, 1981, p. 505.

64. Bye, G. C., and C. R. Howard, *J. Appl. Chem. Biotechnol.* **21**, 319 (1971); **22**, 1053 (1972).

65. Cabradilla, K. E., and S. H. Zeronian, Influence of crystallinity on the thermal properties of cellulose, in F. Shafizadeh, K. V. Karkanen, and D. A. Tillman, Eds., *Thermal Uses and Properties of Carbohydrates and Lignins*, New York: Academic, 1976, pp. 73–96.

66. Cameron, G. G., and J. D. Fortune, *Eur. Polym. J.* **4**, 333 (1968).

67. Capon, A., and F. A. P. Maggs, in D. Dollimore, Ed., *Proc. 1st Eur. Symp. Thermal Analysis*, London: Heyden, 1976, p. 121.

68. Carel, A. B., and D. K. Cabbiness, *Thermochim. Acta* **44**, 363 (1981).

69. Carraher, C. E., *J. Therm. Anal.* **10**, 37 (1976).

70. Carter, M. A., D. R. Glasson, and S. A. A. Jayaweera, *Fuel* **63** 1068 (1984).

71. Chang, F. C., and W. W. Wendlandt, *Thermochim. Acta* **3**, 69 (1971).

72. Charcosset, H., C. Bolivar, R. Frety, R. Gomez and Y. Trambouze, in S. C. Bevan, S. J. Gregg, and N. D. Parkyns, Eds., *Prog. Vac. Microbalance Tech.*, Vol. 2, London: Heyden, 1973, p. 175.

73. Charit, I., H. Harel, S. Fischer, and G. Marom, *Thermochim. Acta* **62**, 237 (1983).

74. Charles, R. G., and A. Langer, *J. Phys. Chem.* **63**, 603 (1959).

75. Charsley, E. L., and H. G. Midgley, personal communication.

76. Charsley, E. L., and A. Kamp, *J. Therm. Anal.* **7**, 173 (1975).

77. Charsley, E. L., and J. G. Dunn, *J. Therm. Anal.* **17**, 535 (1980).

78. Charsley, E. L., and J. G. Dunn, *Plastics and Rubber Processing Applications* **1**, 3 (1981).

79. Charsley, E. L., and J. G. Dunn, *Rubber Chem. Technol.* **55**, 382 (1982).

80. Chatfield, D. A., and I. N. Einhorn, *J. Polym. Sci. Polym. Chem. Ed.* **19**, 601 (1981).

81. Chen, R., *J. Mater. Sci.* **11**, 1521 (1976).

82. Chiu, J., *J. Anal. Chem.* **39**, 861 (1967).

83. Chiu, J., *J. Therm. Anal.* **1**, 231 (1970).

84. Chiu, J., *J. Macromol. Sci. Chem.* **A8**(1), 3 (1974).

85. Cimino, A., B. A. De Angelis, D. Gazzoli, and M. Valigi, *Z. Anorg. Allg. Chem.* **460**, 86 (1980).

86. Cimino, A., D. Gazzoli, G. Minelli, and M. Valigi, *Z. Anorg. Allg. Chem.* **494**, 207 (1982).

87. Clark, R. P., P. K. Gallagher, and B. M. Dillard, *Thermochim. Acta* **33**, 141 (1979).

88. Coats, A. W., and J. P. Redfern, *Analyst* **88**, 906 (1963).

89. Coats, A. W., and J . P. Redfern, *Nature* **201**, 68 (1964).

90. Cobbler, J. G., and C. D. Chou, *Anal. Chem.* **49**, 162R (1977); **51**, 287R (1979).

91. Collepardi, M., M. Corradi, G. Baldini, and M. Pauri, *Cem. Concr. Res.* **8**, 571 (1978); *J. Am. Ceram. Soc.* **62**, 33 (1979).

92. Collepardi, M., S. Monosi, G. Moriconi, and M. Corradi, *Cem. Concr. Res.* **9**, 431 (1979).

93. Collins, L. W., W. W. Wendlandt, and E. K. Gibson, *Thermochim. Acta* **8**, 205 (1974).

94. Collins, L. W., and W. W. Wendlandt, *Thermochim Acta* **11**, 253 (1975).

95. Craig, R. G., J. M. Powers, and F. A. Peyton, *J. Dent. Res.* **50**, 450 (1971).

96. Cullis, C. F., and H. S. Laver, *Eur. Polym. J.* **14**, 575 (1978).

97. Cumming, J. W., and J. McLaughlin, *Thermochim. Acta* **57**, 253 (1982).

98. Czanderna, A. W., and S. P. Wolsky, Introduction and microbalance review, in A. W. Czanderna and S. P. Wolsky, Eds., *Microweighing in Vacuum and Controlled Environments*, Amsterdam: Elsevier, 1980, pp. 1–57.

99. D'Ascenzo, G., R. Curini, G. de Angelis, E. Cardarelli, A. Magri, and L. Miano *Thermochim. Acta* **62**, 149 (1983).

100. Daniels, T., *Thermal Analysis*, London: Kogan Page, 1973.

101. David, D. J., *Insulation*, Nov., 38 (1967).

102. De Keyser, W. L., *Bull. Soc. France Ceram.* **20**, 1 (1953).

103. Derry, D. J., and D. G. Lees, *Corros. Sci.* **16**, 219 (1976).

104. Di Lorenzo, A., S. Masi, and A. Pennacchi, in D. Dollimore, Ed., *Proc. 1st European Symp. Thermal Analysis*, London: Heyden, 1976, p. 37.

105. Dickens, B., *Thermochim. Acta* **29**, 41 (1979).

106. Dollimore, D., D. L. Griffiths, and D. Nicholson, *J. Chem. Soc.* 2617 (1963).

107. Dollimore, D., *Proc. Soc. Anal. Chem.* 167 (1965); Oxysalts, in R. C. MacKenzie, Ed., *Differential Thermal Analysis*, Vol. 1, London: Academic, 1970, pp. 395–422.

108. Dollimore, D., F. E. Jones, and P. Spooner, *J. Chem. Soc. A*, 2809 (1970).

109. Dollimore, D., *Thermochim. Acta* **50**, 123 (1981).

110. Dollimore, D., G. A. Gamlen, and T. J. Taylor, *Thermochim. Acta* **75**, 59 (1984).

111. Donbarand, J., *Thermochim. Acta* **79**, 161 (1984).

112. Doyle, C. D., *J. Appl. Polym. Sci.* **5**, 285 (1961).

113. Doyle, C. D., *J. Appl. Polym. Sci.* **6**, 639 (1962).

114. Doyle, C. D., *Nature* **207**, 290 (1965).

115. Doyle, C. D., Quantitative calculations in thermogravimetric analysis, in P. E. Slade and L. T. Jenkins, Eds., *Techniques and Methods of Polymer Evaluation*, Vol. 1, New York: Dekker, 1966, Chapter 4.

116. Dunn, J. G., and C. E. Kelly, *J. Therm. Anal.* **12**, 43 (1977).

117. Dunn, J. G., and C. E. Kelly, *J. Therm. Anal.* **18**, 147 (1980).

118. Dunn, J. G., *Chem. Aust.* **47**, 281 (1980).

119. Dunn, J. G., and S. A. A. Jayaweera, *Thermochim. Acta* **61**, 313 (1983).

120. Dunn, J. G., and S. A. A. Jayaweera, *Thermochim. Acta* **85**, 115 (1985).

121. Dunn, J. G., S. A. A. Jayaweera, and S. G. Davies, *Proc. Australas. Inst. Min. Metall.* **290**, 75 (1985).

122. Dupuis, T., and C. Duval, *Anal. Chim. Acta* **3**, 191 (1949).

123. Durand-Souron, D., Thermogravimetric analysis and associated techniques applied to kerogen, in B. Durand, Ed., *Kerogen*, Paris: Technip, 1980, pp. 143–161.

124. Dutrizac, J. E., *J. Less-Common Metals* **51**, 283 (1977).

125. Dutta, S., and C. Y. Wen, *Ind. Eng. Chem. Process Res. Dev.* **16**, 31 (1977).

126. Duval, C., *Inorganic Thermogravimetric Analysis*, Amsterdam: Elsevier, 1953.

127. Duval, C., *Inorganic Thermogravimetric Analysis*, 2d ed., Amsterdam: Elsevier, 1963.

128. Duval, C., Thermal methods in analytical chemistry, in G. Svehla, Ed., *Comprehensive Analytical Chemistry*, Vol. 7, Amsterdam: Elsevier, 1976.

129. Einhorn, I. N., D. A. Chatfield, K. J. Voorhees, F. D. Hileman, R. W. Nickelson, S. C. Israel, J. H. Futrell, and P. W. Ryan, *Fire Res.* **1**, 41 (1977).

130. Elder, J. P., *Thermochim. Acta* **52**, 235 (1982).

131. Emmerich, W. D., and V. Balek, in H. G. Wiedemann, Ed., *Thermal Analysis: Proc 3rd ICTA*, Vol. 2, Basel: Birkhaeuser, 1971, p. 475.

132. Era, V. A., and A. Mattila, *J. Therm. Anal.* **10**, 461 (1976).

133. Erdey, L., F. Paulik, and J. Paulik, *Acta Chim. Acad. Sci. Hung.* 80 (1963).

134. Escoubes, M., C. Eyraud, and E. Robens, *Thermochim. Acta* **82**, 15 (1984).

135. Escoubes, M., and R. Blanc, in C. Eyraud and M. Escoubes, Eds. *Prog. Vac. Microbalance Tech.*, Vol. 3, London: Heyden, 1975, p. 388.

136. Farrauto, R. J., *J. Catal.* **41**, 482 (1976).

137. Figueiredo, J. L., and D. L. Trimm, *J. Appl. Chem. Biotechnol.* **28**, 611 (1978).

138. Fischer, S. G., and J. Chiu, *Thermochim. Acta* **65**, 9 (1983).

139. Flynn, J. H., and L.A. Wall, *J. Res. Natl. Bur. Std A* **70**, 487 (1966).

140. Flynn, J. H., and L. A. Wall, *J. Polym. Sci.* B**4**, 323 (1966).

141. Flynn, J. H., and B. Dickens, *Thermochim. Acta* **15**, 1 (1976).

142. Flynn, J. H., Thermogravimetric analysis and differential thermal analysis, in H. H. G. Jellinek, Ed., *Aspects of Degradation and Stabilisation of Polymers*, Amsterdam: Elsevier, 1978, p. 573.

143. Flynn, J. H., *Polym. Eng. Sci.* **20**, 675 (1980).

144. Formes-Marquina, J. M., and N. B. Chanh, *J. Therm. Anal.* **7**, 263 (1975).

145. Freeman, A. G., A. A. Hodgson, and H. F. W. Taylor, *Mineral. Mag.* **35**, 5 (1965); **35**, 445 (1965).

146. Freeman, E. S., and B. Carroll, *J. Phys. Chem.* **62**, 394 (1958).

147. Freeman, E. S., and B. Carroll, *J. Phys. Chem.* **76**, 1474 (1972).

148. Friedman, H. L., *J. Polym. Sci. C* **6**, 183 (1965).

149. Friedman, H. L., The application of thermal analysis in the aerospace industry, in H. Kambe and P. D. Garn, Eds., *Thermal Analysis: Comparative Studies on Materials*, New York: Halsted, 1976, p. 303.

150. Furminsky, E., *Ind. Eng. Chem. Prod. Res. Dev.* **22**, 637 (1983).

151. Gachet, C. G., and J. Y. Trambouze, in C. Eyraud and M. Escoubes, Eds., *Prog. Vac. Microbalance Tech.* Vol. 3, London: Heyden, 1975, p. 144.

152. Gadag, R. V., and M. R. Gagendragad, *J. Indian Chem. Soc.* **55**, 789 (1978).

153. Gajardo, P., P. Grange, and B. Delmon, *Chem. Soc. Faraday Trans. 1* **76**, 929 (1980).

154. Gallagher, P. K., Some Thermoanalytical applications to the chemistry of coordination compounds, in H. Kambe and P. D. Garn, Eds., *Thermal Analysis: Comparative Studies on Materials*, New York: Halsted, 1974, p. 17.

155. Gallagher, P. K., and J. P. Luongo, *Thermochim Acta* **12**, 159 (1975).

156. Gallagher, P. K., and S. St. J. Warne, *Thermochim Acta* **43**, 253 (1981); ibid, *Mater, Res. Bull.*, *16*, 141 (1981).

157. Gallagher, P. K., E. M. Gyorgy, and W. R. Jones, *J. Chem. Phys.* **75**, 3847 (1981).

158. Gallagher, P. K., E. M. Gyorgy, and W. R. Jones, *J. Therm. Anal.* **23**, 185 (1982).

159. Gallagher, P. K., *J. Therm. Anal.* **25**, 7 (1982).

160. Gallagher, P. K., *Therm. Anal. Abstr.* **14**, 105 (1985).

161. Gardner, D. C., and C. H. Bartholomew, *Ind. Eng. Chem. Prod. Res. Dev.* **20**, 80 (1981).

162. Garn, P. D., *Thermoanalytical Methods of Investigation*, New York: Academic, 1965.

163. Garn, P. D., O. Menis, and H. G. Wiedemann, *J. Therm. Anal.* **20**, 185 (1981).

164. Garn, P. D., *Thermochim. Acta* **55**, 121 (1982).

165. Garnaud, J., *J. Therm. Anal.* **8**, 273 (1975).

166. Garner, W. E., Ed., *Chemistry of the Solid State*, London: Butterworths, 1955.

167. Gast, T. H., *J. Phys. E: Sci. Instrum.* **7**, 865 (1974).

168. Gast, T., in C. Eyraud and M. Escoubes, Eds., *Prog. Vac. Microbalance Tech.*, Vol. 3, London: Heyden, 1975, p. 108.

169. George, C. M., *Trans. J. Br. Ceram. Soc.* **79**, 82 (1980).

170. George, T. D., and W. W. Wendlandt, *J. Inorg. Nucl. Chem.* **25**, 395 (1963).

171. Ghodsi, M., R. Derie, and J. P. Lempereur, *Thermochim. Acta*, **28**, 259 (1979).

172. Ghodsi, M., and C. Neumann-Tilte, *Thermochim. Acta* **62**, 1 (1983).

173. Gomes, W., *Nature* **192**, 865 (1961).

174. Gomory, I., *J. Therm. Anal.* **11**, 327 (1977).

175. Gordon, S., and C. Campbell, *Anal. Chem.* **32** 271R (1960).

176. Gouda, M. W., A. R. Ebien, M. A. Moustafa, and S. A. Khalil, *Drug Dev. Ind. Pharm.* **3**, 273 (1977).

177. Grassie, N., M. Zulfigar, and M. I. Guy, *J. Polym. Sci. Polym. Chem. Ed.* **18**, 265 (1980).

178. Green, J., *J. Mater. Sci.* **18**, 637 (1983).

179. Griffin, W. R., Thermogravimetry of elastomers, in H. Kambe and P. D. Garn, Eds., *Thermal Analysis: Comparative Studies on Materials*, New York: Halsted, 1974, p. 132.

180. Groves, A. W., *Silicate Analysis*, 2d ed., London: Allen Unwin, 1981, p. 95.

181. Habersberger, K., *J. Therm. Anal.* **12**, 55 (1977).

182. Habery, F., and A. Kockel, *IEEE Trans. Magn.* **12**, 983 (1976).

183. Hadrich, W., and E. Kaisersberger, in H. G. Wiedeman, Ed., *Thermal Analysis: Proc. 6th ICTA*, Vol. 1, Basel: Birkhaeuser, 1980, p. 511.

184. Haflan, H., and P. Kofstad, *Corros. Sci.* **23**, 1333 (1983).

185. Haglund, B. O., in D. Dollimore, Ed., *Proc. 1st European Symp. Thermal Analysis*, London: Heyden, 1976, p. 415.

186. Haglund, B. O., and T. Luks, in H. G. Wiedemann, Ed., *Thermal Analysis: Proc. 6th ICTA*, Vol. 1, Birkhauser, 1980, p. 207.

187. Hanf, N., and M. Sole, *Trans. Faraday Soc.* **66**, 3065 (1970).

188. Hassel, R. L., *J. Am. Oil Chem. Soc.* **53**, 179 (1976).

189. Hills, A. W. D., in *Heat and Mass Transfer in Process Metallurgy*, London: Inst. of Min. and Met. 1967, p. 39.

190. Hills, A. W. D., *Chem. Eng. Sci.* **23**, 297 (1968).

191. Honda, K., *Sci. Rep. Tohoku Univ.* **4**, 97 (1915).

192. Hornof, V., B. V. Kokta, J. L. Valade, and J. L. Fassen, *Thermochim. Acta* **19**, 63 (1977).

193. Horowitz, E., and T. P. Perros, *J. Inorg. Nucl. Chem.* **26**, 139 (1964).

194. Howard, J. B., *Polym. Eng. Sci.* **13**, 429 (1973).

195. Hughes, M. A., *J. Therm. Anal.* **8**, 99 (1975).

196. Hughes, R., and H. M. Zadeh, in D. Dollimore, Ed., *Proc. 1st European Symp. on Thermal Analysis*, London: Heyden, 1976, p. 131.

197. Hughes, R., E. K. T. Kam, and H. Mogadam-Zadeh, *Thermochim. Acta* **59**, 367 (1982).

198. Humphries, W. T., and R. H. Wildnauer, *J. Macromol. Sci. Chem.* **A8**, 65 (1974).

199. Hunt, P. J., and K. N. Strafford, *J. Electrochem. Soc. Solid State Sci. Technol.* **128**, 352 (1981).

200. Ingraham, T. R., and P. Marier, *Trans. Met. Soc. AIME* **242**, 2039 (1968).

201. Inui, T., K. Ueda, M. Funabiki, M. Suehiro, T. Sezume, and Y. Takegami, *J. Chem. Soc. Faraday Trans. 1* **75**, 1495 (1979).

202. Inui, T., T. Ueda, and M. Suehiro, *J. Catal.* **65**, 166 (1980).

203. Jehn, H., *Thermochim. Acta* **29**, 229 (1979).

204. Jha, K. N ., *ACS Div. Pet. Chem. Preprints* **28**, 1176 (1983).

205. Joesten, B. L., and N. W. Johnston, *J. Macromol. Sci. Chem.* **A8**, 83 (1974).

206. Johnson, D. W., and P. K. Gallagher, *J. Phys. Chem.* **76**, 1474 (1972).

207. Johnson, B. B., and J. Chiu, *Thermochim. Acta* **50**, 57 (1981).

208. Jona, E., T. Sramko, and J. Gazo, *J. Therm. Anal.* **7**, 551 (1975).

209. Joseph, J., and T. D. Nair, *J. Therm. Anal.* **14**, 271 (1978).

210. Juranic, N., D. Karaulik, and D. Vucelic, *J. Therm. Anal.* **8**,, 417 (1975).

211. Kambe, H., and P. D. Garn, *Thermal Analysis: Comparative Studies on Materials*, New York: Halsted, 1974.

212. Kamishita, M., O. P. Mahajan, and P. L. Walker, Jr., *Fuel* **56**, 444 (1977).

213. Kanamaru, F., M. Shimada, and M. Koizumi, *J. Phys. Chem. Solids* **33**, 1169 (1972).

214. Keattch, C. J., and D. Dollimore, *An Introduction to Thermogravimetry*, 2d ed., London: Heyden, 1975.

215. Kennedy, T., and B. T. Sturman, *J. Therm. Anal.* **8**, 329 (1975).

216. Kofstad, P., and G. Akesson, *Oxidn Metals* **14**, 301 (1980).

217. Kohli, R., and K. I. Vasu, *Indian J. Technol.* **16**, 105 (1978).

218. Krajewski, W., M. Kusi-Mensah, and W. Winterhager, *Thermochim. Acta* **20**, 93 (1977).

219. Kulp, J. L., and J. N. Perfetti, *Mining Mag.* **29**, 239 (1950).

220. Levy, J. H., and W. I. Stuart, *Thermochim. Acta* **74**, 227 (1984).

221. Liptay, G., E. Papp-Molnar, and K. Burger, *J. Inorg. Nucl. Chem.* **31**, 253 (1969).

222. Liptay, G., *The Atlas of Thermoanalytical Curves*, Budapest: Akademiai Kiado and London: Heyden, Vol. 1 (1971), Vol. 2 (1973), Vol. 3 (1974), Vol. 4 (1975), and Vol. 5 (1976).

223. Lochr, A. A., and P. F. Levy, *Am. Lab.* **4**, 11 (1970).

224. Loft, B. C., *J. Polym. Sci. Symp. No. 49*, 127 (1975).

225. Lucchini, E., D. Minichelli, and G. Sloccari, *J. Am. Chem. Soc.* **57**, 42 (1974).

226. Lumme, P., and J. Korvola, *Thermochim. Acta* **9**, 109 (1974).

227. MacCallum, J. R., and J. Tanner, *Eur. Polym. J.* **6**, 907 (1970).

228. MacKenzie, K. J. D., *J. Inorg. Nucl. Chem.* **32**, 3731 (1970).

229. MacKenzie, K. J. D., Applications of thermogravimetry to inorganic chemistry, in T. S. West, Ed., *Reviews in Analytical Chemistry*, Vol. 7, 1983, pp. 193–296.

230. MacKenzie, K. J. D., *Rev. High Temp. Mater.* **4**, 251 (1984).

231. Mackenzie, R. C., *Ber. dt. Keram. Ges.* **41**, 696 (1964).

232. Mackenzie, R. C., and S. Caillere, Thermal characteristics of soil minerals and the use of these characteristics in the qualitative and quantitative determination of clay minerals in soils, in J. E. Gieseking, Ed., *Soil Components*, Vol. 2, New York: Springer, 1975, pp. 529–571.

233. Mackenzie, R. C., *Thermochim. Acta* **28**, 1 (1979).

234. Mackenzie, R. C., Nomenclature in thermal analysis, in P. J. Elving, C. B. Murphy, and I. M. Kolthoff, Eds., *Treatise on Analytical Chemistry*, 2d ed., Pt. I, Vol. 12, New York: Wiley, 1983, pp. 1–16.

235. McAdie, H. G., *Anal. Chem.* **39**, 543 (1967).

236. McGhie, A. R., J. Chiu, P. G. Fair, and R. L. Blair, *Thermochim. Acta* **67**, 241 (1983).

237. McIntosh, R. M., J. H. Sharp, and F. W. Wilburn, *Thermochim Acta.* 165, 281 (1990); ibid, *J. Therm. Anal.*, 37, 2021 (1991).

238. McIntosh, R. M., J. H. Sharp, F. W. Wilburn, and D. M. Tinsley, *J. Therm. Anal.*, 37, 2003 (1991).

239. McKee, D. W., and G. Romeo, *Met. Trans.* **5**, 1127 (1974).

240. McNeil, I. C., The application of thermal volatilisation analysis in studies in polymer degradation, in N. Grassie, Ed., *Development in Polymer Degradation*, Vol. 2, London: Applied Science, 1977, pp. 43–66.

241. Macurdy, L. B., Measurement of mass, in I. M. Kolthoff and P. J. Elving, Eds., *Treatise on Analytical Chemistry*, Part 1, Vol. 7, Chapter 74, 1967, p. 4247.

242. Majumdar, A. K., and A. K. Mukherjee, *J. Inorg. Nucl. Chem.* **26**, 2177 (1964).

243. Mandal, S. K., and J. Das, *J. Indian Chem. Soc.* **54**, 951 (1977).

244. Mande, C. R., and F. R. Sale, *Trans. Inst. Min. Met.* June, C82 (1977).

245. Martinez, E., *Trans. Am. Inst. Min. Eng.* **238**, 172 (1967).

246. Massen, C. H., and J. A. Poulis, Sources of error in microweighing in controlled atmospheres, in A. W. Czanderna and S. P. Wolsky, Eds., *Microweighing in Vacuum and Controlled Atmospheres*, Amsterdam: Elsevier, 1980, pp. 95–125.

247. Maurer, J. J., *Rubber Age* **102**(2): 47 (1970).

248. Maurer, J. J., *Nat. Bur. Stds. (USA) Spec. Publ. No. 338*, 165 (1973).

249. Maurer, J. J., *J. Macromol. Sci. Chem.* **A8**, 73 (1974).

250. Maurer, J. J., Elastomers, in E. A. Turi, Ed., *Thermal Characterization of Polymeric Materials*, New York: Academic, 1981, Chapter 6.

251. Maycock, J. N., and V. R. Pai Verneker, *Thermochim. Acta* **1**, 191 (1970).

252. Melnick, A. M., and E. J. Nolan, *J. Macromol. Sci. Chem.* **A3**, 641 (1969).

253. Mercier, J. G., in I. Buzas, Ed., *Thermal Analysis: Proc 4th ICTA*, Vol. 3, Budapest: Akademiai Kiado, 1975, p. 1041.

254. Metrot, A., in C. Eyraud and M. Escoubes, Eds., *Prog. Vac. Microbal. Tech.*, Vol. 3, London: Heyden, 1975, p. 153.

255. Midgley, H. G., *Trans. Br. Ceram. Soc.* **66**, 161 (1967).

256. Midgley, H. G., and A. Midgley, *Mag. Concr. Res.* **27**, 59 (1975).

257. Midgley, H. G., *Cem. Concr. Res.* **9**, 77 (1979).

258. Milner, O. J., and L. Gordon, *Talanta* **4**, 115 (1960).

259. Mimura, H., Imai, Sugawara, and Suzuki, *Nippon Kogyo Kaishi* **82**, 26 (1966).

260. Mitchell, J., and J. Chiu, *Anal. Chem.*, alternate years to 1975.

261. Monteil, J. B., L. Radel, and J. C. Bernier, *J. Solid State Chem.* **25**, 1 (1978).

262. Morgan, P., and W. W. Wright, in D. Dollimore, Ed., *Proc 1st European Symp on Thermal Analysis*, London, Heyden, 1976, p. 164.

263. Muller, A. C., P. A. Engelhard, and J. E. Weisang, *J. Catal.* **56**, 65 (1979).

264. Murphy, C. B., and C. D. Doyle, *Appl. Polym. Symp.* **2**, 77 (1966).

265. Murphy, C. B., *Anal. Chem.*, alternate years from 1964 to 1980.

266. Naumann, A. W., and A. S. Behan, Jr., *J. Catal.* **39**, 432 (1975).

267. Neumeyer, J. P., J. I. Wadsworth, N. B. Knoepfler, and C. H. Mack, *Thermochim. Acta* **16**, 133 (1976).

268. Newkirk, A. E., *Anal. Chem.* **32**, 1558 (1960).

269. Newkirk, A. E., and E. L. Simons, *Talanta* **13**, 1401 (1966).

270. Newkirk, A. E., *Thermochim. Acta* **2**, 1 (1971).

271. Niepce, J.-C., and G. Watelle, *J. Mater. Sci.* **13**, 149 (1978).

272. Niepce, J.-C., G. Watelle, and N. H. Brett, *J. Chem. Soc. Faraday Trans. 1* **74**, 1530 (1978).

273. Nieschlag, H. J., J. W. Hagemann, and J. A. Rothfus, *Anal. Chem.* **46**, 2215 (1974).

274. Nikalaev, A. V., and V. A. Logvinenko, *J. Therm. Anal.* **13**, 253 (1978).

275. Noel, F., and G. E. Cranton, *Am. Lab.* **11**, 27 (1979).

276. Norem, S. D., M. J. O'Neill, and A. P. Gray, *Thermochim. Acta* **1**, 29 (1970).

277. O'Gorman, J. V., and P. L. Walker, *Fuel* **52**, 71 (1973).

278. Ocone, L. R., J. R. Soulen, and B. P. Block, *J. Inorg. Nucl. Chem.* **15**, 76 (1960).

279. Oswald, H. R., and H. G. Wiedemann, *J. Therm. Anal.* **12**, 147 (1977).

280. Ozawa, T., *Bull. Chem. Soc. Jpn.* 38, 1881 (1965).

281. Pampuch, R., *Prace Mineral.* **6**, 53 (1965).

282. Parker, K. M., and J. H. Sharp, *Trans. J. Br. Ceram. Soc.* **81**, 35 (1982).

283. Paulik, F., and L. Erdey, *Acta Chem. Hung. Acad. Sci.* **13**, 132 (1958).

284. Paulik, F., E. Buzagh, L. Polos, and L. Erdey, *Acta Chem. Acad. Sci. Hung.* **38**, 311 (1963).

285. Paulik, F., and J. Paulik, *Thermochim. Acta* **3**, 13 (1971).

286. Paulik, F., and J. Paulik, in I. Buzas, Ed., *Thermal Analysis: Proc 4th ICTA*, Vol. 3, Budapest: Akademiai Kiado, 1975, p. 779.

287. Paulik, J., and F. Paulik, in I. Buzas, Ed., *Thermal Analysis: Proc 4th ICTA*, Vol. 3, Budapest: Akademiai Kiado, 1975, p. 789.

288. Paulik, F., and J. Paulik, *Analyst* **103**, 417 (1978).

289. Pautrat, R., B. Metivier, and J. Marteau, *Rubber Chem. Technol.* **49**, 1060 (1976).

290. Pearce, E. M., Y. P. Khanna, and D. Raucher, Thermal analysis in polymer flammability, in E. A. Turi, Ed., *Thermal Characteristics of Polymeric Materials*, New York: Academic, 1981, Chapter 8.

291. Pekenc, E., and J. H. Sharp, in I. Buzas, Ed., *Thermal Analysis: Proc 4th ICTA*, Vol. 2, Budapest: Akademiai Kiado, 1975, p. 585.

292. Peters, K. R., D. P. Whittle, and J. Stringer, *Corros. Sci.* **16**, 791 (1976).

293. Piehl, D. H., *J. Therm. Anal.* **6**, 221 (1974).

294. Pope, D., D. S. Walker, and R. L. Moss, *J. Catal.* **47**, 33 (1977).

295. Poppl, L., M. Gabor, J. Wajand, and Z. G. Szabo, in D. Dollimore, Ed., *Proc 1st European Symp Thermal Analysis*, London: Heyden, 1976, p. 332.

296. Porter, J. R., and L. C. de Jonghe, *AIME Metall. Trans. B* **12**, 299 (1981).

297. Poulis, J. A., and J. M. Thomas, in K. H. Berndt, Ed., *Vacuum Microbalance Techniques*, Vol. 3, New York: Plenum, 1963, p. 1.

298. Radecki, A., and M. Wesolowski, *Thermochim. Acta* **17**, 217 (1976).

299. Radecki, A., and M. Weselowski, *J. Therm. Anal.* **9**, 29 (1976); **9**, 357 (1976); **10**, 233 (1976); **11**, 39 (1977).

300. Radecki, A., and M. Wesolowski, *Talanta* **27**, 507 (1980).

301. Rajeshwar, K., and J. B. Dubow, *Thermochim. Acta* **54**, 71 (1982).

302. Rajeshwar, K., *Thermochim. Acta* **63**, 97 (1983).

303. Rajeshwar, K., R. J. Rosenvold, and J. B. Dubow, *Thermochim. Acta* **66**, 373 (1983).

304. Ramachandran, V. S., *Applications of Differential Thermal Analysis in Cement Chemistry*, Chemical Publishing Co., New York (1969).

305. Rankin, W. J., and J. S. J. Van Deventer, *J. South African IMM* 239 (1980).

306. Rawe, R. L., and C. J. Rosa, *Oxid. Metals* **14**, 549 (1980).

307. Reed, R. L., L. Weber, and B. S. Gottfried, *Ind. Eng. Chem. Fundam.* **4**, 38 (1965).

308. Rees, L. V. C., in D. Dollimore, Ed., *Proc 1st European Symp Thermal Analysis*, London: Heyden, 1976, p. 310.

309. Reich, L., *J. Polym. Sci. B* **2**, 621 (1964); *J. Appl. Polym. Sci.* **9**, 3033 (1965).

310. Rilling, J., and D. Balesdent, in C. Eyraud and M. Escoubes, Eds. *Prog. Vac. Microbalance Technol.*, Vol. 3, London: Heyden, 1975, p. 182.

311. Robens, E., in H. G. Wiedemann, Ed., *Thermal Analysis: Proc 6th ICTA*, Vol. 1, Basel: Birkhauser, 1980, p. 213.

312. Robens, E., C. Eyraud, and M. Escoubes, *Thermochim. Acta*, **82**, 23 (1984).

313. Rode, E. Ya, *Trans. 1st. Conf. on Thermal Analysis*, Kazan, 1955.

314. Rose, A. G., and C. Woodfine, *Br. J. Urol.* **48**, 403 (1976).

315. Rosenvold, R. J., K. Rajeshwar, and J. B. Dubow, *Thermochim. Acta* **57**, 1 (1982).

316. Rouquerol, J., and M. Gauteaume, *J. Therm. Anal.* **11**, 201 (1977).

317. Sagi, S. R., K. V. Ramana, and M. S. Prasada Rao, *Indian J. Chem.* **16A**, 1115 (1978).

318. Salvador, P., and M. L. G. Gonzalez, *J. Colloid Interface Sci.* **56**, 577 (1976).

319. Satava, V., and F. Skvara, *J. Am. Ceram. Soc.* **52**, 591 (1969).

320. Saxby, J. D., *Thermochim. Acta* **47**, 121 (1981).

321. Scheve, J., and K. Heise, in H. Wiedemann, Ed., *Thermal Analysis: Proc 3rd ICTA*, Vol. 3, Basel: Birkhauser, 1971, p. 71.

322. Schreiber, H. P., and Y. B. Tewari, *Ind. Eng. Chem. Prod. Res. Dev.* **17**, 27 (1978).

323. Schubart, B., and E. Knothe, in T. Gast and E. Robens, Eds., *Prog. Vac. Microbal. Technol.*, Vol. 1, London: Heyden, 1972, p. 207.

324. Schwartz, N. V., and D. W. Brazier, *Thermochim. Acta* **26**, 349 (1978).

325. Schwoebel, R. L., Beam microbalance design, construction and operation, in A. W. Czanderna and S. P. Wolsky, Eds., *Microweighing in Vacuum and Controlled Environments*, Amsterdam: Elsevier, 1980, pp. 59–93.

326. Sekhar, M. V. C., and M. Ternan, *ACS Div. Pet. Chem. Preprints* **23**, 208 (1978).

327. Sestak, J., *Talanta*, **13**, 567 (1966).

328. Sestak, J., in F. Schwenker and P. D. Garn, Eds., *Thermal Analysis: Proc. 2nd ICTA*, Vol. 2, New York: Academic, 1969, p. 1085.

329. Sestak, J., in H. G. Wiedemann, Ed. *Thermal Analysis: Proc 3rd ICTA*, Vol. 2, Basel: Birkhauser, 1972, p. 3.

330. Sestak, J., and V. Satava, *Thermochim. Acta* **27**, 383 (1978).

331. Shafizadeh, F., K. V. Karkanen, and D. A. Tillman, Eds., *Thermal Uses and Properties of Carbohydrates and Lignins*, New York: Academic, 1976, pp. 429.

332. Sharp, J. H., G. W. Brindley, and B. N. N. Achar, *J. Am. Ceram. Soc.* **49**, 379 (1966).

333. Sharp, J. H., and S. A. Wentworth, *Anal. Chem.* **41**, 2060 (1969).

334. Sharp, J. H., Reaction kinetics, in R. C. Mackenzie, Ed., *Differential Thermal Analysis*, Vol. 2, London: Academic, 1972, pp. 45–75.

335. Sharp, J. H., *Thermal Analysis Abstracts*, Vols. 2–12, Heyden, UK: Wiley, 1973–1983; R. C. Mackenzie, Vols. 13–14, (1984–1986); G. M. Clark, Vol. 16–20, (1987–1991).

336. Sharp, W. B. A., and D. Mortimer, *J. Sci. Instrum.* (*J. Phys. E*), Series 2 **1**, 843 (1968).

337. Sharp, W. B. A., and D. Mortimer, in T. Gast and H. Robens, Eds., *Prog. Vac. Microbal. Technol.*, Vol. 1, London: Heyden, 1972, p. 101.

338. Shigematsu, T., and K. Utsumomiya, *Anal. Chim. Acta* **58**, 411 (1972).

339. Shimokawabe, M., R. Furuichi, and T. Ishii, *Thermochim. Acta* **21**, 273 (1977).

340. Shimokawabe, M., R. Furuichi, and T. Ishii, *Thermochim. Acta* **28**, 287 (1979).

341. Simons, E. L., and A. E. Newkirk, *Talanta* **11**, 549 (1964).

342. Simons, E. L., and W. W. Wendlandt, *Anal. Chim. Acta* **35**, 461 (1966).

343. Sircar, A. K., and T. G. Lamond, *Rubber Chem. Technol.* **48**, 301 (1975).

344. Smallwood, T. B., and C. D. Wall, *Talanta* **28**, 265 (1980).

345. Smeggil, J. G., and N. S. Bornstein, *J. Electrochem. Soc. Solid State Sci. Technol.* **125**, 1283 (1978).

346. Smith, M. L., and B. Topley, *Proc. R. Soc. A* **134**, 224 (1931); *J. Chem. Soc.* 321 (1935).

347. Soboleva, T. N., L. A. Rudnitsky, and A. M. Alekseyev, *J. Therm. Anal.* **18**, 517 (1980).

348. Solymosi, F., *Structure and Stability of Salts of Halogen Oxyacids in the Solid Phase*, New York: Wiley, 1977.

349. Sosa, J. M., *J. Polym. Sci. Poly. Chem. Ed.* **13**, 2397 (1975).

350. Spacsek, K., A. Somlo, and I. Soo's, *J. Therm. Anal.* **11**, 211 (1977).

351. Sramko, T., G. Liptay, and E. Jona, *J. Therm. Anal.* **12**, 217 (1977).

352. Steinheil, E., in H. G. Wiedemann, Ed., *Thermal Analysis: Proc 3rd ICTA*, Vol. 1, Basel: Birkhauser, 1971, p. 187.

353. Still, R. H., The use and abuse of thermal methods of stability assessment, in N. Grassie, Ed., *Developments in Polymer Degradation*, Vol. 2, London: Applied Science, 1977, pp. 1–42.

354. Stone, R. L., *Anal. Chem.* **32**, 1582 (1960).

355. Strafford, K. N., and B. A. Nagaraj, *Oxid. Metals* **14**, 109 (1980).

356. Suzuki, M., D. M. Misic, O. Koyama, and K. Kawazoe, *Chem. Eng. Sci.* **33**, 271 (1978).

357. Swarin, S. W., and A. M. Wims, *Rubber Chem. Technol.* **47**, 1193 (1974).

358. Szendrei, T., and P. C. Van Berge, *Thermochim. Acta* **44**, 11 (1981).

359. Tanabe, S., R. A. Earle, A. N. Pick, and J. H. Sharp, unpublished results.

360. Tang, W. K., Forest products, in R. C. Mackenzie, ed., *Differential Thermal Analysis*, Vol. 2, London: Academic, 1972, p. 532.

361. Teychenné, D. C., *Concrete* **9**, 24 (1975).

362. Tidd, B. K., *Plastics and Rubber: Materials and Applications*, Aug., 100 (1977).

363. Till, L., *J. Therm. Anal.* **3**, 177 (1977).

364. Tinsley, D. M., and J. H. Sharp, *J. Therm. Anal.* **3**, 43 (1971).

365. Todor, D. N., *Thermal Analysis of Minerals*, Tonbridge Wells, UK: Abacus, 1976.

366. Tomassetti, M., L. Campanella, P. Cignini, and G. D'Ascenzo, *Thermochim. Acta* **84**, 295 (1985).

367. Tomlinson, W. J., and J. Yates, *J. Electrochem. Soc. Solid State Sci. Technol.* **125**, 803 (1978).

368. Toop, D. J., *IEEE Trans. Elect. Insul.* **EI-6**, 2 (1971).

369. Turi, E. A., Ed. *Thermal Characteristics of Polymeric Materials*, New York: Academic, 1981.

370. Unmuth, E. E., L. H. Schwartz, and J. B. Butt, *J. Catal.* **61**, 242 (1980).

371. Van Krevelen, D. W., C. Van Heerden, and F. J. Huntjens, *Fuel* **30**, 253 (1951).

372. Van Luik, F. W., and R. E. Rippere, *Anal. Chem.* **34**, 1617 (1962).

373. Vasantasree, V., and M. G. Hocking, *Corros. Sci.* **16**, 261 (1976).

374. Verneker, P. V. R., M. McCarty, and J. N. Maycock, *Thermochim. Acta* **3**, 37 (1971).

375. Vogel, E., and D. W. Johnson, *Thermochim. Acta* **12**, 49 (1975).

376. Waters, P. L., *J. Sci. Instrum.* **35**, 41 (1958); *Anal. Chem.* **32**, 852 (1960).

377. Webber, J., *Corros. Sci.* **16**, 499 (1976).

378. Weltner, M., *Acta Chim. Hung.* **31**, 449 (1962).

379. Weltner, M., *Acta Chem. Hung.* **47**, 311 (1966).

380. Wendlandt, W. W., and G. R. Horton, *Analyt. Chem.* **34**, 1098 (1962).

381. Wendlandt, W. W., and J. P. Smith, *The Thermal Properties of Transition Metal Ammine Complexes*, Amsterdam: Elsevier, 1967.

382. Wendlandt, W. W., *Thermal Methods of Analysis*, 3rd ed. New York: Wiley, 1986.

383. Wendlandt, W. W., and L. W. Collins, *Anal. Chim. Acta* **71**, 411 (1974).

384. Wendlandt, W. W., and L. W. Collins, *Benchmark Papers in Analytical Chemistry*, Vol. 2, *Thermal Analysis*, Stroudsbourg, PA: Dowden, Ross & Hutchison, 1976.

385. Wendlandt, W. W., *Thermochim. Acta* **21**, 295 (1977).

386. Wendlandt, W. W., and P. K. Gallagher, Instrumentation, in E. Turi, Ed., *Thermal Characterization of Polymeric Materials*, New York: Academic, 1981, Chapter 1.

387. Wentworth, S. A., and J. H. Sharp, *Anal. Chem.* **42**, 1297 (1970).

388. White, A. D., and J. H. Sharp, *J. Chem. Soc. A* 3062 (1971).

389. Wiedemann, H. G., *Thermochim. Acta* **3**, 355 (1972).

390. Wiedemann, H. G., *Thermochim. Acta* **6**, 257 (1973).

391. Wiedemann, H. G., and G. Bayer, in C. Eyraud and M. Escoubes, Eds., *Prog. Vac. Microbalance Technol.*, Vol. 3, London: Heyden, 1975, p. 103.

392. Wiedemeier, H. G., and F. J. Csillag, *Thermochim. Acta* **34**, 257 (1979).

393. Wilburn, F. W., C. J. Keattch, H. G. Midgley, and E. L. Charsley, *Recommendations for the Testing of High Alumina Cement Concrete Samples by Thermoanalytical Techniques*, London: Thermal Methods Group, Analytical Division of the Chemical Society, 1975.

394. Williams, J. R., E. L. Simons, and W. W. Wendlandt, *Thermochim. Acta* **5**, 101 (1972).

395. Wilson, D. E., and F. M. Hamaker, in R. F. Schwenker and P. D. Garn, Eds., *Thermal Analysis: Proc 2nd ICTA*, Vol. 2, New York: Academic, 1969, p. 1251.

396. Wirth, H. O., and H. Andreas, *Pure Appl. Chem.* **49**, 627 (1977).

397. Wright, W. W., Application of thermal methods to the study of the degradation of polyimides, in N. Grassie, Ed., *Developments in Polymer Degradation*, Vol. 3, London: Applied Science, 1981, pp. 1–26.

398. Young, D. J., W. E. Smeltzer, and J. S. Kirkaldy, *Met Trans.* **6A**, 1205 (1975).

399. Zsako, J., *J. Phys. Chem.* **72**, 2406 (1968).

ADDENDUM TO THERMOGRAVIMETRY

It is inevitable that during the production of this book further advances have taken place. The literature has grown by the publication of four important texts (1, 3, 5, 8), and the review articles in Analytical Chemistry have been continued by Dollimore (4).

Advances in instrumentation have included the addition of automatic sampling devices which permit unattended operation of the thermobalance to examine multiple samples. More attention has been given to the technique of Controlled Rate Thermal Analysis, in which the programmed temperature is controlled by the rate of reaction.

Clearly there have been further studies in all the various fields of application, but the most significant new area of application has been to the recently discovered superconducting ceramics. Thermogravimetry is particularly important because it enables the determination of the variable oxygen content of these materials with temperature and partial pressure of oxygen (2, 6, 7).

REFERENCES

1. Brown, M. E., *An Introduction to Thermal Analysis*, Chapman and Hall, London (1988).

2. Button, T. W., B. Rand, P. J. Ward, E. A. Harris, J. H. Sharp and P. F. Messer, *Br. Ceram. Trans. J.*, **86**, 166 (1987); ibid, *Br. Ceram. Proc.*, **40**, 93 (1988).

3. Dodd, J. W. and K. H. Tonge, *Thermal Methods*, Wiley, Chichester (1987).

4. Dollimore, D., *Analyt. Chem.*, 1988, 1990, 1992.

5. Earnest, C. M., *Compositional Analysis by Thermogravimetry*, ASTM STP 997 (1988).

6. Gallagher, P. K., H. M. O'Bryan, S. A. Sunshine and D. W. Murphy, *Mat. Res. Bull.*, **22**, 995 (1987).

7. Strobel, P., J. J. Capponi, C. Chaillout, M. Marezio and J. Tholence, *Nature*, **327**, 306 (1987).

8. Wunderlich, B., *Thermal Analysis*, Academic Press, New York (1990).

Chapter 4

THE APPLICATION OF THERMODILATOMETRY TO THE STUDY OF CERAMICS

By Moshe Ish-Shalom, *Israel Ceramic and Silicate Institute, Technion City, Haifa, Israel*

Contents

I. INTRODUCTION

The term *ceramics* in this chapter is used in the broad sense, covering inorganic nonmetallic materials, the manufacture or the use of which involves heat. It thus includes conventional and advanced ceramics, refractories, glass, glass–ceramics, and cements (cements are not reviewed in this chapter).

Since heating of the material is almost always used in the processing of ceramics, and quite often in their application, it is natural to have thermoanalytical equipment in a ceramic research and development laboratory, as well as in a control laboratory. In such a laboratory a thermodilatometer is an almost indispensable piece of equipment, and the results are capable of far wider interpretation than the mere "trivial" measurement of change in length (37). This chapter attempts to demonstrate to the thermal analyst the wider contribution of thermodilatometry (TDA) studies to ceramic science and technology; and to the ceramicist, the importance of the use of TDA especially in conjunction with other thermoanalytical techniques, as well as various other methods. This is achieved mainly by presenting examples of various applications of TDA to different types of problems or systems.

Treatise on Analytical Chemistry, Part 1, Volume 13,
Second Edition: Thermal Methods, Edited by James D. Winefordner.
ISBN 0-471-80647-1   © 1993 John Wiley & Sons, Inc.

However, that it is not intended to present an exhaustive literature survey, but rather selected examples.

Even in the narrow sense of the term, thermal analysis is a useful tool in the analysis of samples, through the ability to identify phases present and if suitably calibrated to determine their concentration. However, thermogravimetric analysis (TG), differential thermal analysis (DTA), and differential scanning calorimetry (DSC) have been applied to a much broader range of problems, such as the determination of the kinetics of chemical reactions, phase changes, melting, and many other physical processes in organic and inorganic systems, including ceramics. DTA and DSC can be operated only in a dynamic mode, and measure the result of a set heating rate on the sample, with the slowest practical heating rate being determined by the versatility of the instrument in use. TG can be used in a dynamic as well as in a static, isothermal mode. The derivative DTG curve representing the rate of these changes may be compared with the DTA or DSC curves.

In dilatometric measurements the source of the signal is the density change in the sample with time and temperature. As in TG, the measurement can be made with the temperature program in a dynamic or isothermal mode. In the actual implementation sometimes volume, but mostly linear, changes are being measured. The density usually changes continuously with temperature, although sharp changes in density or in the thermal expansion coefficient (TEC) can occur due to first- or second-order transitions or other chemical reactions. The derivative of the dilatometer curve can be helpful in its interpretation and comparison with DTA and TG curves.

Another important distinction between dilatometry and the other thermal methods discussed so far is related to the physical form of the sample. In other methods the properties of a sample, whatever its form, are measured, whereas in dilatometry the dimensions of a solid sample or a compacted powder that can support its shape during the experiment are measured. In the case of a solid, the inherent properties of the sample are sensed; but in the latter, in addition to the inherent properties large dimensional changes may be observed due to changes in the special distribution or packing arrangement of the compacted particles. For example, large shrinkages accompany the removal of porosity during sintering, even without any chemical reactions. While the inherent properties are of interest in characterizing the ceramic materials and products, the latter type of changes are of importance in the study of the processing of ceramics, for which thermal dilatometry is a unique tool. Of course, in many cases the situation is complicated because the two types of changes occur simultaneously and are difficult to separate using TDA alone. This possibility of more complex behavior of a sample may make this technique less attractive as an analytical tool for chemical identification or quantitative estimation. Clearly, like many other tools, it is not a stand-alone tool, but needs the support of other techniques. Similarly, TDA can provide support for other techniques.

In modern ceramic processing sintering with or without application of mechanical pressure is used. Therefore, the scope of this chapter has been extended to include examples where mechanical pressure has been used as an additional parameter. It thus includes the measurement of the viscosity of glass, sintering under axial load, and hot pressing (under 30–40 MPa). It would have been justified to include higher pressure experiments, such as hot-isostatic pressing (at 200–400 MPa), or even at higher pressures in the GPa range, which are used in the manufacture of synthetic diamonds or similar products, but there is a lack of published examples of such studies.

There is an extensive literature on research into the development, production and application of ceramics, as indicated by the large number of papers and books appearing, and the regular presentation of papers at international conferences (7, 23, 24).

II. EQUIPMENT AND METHODOLOGY

For most ceramic applications pushrod dilatometers are used. These are available commercially, and are described in several sources (76, 79, 80). Touloukian et al. (76) described the principles of a variety of methods of measurement useful for different applications, such as methods for high precision and sensitivity at very low temperatures or for the measurement of very low TECs, moderate temperature industrial applications where automation and fast response are important, or where knowledge of the lattice spacings are needed to understand the behavior of the materials. These include pushrod dilatometry, twin telemicroscopes, interferometry, and x-ray diffraction (XRD).

In the following a brief description of pushrod dilatometers is given, with some considerations of their design and applications, as well as some modifications used in examples given in the text. More details can be found in the literature cited as well as in commercial sources. In the examples in the text, commercial equipment used is identified where possible.

Figure 1 is a schematic representation of a single- and double-pushrod dilatometer. In most modern dilatometers a linear variable differential transformer (LVDT) is used as the displacement sensing element. In it the motion of the core relative to the coil gives rise to the linear displacement signal. This signal, after proper conditioning, can be used as an analog outout signal to a recorder, or can be converted to a digital signal and fed to a microcomputer, with possibilities of simultaneous viewing, later viewing, manipulation of the data, and analysis. In a single-pushrod unit the rod pushes the core while the supporting tube leans against the structure, to which the LVDT coil is also fixed. A reference sample of known expansion is needed to calibrate the instrument. For moderate temperature dilatometers (up to 1100°C) fused silica tubes and rods are used because of their very low TEC. In a dual-

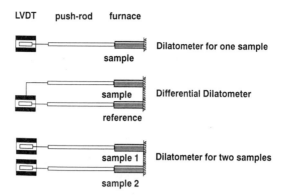

Fig. 1.   Schematic diagrams of single- and double-pushrod dilatometers.

pushrod instrument one of the rods pushes the core while the other pushes the coil. Thus only the difference between the two samples in the dilatometer is actually measured, one of them serving as a reference, and this configuration is referred to as a differential dilatometer. For higher temperatures the support and pushrod are made from alumina (up to about 1700°C).

The temperature of the sample is normally monitored by a thermocouple (TC) located as close as practicable to the sample. For optimization of control of the furnace temperature, another TC is located close to the heating element.

In some commercial dilatometers a vertical and in others a horizontal configuration is adopted (12). Positive contact between the sample and pushrod is needed, but the axial pressure on the sample should be minimized. Weak spring mounting is used in the horizontal configuration, and spring flotation is used in the vertical one. For the cases of larger length changes, such as in sintering studies in a horizontal configuration, the spring is replaced by small weights to ensure contact under constant minimum axial stress. In any case, the control and knowledge of this minimum load is a desirable feature and is available in some commercial equipment.

It is sometimes more convenient, or desirable, to measure directly the volume change with temperature, for example, in the cases of odd-shaped samples or powders, which cannot be easily compacted. A suitable adaptor for a commercial dilatometer is shown in Fig. 2 (80). Since a very low-expansion container is needed, fused quartz is used, which of course limits the useful temperature range.

Dilatometer measurements have been made under controlled axial load using weights in a vertical configuration (21) to measure the viscosity of glass, and in the horizontal configuration to study sintering, using gas pressure on a modified commercial instrument (13). In another study, simultaneous axial and radial strains were measured (6, 78).

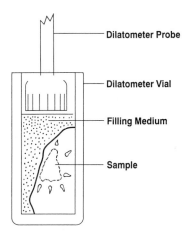

Dilatometer Probe

Dilatometer Vial

Filling Medium

Sample

Fig. 2. Apparatus for the determination of volume change with temperature for irregular shaped samples.

Some sintering of modern ceramics is done by die pressing in a hot press. It should be noted that in this case, as distinct from dilatometry under applied axial load, radial expansion is restrained by the die. Hot-pressing experiments recording the length changes are considered here as special cases of dilatometry.

Although it is not the intention here to analyze the considerations of the accuracy and precision of the instrumentation, it may be useful to bring out some points related to the use of pushrod dilatometers. These relate to sample size, thermal properties, and heating rate.

For economical reasons, as well as better temperature resolution, the smallest size sample is desirable, compatible with proper representation of the material and limited by the sensitivity of the sensing element in the instrument. In DTA, DSC, and TG the sensitivity of the equipment can usually be very high, so that proper representation becomes the determining factor. In pushrod dilatometers, the reverse is normally true, and the minimum size sample is prescribed by the sensitivity of the sensor, except for very coarse-grain materials, such as refractory bricks, which may contain grains of a few millimeters, with sometimes relatively large-scale inhomogeneities. Clusner (10) refers to international standards requiring sample length of 5.10^4 times the sensitivity of the sensor, so that for a sensitivity of about 0.1 μm a 1-cm-length sample should be adequate. The highest allowable heating rate compatible with temperature uniformity in a sample depends upon the geometry of the sample and its thermal diffusivity (K), which is the thermal conductivity divided by density multiplied by the specific heat. Assuming a cylindrical furnace with sufficient axial temperature uniformity, a small practical radial dimension of the sample is desirable (a diameter of about $\frac{1}{4}$ to $\frac{1}{8}$ of the length is considered adequate). The lower the thermal diffusivity, the lower is the permissible heating or cooling rate. Clusner (10) suggests, for instance, that for a glass sample of 3 mm diameter a heating rate of about

8°C/min is in order; for thermal insulators (K of less than 1) a heating rate of 0.1°/min is recommended, and for Pyrex or quartz glass ($K = 1$–10) 1–5°C/min is appropriate. Heckman and Elrick (22) have also examined in some detail the errors in a single-pushrod dilatometer related to heating rate and thermal properties. Placement of the TC bead in very good contact with the sample or even in a well in it is highly desirable, especially at higher heating rates, to avoid sizable errors in temperature readings of the sample.

Plummer (49) considered the advantages and applications of the differential dilatometer. A clear distinction in terminology should be made between the differential dilatometric curve, which is simply the difference in the length change of the two samples in the double-pushrod dilatometer, and the derivative curve, which is a plot of the first derivative of the dilatometric curve. Plummer evaluated the precision and accuracy of the measurements and, among other things, the effect of heating rate. When using a reference and sample with similar thermal properties two advantages can be achieved: higher sensitivity to the difference in extension and reduced sensitivity to heating rate errors. By matching not only the TECs of the sample and reference, but also the thermal diffusivity, rapid heating can be employed with only small sacrifice in precision and accuracy.

In a linear measurement in a dilatometer a plot of length change with temperature is obtained, $d(L/L_0)$, normalized by the original length L_0 at room temperature T_0. We then have

$$d\frac{L}{L_0} = f(T) \tag{1}$$

where $f(T_0) = 0$ and

$$\alpha(T) = \frac{d}{dT(dL/L_0)} = \frac{df(T)}{dT} \tag{2}$$

where $\alpha(T)$ is the TEC at temperature T and is obtained from the slope of the plot of $f(T)$ at temperature T.

It is customary, however, to report the mean TEC between two temperatures T_1 and T_2 or between room temperature T_0 and temperature T, as per Eqs. 3 and 4, respectively:

$$\alpha(T_1 - T_2) = \frac{f(T_2) - f(T_1)}{T_2 - T_1} \tag{3}$$

$$\alpha(T_0 - T) = \frac{f(T) - f(T_0)}{T - T_0} \tag{4}$$

Although for many practical applications the mean TEC is a useful parameter, $\alpha(T)$ is a more physically meaningful function. Second-order transformations or other small changes with temperature are also more visible in this function than in the mean TEC.

In many cases the interest is indeed in the linear strains, but in others it is the volume changes that are of interest. The correlation between the two is normally taken as

$$\frac{dV}{V_0} = \frac{V - V_0}{V_0} = \frac{3\,dL}{L_0} \tag{5}$$

It is worth noting the assumption and approximations involved. For a cube with initial side L_0 and initial volume $V_0 = L_0^3$ expanding isotropically to a cube of side L and volume V, we have

$$dV = (L_0 + dL)^3 - L_0^3$$

$$dV = 3L_0^2(dL)\left[1 + \frac{dL}{L_0} + \frac{(dL)^2}{3L_0^2}\right] \tag{6}$$

Equation 5 applies for small linear strains, i.e., for $(dL/L_0) < 1$. For ordinary TEC measurements the strains are small enough and (5) is valid. In other cases, as for instance for sintering studies, the linear strain may be as high as 10 or 20% and a better approximation for the computation of volume changes may be obtained from the use of Eq. 6. Moreover, in some cases the sintering shrinkage is not isotropic and the volume changes should be computed using linear changes in an additional direction, e.g., in the radial direction.

A final comment in this section relates to computerization. Most modern commercial dilatometers use computers to various degrees. They are used to program, set up the experiment, collect the data, and display and store it. Different software packages are offered for data viewing, manipulation, analysis, and reporting, which are dealt with in the manufacturer's instruction manuals. Besides the many obvious advantages of the application of the computer, attention is directed to two seemingly simple points:

1. The possibility of magnifying small sections of the TDA record to better examine the details of the data.

2. The facility of obtaining the derivative dilatometer curve, which is very helpful in interpreting the DTA curves.

III.   GLASS

Glass has been defined as an inorganic product of melting that has been cooled to a rigid state without crystallization. Since other methods can be used to produce rigid noncrystalline solids with structure and properties similar to glass without going through the melting, the other products can be referred to as noncrystalline or amorphous solids, which can include glass obtained by melting.

Commercial glasses can be considered to be highly cross-linked polymers, consisting of silica tetrahedra connected by $-Si-O-Si-$ (siloxane) bonds, without long-range periodic order, with possible substitutions of other high-valence ions for Si in the network (e.g., Al, P, B). In glasses containing alkali or alkaline earth ions part of the network bonds are depolymerized to give negatively charged oxygen ends, which are locally neutralized by the low valence ions. This simplified model of the structure of silicate glasses can account for the main features of these glasses; viz.,

1.   The lowering of the viscosity of glasses containing these low valence cations, the presence of which is associated with a lower degree of polymerization. This effect is more significant for alkali than for alkaline earth cations.

2.   The cation exchange capability of the alkali cation.

3.   Increased ionic conductivity at elevated temperature.

Due to its rigid but noncrystalline nature, glass provides a convenient starting point for the discussion of applications of TDA. Its presence in and its influence on the thermal behavior of many ceramic systems amply justifies its discussion at this stage.

A typical dilatometer curve of a glass sample is shown in Fig. 3. The two parts of the curve, below and above T_g, are readily distinguishable by the slope of the curve, which is related to the TEC of the sample: at temperatures below T_g the TEC is low and above T_g it is much higher. The apparent shrinkage shown at the upper part of the curve starting at the point M_g is in fact not shrinkage of the glass, but a result of its deformation under its own weight in a vertical dilatometer or under the slight retaining load on the sample in a horizontal one. We shall refer to this point later.

T_g designates the glass transition point of the sample, above which it is a highly viscous supercooled liquid and below which it is in its glassy state. T_g is normally determined by linear extrapolation, as shown in Fig. 3, although the two parts of the curve are not necessarily linear. The dilatometer curve was obtained from a well-annealed sample, otherwise the curve may look quite different for reasons to be given later. T_g can also be obtained from a cooling curve. This procedure can be considered an operational definition of T_g.

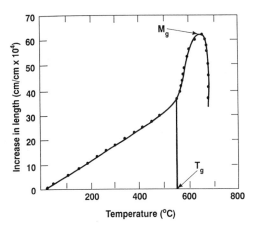

Fig. 3.   Typical thermal expansion curve of a glass, showing the glass transition temperature T_g.

However, it turns out experimentally that the value of T_g is not quite unique, but varies somewhat, depending on the rate of cooling or heating. As will be explained shortly, this is not an experimental shortcoming, but is in the nature of the phenomenon involved.

In Fig. 4a are shown schematic cooling curves of a liquid, crystal, and glass. Under normal laboratory cooling conditions (in a matter of days) the sample will not crystallize when going through its melting point, but will keep on shrinking along the same line into the supercooled liquid zone. A relatively sharp change in the slope will occur at a much lower temperature, from which T_g can be determined. From Fig. 4b it may be seen that T_g is higher for higher cooling rates. It may also be observed that the slope of the sample in its glassy state is similar to that of the crystalline material. This implies that in the glassy state, even though it is noncrystalline, the mechanism of thermal shrinking or expansion is similar to that in a crystalline solid. In a crystalline solid the thermal expansion is a result of the change in average distance between pairs of atoms vibrating in their free energy-distance trough when the thermal energy level is raised. The thermal expansion is higher the higher the asymmetry of this trough. Presumably, in the glassy state, below T_g, the same mechanism takes place within the short range order in the material.

Above the glass transition point there must be an additional mechanism of expansion with a much higher contribution. This is a process of reconfiguration of the structure of the liquid by longer range movement of atoms, ions, or larger volume units. By its nature this process is very much slower than the interatomic vibrations and is very strongly dependent on temperature as expressed by its viscosity. Thus, when the liquid is being cooled its viscosity rises sharply and so does the relaxation time t_{ou}, i.e., the time required by the system to reach equilibrium volume under the new, lower temperature

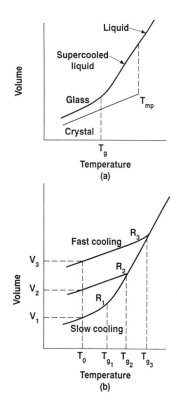

Fig. 4. Schematic specific volume–temperature relationships: (a) for liquid glass and crystal; (b) glasses formed at different cooling rates $R_1 < R_2 < R_3$.

conditions. For a given material at a given cooling rate a temperature T_g may be reached at which the time scale of the cooling t_s becomes much larger than the relaxation time t_{ou} such that the mechanism of contraction by reconfiguration of the structure is practically frozen in and only the interatomic vibrational mechanism remains operative. It may thus be suggested that the glass transition point T_g actually reflects a relaxation phenomenon such that

Above T_g, $t_s/t_{ou} > 1$, both mechanisms occur.
Below T_g, $t_s/t_{ou} < 1$, only the vibrational mechanism operates.
Around T_g, $t_s/t_{ou} \approx 1$, transition zone; relaxation phenomena.

Since t_s is inversely related to the rate of cooling, T_g will vary with it too. Therefore, the precise value of T_g will vary somewhat with the rate of cooling, as shown in Fig. 4b.

Glasses of the same composition but which have been differently annealed show dissimilar dilatometric properties. Less annealed glasses, when heated slowly enough, contract when heated to relax the frozen higher temperature configuration to the lower temperature one. If these reheating experiments had been carried out very rapidly up to the T_g then no shrinkage should have been observed.

As noted before, in the transition zone the behavior of the glass should be strongly time-dependent, reflecting the relaxation of the glass, i.e., the rate of approach of the system to equilibrium. Some examples of studies of relaxation phenomena in glass are those by Tool (75), Collier (11), Ritland (52), and Scherer (55).

In Fig. 3 attention was drawn to the point M_g on the dilatometer curve of glass. The apparent contraction following this point turns out to be the beginning of deformation of the sample even under the minimum applied axial load on the sample either under its own weight or the light spring loading in the dilatometer. This point, for normal heating rates in a commercial dilatometer represents the temperature at which the viscosity of the glass is 10^{11}–10^{12} P. This is expressed as a range since the loading conditions are not well defined at this sensitive temperature range. To obtain further information on the temperature-dependent flow properties of glass, a natural extension of dilatometer measurements is to apply a controlled load on a glass sample and observe its deformation. In a vertical parallel plate viscometer (21, p. 358) a small size sample is used, of 2–4 mm thick and 3–6 mm diameter, with a known vertical load to apply uniaxial compression. Viscosities in a broad range can be measured (ca. 10^8–10^{15} P), which include the annealing and strain points of $10^{13.4}$ and $10^{14.5}$ P, respectively. It should be noted that the currently adopted ASTM standard definitions avoid using actual viscosity values for these, but rather use standard samples with a specified operational definition specifying the method of measurement. The parallel plates are made of inconel and the pushrod of fused silica to eliminate the need for TEC corrections. The rate of change in height with time and temperature is measured and used to calculate the viscosity η in the following equation:

$$\eta = \frac{2\pi Mqh^5}{3Ve(2\pi h^3 + V)} \tag{7}$$

where M is the applied load (g), h is the height of the specimen at the particular time of interest (cm), V is the volume of the specimen (cm^3), and e is the rate of change of specimen height at the time of interest (cm/s).

As seen from the equation the precision of the height measurement is critical. Hagy (21) reports that he applied this instrument to study the dynamic behavior of pellets made of devitrifying frits, although he does not claim that these are actual viscosity values, but rather viscosity related values. He interprets the results in terms of the upturn of the "viscosity" at around 450–460°C as an indication of devitrification of the frits. An alternative interpretation of these data is the possible reversion of the unannealed frits (which are often made by quenching the molten glass in cold water to ease its milling) to their lower density upon heating, when approaching their glass transition temperature. Once so "annealed" the samples should then expand with temperature. Other methods of determining the viscosity of glass measure the rate of extension of a glass fiber under its own weight (softening point, 3, 21), or under an additional weight (annealing and strain points 2, 21), or the rate of bending of a beam under a known load (bending beam, 20, 21).

Practical experience points to the sensitivity of glass and other ceramic objects to thermal shock (abrupt heating or cooling) and to some specific products that better withstand thermal shock. This is due mainly to the brittleness of ceramics and is strongly dependent on the actual values of TEC, which is the dominant variable parameter in internal stresses of ceramic materials.

Abrupt cooling or heating of a glass slab will induce stresses by internal restraints due to differences in thermal contraction of adjacent layers, resulting from differences in their transient temperatures. The slab will shatter due to this thermal shock when the critical thermal stress is exceeded. The stresses are related to the properties of the glass, temperature difference, and geometry, expressed in a general form as

$$\sigma = \frac{AE(TEC)\,dT}{(1-\mu)k} \tag{8}$$

where σ is the stress developed, E is Young's modulus, dT is the initial temperature difference between the slab and the cooling medium, μ is the Poisson ratio, k is the heat conductivity, and A is a combined parameter including a geometrical constant and other material constants. This parameter is only weakly dependent on composition, as are E and μ. The value of TEC is the only material constant that is strongly dependent upon composition and varies in a wide range between 6 and 100×10^{-7}, as seen in Table 1 (more data and discussion on the TEC of glasses can be found in Refs. 19, 26, 32, 33, 36). Data from the table help to explain the significance of a low TEC for thermal shock resistance. Examination of the table shows why Pyrex glass is so useful for chemical labware or ovenware, why quartz glass can withstand being plunged red hot into cold water, and why glass ceramics based on cordierite and spodumene can be used for ovenware and flameware.

The phenomenon of induced stresses is being used to produce thermally toughened glass, by controlled abrupt cooling of the glass object, using air jets or a liquid cooling medium. The compressive stress imposed on the external layer of the object increases the tensile stresses required to break the object. If this is overdone, the excessive tensile stress on the internal layer

TABLE 1
Physical Properties of Some Commercial Glasses

Type of Glass	Thermal Expansion $\alpha \times 10^7 (K^{-1})$	Thermal Conductivity $(W/cm\ K^1)$	Specific Heat 20–300°C $(J/g\ K^1)$	Density (g/cm^3)	Refractive Index n_d
Sodium–potassium crystal glass	90–96			2.5	1.52
Lead glass (50 wt% PbO)		0.0071		3.8	1.64
Sheet and glass container	80–90	0.0096	0.977	2.5	1.51
Soft instrument glass (Unihost)	98			2.49	
Thermometer glass (PN)	82			2.61	
Simax (Pyrex type)	33	0.0105		2.23	1.47
Low-alkali fiberglass (Eutal)	52			2.68	
Quartz glass	6.7	0.0138	0.887	2.20	1.46

can cause spontaneous shattering of the object. Such treatment can be performed only with glasses of sufficiently large TEC.

Another commercial application of induced thermal stresses to strengthen glass objects uses glass layers of different TEC: a sandwich of a middle glass layer is produced with two outside thinner layers of lower TEC glass. When cooled, compressive stresses are developed in the external layers, causing increased strength.

In other applications, such as glass-to-metal seals, matching of TECs or graded TECs are required. In yet others, extremely low TECs are required per se, for example, for large telescope mirrors, in which very sensitive measurements of TEC are needed. Plummer (49) describes the application of a differential dilatometer of his own and a commercial one. Figure 5 shows the absolute and differential measurements of thermal expansions of a pair to be matched—a Kover metal and a glass. The advantage of the direct differential measurement is evident, compared to deducing differences from integral expansion. This has been taken advantage of for quality control measurements of various products, such as gas turbine components, emission control substrates, and glass ceramics for electric range cooking surfaces. When a proper reference substance is selected, a differential measurement will provide a much more sensitive measurement. An added advantage of differential measurement can be obtained if the reference material is chosen to have thermal properties similar to those of the examined sample in that the sensitivity of the expansion difference is less sensitive to the heating rate.

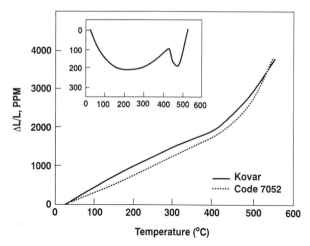

Fig. 5.   Absolute and differential expansion curves for KOVAR and Corning 7052 (49).

It is worth mentioning here the polarizing microscope can be a powerful tool for the study and evaluation of internal stresses in glass.

Elmer (16) has studied the sintering of a porous glass sample. In the process of manufacture of Vycor glassware of 96% SiO_2 content, this porous glass represents an intermediate stage with high open interconnected porosity. The final stage entails heat treatment of the body to shrink it to full density. Incidentally, this porous glass has many interesting applications that take advantage of its high specific surface area. In this experimental study the shrinkage was measured and in parallel the changes in weight, surface area, and equilibrium water absorption as a function of temperature. The composition of the porous glass was (in wt%): SiO_2, 96; B_2O_3, 3; (R_2O_3 + RO_2), 0.4 (chiefly Al_2O_3 and ZrO_2); and traces of Na_2O and As_2O_3. Its specific surface area was approximately 200 M^2/g, with an internal pore volume of 28%, and a very narrow pore-size distribution with about 96% at a radius of 0.5 ± 0.03 nm. Shrinkage measurements were made on samples that were heated at 100°C intervals in a stream of air having a dew point of −43°C. The shrinkage was measured on the cooled samples using a traveling microscope. The main features observed were the following:

- A slow limited shrinkage up to 900°C, with an increasing rate above that temperature. At 1100°C the glass is essentially consolidated into a clear, impervious material.
- After loss of the physical water up to 100°C there is a continuous slow loss of water of constitution to 1100°C.
- The surface area is virtually constant up to 900°C followed by a rapid loss, accompanied with a rapid shrinkage.

- The equilibrium water absorption, representing the pore volume, does not change much up to 500–600°C, followed by a slow but increasing loss up to 900°C, followed by a rapid loss.

A model of the porous structure is presented and the interpretation of the results is given in terms of removal of physical water first, followed by gradual dehydration of neighboring–SiOH or –BOH on the surface of the walls of the open pores. This is followed, in the final stage at temperatures above 900°C, by a condensation reaction between silonal groups on adjacent pore walls, expelling additional water and assisting in the closure of the residual porosity. The high shrinkage involves viscous flow driven by the reduction in surface energy.

IV. CRYSTALLINE MATERIALS

In crystalline solids, the shape, dimensions, and chemical content of the unit cell, representing the repetitive unit in the structure, determine the specific volume or density of the material. This is the true density, sometimes referred to as XRD density, since it can be determined by XRD. The average atomic distances in the thermal vibrations, represented by a potential energy well, determine the linear dimensions of the unit cell, and its asymmetry gives rise to a shift in this average distance with temperature, leading normally to thermal expansion. In metals and organic materials this potential energy well is relatively shallow, with larger asymmetry as compared to ceramic materials. Therefore, the TEC values of the former are relatively high compared to those of the latter. However, despite the moderate values of the TEC of the ceramic materials, their actual values are of great significance in determining their usefulness and applications. Typical values of TECs of crystalline materials are given in Table 2.

TABLE 2
Mean Thermal Expansion Coefficients for a Number of Materials

Material	Linear Expansion Coefficient, 0–1000°C (in./in.°C $\times 10^6$)	Material	Linear Expansion Coefficient, 0–1000°C (in./in.°C $\times 10^6$)
Al_2O_3	8.8	ZrO_2 (stabilized)	10.0
BeO	9.0	Fused silica glass	0.5
MgO	13.5	Soda–lime–silica glass	9.0
Mullite	5.3	TiC	7.4
Spinel	7.6	Porcelain	6.0
ThO_2	9.2	Fire-clay refractory	5.5
UO_2	10.0	Y_2O_3	9.3
Zircon	4.2	TiC cermet	9.0
SiC	4.7	B_4C	4.5

TABLE 3
Thermal Expansion Coefficients for Some Anisometric Crystals ($a \times 10^6/°C$)

Crystal	Normal to c Axis	Parallel to c Axis
Al_2O_3	8.3	9.0
Al_2TiO_5	-2.6	$+11.5$
$3Al_2O_3 \cdot 2SiO_2$	4.5	5.7
TiO_2	6.8	8.3
$ZrSiO_4$	3.7	6.2
$CaCO_3$	-6	25
SiO_2 (quartz)	14	9
$NaAlSi_3O_8$ (albite)	4	13
C (graphite)	1	27

Dilatometeric measurements of crystalline materials, where the change in linear dimensions is being measured, requires that anisotropy be considered. Only cubic crystals are isotropic in their unit cell dimensions and their TEC, while the lower symmetry crystals are not. The need to ensure definition of crystal direction in measurement of single crystals (which is seldom done) or to ensure randomization of crystal orientation of powder samples is evident. Examples of such anisotropy are shown in Table 3. It is worth noting the cases of negative TEC in certain directions in some crystals. Touloukian et al. (76) list several sources of information on thermal expansion data of ceramic materials (31, 48, 60, 81). Taylor has published a review of thermal expansion data of different types of crystalline materials (61–73), giving the values along the crystal axes.

Phase transformations are frequently encountered in ceramics, and are accompanied by a sharp change in specific volume and thermal properties. An example of interest and technical importance in the processing of conventional ceramics and refractories is the case of silica. In Fig. 6 are shown the dilatometeric curves of the different polymorphic forms of SiO_2. The following simplified chart shows the sequence of these transformations.

$$\text{quartz} \xleftrightarrow{870°C} \text{tridymite} \xleftrightarrow{1470°C} \text{cristobalite} \xleftrightarrow{1728°C} \text{melt}$$

$$\beta \xleftrightarrow{573°C} \alpha, \quad \beta \xleftrightarrow{117-163°C} \alpha, \quad \beta \xleftrightarrow{220-270°C} \alpha$$

$$\text{berlinite} \xleftrightarrow{815°C} \text{tridymite form} \xleftrightarrow{1025°C} \text{cristobalite form} \xleftrightarrow{1600°C} \text{melt}$$

$$\beta \xleftrightarrow{586°C} \alpha, \quad \beta_2 \xleftrightarrow{93°C} \beta_1 \xleftrightarrow{130°C} \alpha, \quad \beta \xleftrightarrow{210°C} \alpha$$

The transformations shown in the horizontal line in this chart indicate reconstructive transformations, involving reorganization of the whole crystal structure, which are very sluggish. The vertical ones represent displacive transformations, which involve only changing bond angles and are thus very fast. The latter are reversible in the dilatometer experiments, while the former, once formed at high temperatures, will, in practice, not revert back

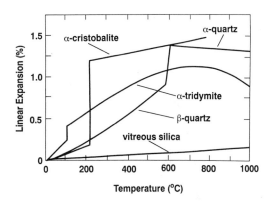

Fig. 6.   Thermal expansion of silica polomorphs.

to their equilibrium polymorphic form upon normal or even slower cooling. The crystal systems and densities of the polymorphic forms of silica are shown in Table 4. From the table and the graph it may be observed that the high-temperature allotropic forms are less dense and also have very low or even negative TEC values.

It is interesting to note the similarity between this system and the isostructural system of $Al(PO_4)$ (54). The density of Berlinite is 2.65 and its XRD pattern is very much like that of quartz. Taylor (61–73), in his compilations of thermal expansion data of inorganic crystalline substances, tried to correlate thermal expansion data of similar structures. Applying the "rule of corresponding states," he was able to draw a reasonably good master plot of TECs of a group of inverse spinels against the Tamen temperature (66) (the ratio of the temperature to the melting temperature, in K). A similarity in thermal expansion was observed also in the group $BeAl_2O_4$, Al_2O_3, and BeO, where the oxygen atoms are close packed and the unit dimensions are similar. However, in general, no good correlations were found to exist (76). Second-order transformations are generally characterized by the disappearance of ordered structures, resulting in a discontinuous or abrupt change in the TEC (76), which can be readily observed using the derivative dilatometer curve.

TABLE 4
Polymorphic Forms of Silica

Modification	Crystal System	Density (g/cm³)
β-Quartz	Trigonal	2.65 (20°C)
α-Quartz	Hexagonal	2.53 (600°C)
γ-Tridymite	Orthorhombic	2.26 (20°C)
β-Tridymite	Hexagonal	—
α-Tridymite	Hexagonal	2.22 (200°C)
β-Cristobalite	Tetragonal	2.32 (20°C)
α-Cristobalite	Cubic	2.20 (500°C)

The thermal expansion of particulate composite materials does not obey the simple rule of mixtures, except for loose powders, due to internal stresses that result from the thermal mismatch between the crystallites of the different phases. Even in a single-phase, anisotropic polycrystalline material, internal stresses develop with change in temperature. Therefore, to calculate the TEC of a composite, the elastic properties must be taken into account and different results have been obtained depending on the model used (29, 76), assuming that the internal stresses were not relieved by microcracking.

The larger the anisotropy, the larger are the internal stresses. Therefore, in polycrystalline materials with large differences in thermal expansion in the different crystal directions or in mixtures with large differences in thermal expansion, the stresses that develop may be sufficient to cause microcracking. These are important in understanding the behavior of the ceramics and they show up as hysteresis in heating and cooling dilatometer experiments. For example, when titania is cooled from the firing temperature, microcracks develop between the crystals and the sample shrinks less than the individual crystals (29, p. 608). Upon heating, these cracks tend to close and the expansion is lower at the lower temperature. An extreme example of this effect is exhibited in the measured TEC of graphite. Although the TEC normal and parallel to the c plane are 1×10^{-6} and 27×10^{-6}, respectively, the measured TEC of a polycrystalline graphite sample was only $1-3 \times 10^{-6}$. Grain boundary cracking and thermal expansion hysteresis occurred predominantly in large grain samples. Such cracking can occur upon heating of this type of body, leading, upon cooling, to residual expansion. A simple model has been proposed to show schematically both types of hysteresis (18, p. 66).

Redox reactions in iron oxide provide another example of hysteresis in a dilatometer curve for a sintered hematite (Fe_2O_3) compact that was heated and cooled in air at $10°C/min$ (18, p. 68). The dissociation of rhombohedral hematite to cubic (spinel type) magnetite occurs with loss of oxygen when heated to a temperature that depends on the prevailing oxygen pressure. The transition temperature is $1450°C$ in oxygen and $1323°C$ in air. In air the product is a nonstoichiometric spinel with an O/Fe atomic ratio of 1.373, instead of the 1.333 in Fe_3O_4. The transition is reversible, hematite being formed on cooling, gaining all the oxygen that was lost. In a TDA plot the marked expansion accompanying the transition to magnetite is evident, but the contraction produced by the reoxidation on cooling is appreciably less, resulting in a net growth in size. If the cycle is repeated between 1270 and $1430°C$, sintering as well as growth mechanisms operate at high temperature; after eight cycles, shrinkage sets in, probably as a result of the increase in sintering rate induced by the higher surface area of the more porous test piece. These results have also been suggested as an explanation of the observations that in iron oxide-containing bodies, growth and friability are induced by thermal and atmospheric cycling. They occur as a result of changes in the state of oxidation of iron. In pure iron oxide this was difficult to show because of the high transformation temperature, but in iron oxide containing other oxides this transformation could be lowered to as much as 700°C, faciliting the study.

Aluminum titanate Al_2TiO_5 (AT), or tialite, provides a unique example of thermal expansion behavior and is of technological interest in heat engine applications due to its very low TEC and low heat of conductivity. It has been successfully tested for portlines and piston inserts. The thermal expansion of tialite is different in its three axial directions, with TEC values of 11.2, 19.4, and −2.6 ppm for the a, b, and c axes, respectively (15). The large expansion anisotropy accounts for extensive microcracking, resulting in thermal hysteresis and low thermal expansion. As a consequence, the material has very low strength, which needs to be considered in its anticipated applications.

AT is prepared by reaction sintering, i.e., by heating a compact of a mixture of the reactants (alumina and titania) to above the reaction temperature. In Fig. 7 the dilatometer curve obtained using a compact of an equimolar mixture is shown. The mixture is seen to expand normally then begins to sinter and shrink, thus temporarily offsetting the expansion caused by the formation of tialite from the reactants. This expansion takes over above 1300°C. The final slow expansion is that of the product. On cooling from 1600°C, internal stresses are induced due to the anisotropic contraction of tialite, with final expansion upon cooling due to the relief of these stresses by microcracking. Almost no expansion occurred upon reheating of the cracked body to 400°C.

ZrO_2 affords an example of interest from the viewpoint of volume changes upon heating, especially because of its applications in advanced ceramics. Pure zirconia has a melting point of 2710°C and thus could presumably be used as a high-temperature refractory. However, it exists in three polymorphic forms, monoclinic below 1190°C, tetragonal between 1130 and 2350°C, and cubic above 2350°C. The tetragonal to monoclinic transformation involves a 5% volume increase, rendering pure zirconia practically useless as a

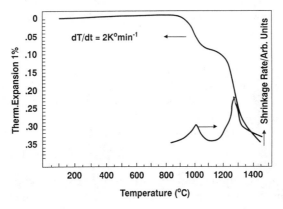

Fig. 7.   Shrinkage behavior of an Al_2O_3/TiO_2 mixture.

high-temperature refractory due to the disruptive effect of this volume change. Fully stabilized zirconia refractories have been produced and used successfully by addition of calcia or magnesia, which form solid solutions with the cubic phase.

Partially stabilized zirconia (PSZ), in which smaller additions of MgO, Y_2O_3 or other additives are used, provides in advanced ceramics an example of the concept of toughening ceramics for high-temperature mechanical applications. The principle in this technology is to obtain in a stable matrix, a metastable dispersed phase, which is prevented from transforming to the stable form because the transition is accompanied by an expansion that is unable to take place because of the constraint imposed by the matrix. For example, in a typical composition, containing 3.2% MgO fired in the single-phase region of the ZrO_2–MgO diagram (14) and cooled quickly, a large amount of homogeneously distributed t-ZrO_2 precipitate was formed. The matrix was the unchanged cubic phase. From the standpoint of toughening, the important variables are the volume fraction of the precipitate formed and the size and mean separation of the dispersed particles. These are controlled by the composition and thermal history and the proportion of t-ZrO_2 retained due to the constraint on the t-m expansion by the matrix, which is strongly dependent on particle size.

In practice it was found that the toughening can be obtained in one of two ways:

1. In PSZ in which the t form has been largely retained, a propagating crack, induced by relieving pressure on the t phase particles in the vicinity of the crack tip, enables the t-m transition to proceed. This in turn sets up compressive stresses within a zone enclosing the tip, obstructing further crack growth.

2. In PSZ in which sufficient transformation to the m form has already occurred, the resultant microcracking causes the propagation to branch, increasing its energy requirements. This mechanism is found to operate when the size of the dispersed particles exceeds a critical value.

It has been found that partially stabilized zirconia can be used to toughen other oxides, such as alumina, mullite, and spinel (15). For example, an additive content of 10–15 vol% and particle size range of 0.2–0.6 μm to alumina was found to be an optimal composition.

Dworak et al. (14, 15) describe the dilation behavior and properties of commercial MgO–PSZ and a MgO–Y_2O_3–PSZ material. The dilation behavior of the products was correlated with the composition and thermal treatment of the reactants and the resulting phase composition and microstructure, on the one hand, and with the loss of strength upon aging at various temperature levels on the other. Samples ZN40 and ZN50 had the same composition of 3.2 wt% MgO in ZrO_2. ZN40 was fired in the single-phase region and cooled quickly; ZN50 was treated at 1000–1200°C for 3–100 h, where the cubic phase decomposed to t-ZrO_2 and MgO, and cooled. ZN40 had a large amount of homogeneously distributed t-ZrO_2 precipitate (40–50%), while ZN50 had only 20–30% of the t phase, predominantly located on the grain boundaries. The ZN40 TDA curve, after the first cycle, is repeatable with no apparent change in length, whereas that of ZH50 showed large hysteresis loops with continued change in length. Room temperature flexural strength measurements after aging of ZN40 at different temperatures correlate well with the dilatometric measurements, and the strength is retained up to 10000 h at 850°C. Partial substitution of MgO by Y_2O_3 in this system gave improved retention of flexural strength against PSZ-Y or PSZ-M formulations, with the ternary mixture showing no loss of flexural strength after aging for 2000 h at 1250°C. Moreover, the partial substitution diminishes the difference between the crystal parameters of the cubic and tetragonal precipitates.

In a study of tetragonal zirconia polycrystals (TZP), Claussen et al. (9) examined the thermal expansion and oxidation behavior of hot-pressed TZP and a composite of TZP-30 vol% SiO_2 whiskers in air and Ar atmospheres up to 1400°C. Up to 1000°C the curve for the TZP is as expected. Above 1200°C in air, both samples swell as a result of reoxidation, which is characteristic of oxide samples that have been hot pressed in a C-rich atmosphere. Running the dilation experiment of the composite in Ar did not produce any swelling. The small observed shrinkage is probably due to mismatch stress relaxation.

In a study of the sintering behavior of aluminum titanate the effect of the powder preparation on the sintering shrinkage has been observed. Roosen and Hausner (53) have carried out a detailed study of the sintering behavior of CaO-stabilized ZrO_2 (13 mol% CaO) and the effect of characteristic properties of the powders and their compacts on the sintering behavior. They compared two precipitated powders with a mechanically mixed material, labeled MM. This was prepared from a mixture of ZrO_2 with $CaCO_3$, wet milled, dried at 120°C, and calcined at 1300°C for 60 min (94% cubic 6% monoclinic). The coprecipitated powders were washed and dried, one in an oven at 120°C (120) and the other was freeze-dried (FD). Both were then wet milled and dried at 120°C and calcined at 700°C for 30 min. All samples were then wet milled for 48 h, dried at 120°C, then compacted (350 N/mm) for sintering. Table 5 shows the characteristics of the calcined powders after milling, of the consolidated compacts, and of the sintered samples. The sintering was studied on pressed cylinders (5.5 mm high × 4 mm diameter) in a furnace for 3 h at temperatures of 1100 and 1580°C and on 10 mm high × 10 mm diameter cylinders in a dilatometer, recording shrinkage and shrinkage rate.

Examination of the various characteristics of the powders and compacts suggested that the best correlation was between shrinkage rate and pore size

TABLE 5A
Characteristics of Calcined Powders After Milling

Powder	Tap Density (% Theor. Dens.)	Specific Surface Area (m²/g)	$d_{50}{}^{a}$ (nm)	$d_{ads}{}^{b}$ (nm)	$d_x{}^{c}$ (nm)	Aggl. Parameter d_{50}/d_{ads}
MM	29.6	2.5	1700	423	48.2	4
120	21.5	68	1900	16	12.2	127
FD	15.6	68	700	16	12.2	44

[a]Average grain size (MSA centrifuge).
[b]Equivalent crystallite size (N_2 adsorption).
[c]Crystallite size (X-ray line broadening measurement).

TABLE 5B
Characteristics of Consolidated Compacts
and Sintered Samples

Powder	Green Density (% Theor. Dens.)	Specific Surface Area (m²/g)	Maximum Pore Radius (nm)	Temperature of Maximum Shrinkage Rate (°C)	Sintered Density (1250°C, 3 h) (% theor. dens.)
MM	56.0	2	100	1470	63.0
120	40.8	68	17 a. 80	1195 a. 1370	76.8
FD	39.7	54	9	1185	94.5

distribution of the compacts, which themselves were dependent on the processing route of the powders. The correlation can be seen in Fig. 8. The shrinkage curves (a) show clearly the shrinkage onset temperature and the extent of shrinkage of the three samples and indications of a two-step shrinkage for both precipitated samples. While the extent of shrinkage can be seen best in this plot, onset and rate can be better seen in (b). The onset of shrinkage of the samples corresponds to the precipitation temperatures of the powders. The distinction between the coprecipitated powders (lower temperature double peak) and the mechanically mixed powder MM (single higher temperature peak) is very clear. The difference between the two precipitated powders is mainly in the height of the two peaks. The pore size frequency distribution of the compacts (c) seems to bear a considerable resemblance to the shrinkage rates in (b): MM has most porosity in the high end (100 nm), corresponding mostly to the interparticle distances between the hard agglomerates with very little below 10 nm; FD is mostly concentrated around 9 nm, corresponding to intraparticle distances, with some pores in between, probably interparticle ones; 120 is intermediate. Thus, the first rate peak appears to correspond to the intraparticle sintering for the two precipitated powders, and the second of these and practically all of MM to the interparticle pore elimination, which requires a higher temperature. The

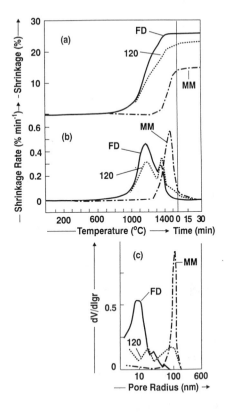

Fig. 8. Pore size frequency distribution of oxide powder compacts (Hg porosity) (53).

rate curves support this interpretation. Changing the heating rate only moves the peaks to higher temperatures, but does not affect the essence of their shape.

V. WHITEWARES AND REFRACTORIES

In ceramic practice a dilatometer can be used for various tasks, such as the following:

- Routine control of products and raw materials
- Development of new or modified compositions and formulations of raw materials and their effect on the firing behavior and properties of products
- Development of suitable firing cycles, optimized from the technological and economic points of view.

Figure 9 shows the dilatometer curves of several substances, and can be used to summarize several points: (1) the difference in behavior of silica, that is,

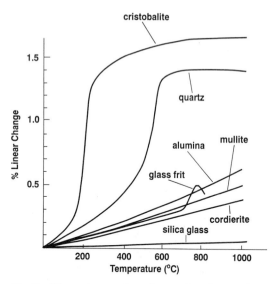

Fig. 9.   Thermal expansion of some ceramic materials.

the extremely low expansivity of fused silica throughout the temperature range, as opposed to the large abrupt charge in quartz and cristobalite due to their low–high phase transformations; (2) the typical behavior of a glass, showing the glass transition range and the flow point, the actual values of which can vary in a wide range with composition; (3) the typical behavior of some ceramic materials with their relative expansivities. Figure 10 gives a schematic overview of the range of expansion of different materials and ceramic products, emphasizing the target properties of the better thermal shock-resistant products.

Dilatometric curves can be routinely used in the quality control of production. For more precise control a differential dilatometer can be used with selected samples as references, as was used by Plummer (49) for low TEC glass or glass ceramics. Most of the conventional ceramic compositions comprise quartz, clay, and feldspar. In ceramic parlance they are referred to as "triaxial" bodies, since their composition can be described on a triangular plot. Reference dilatometric curves of the silica and clay raw materials can be used, among other control methods, and typical curves are also available in the literature and may be furnished by the suppliers. However, to better understand the firing behavior of the ceramic bodies many studies have been conducted on the individual components and partial systems as well.

The thermal behavior of silica has been shown already. Pure quartz normally shows a reversible dilatometer curve up to above 1400°C. Other sources of natural silica, such as flint, having a different mineralogical composition, may have a more complex behavior (40, p. 248). In contrast, clay minerals undergo irreversible changes upon heating. There are various types

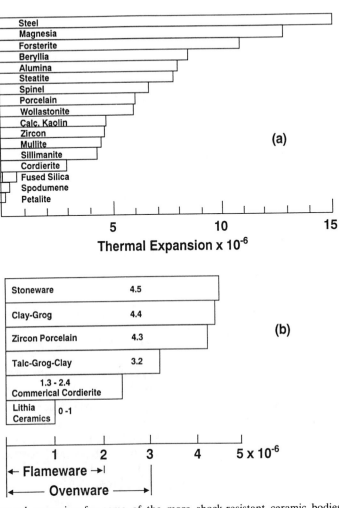

Fig. 10. Thermal expansion for some of the more shock-resistant ceramic bodies with the thermal expansion range desirable for ovenware and flameware.

of clay minerals and, being natural materials, they contain impurities that affect their thermal behavior. Kaolinite, which is now commercially available after various careful pretreatments, has been studied in detail. On heating, the major endothermic peak is the result of dehydroxylation, and any exothermic peaks are the result of recrystallization. Figure 11 shows the shrinkage steps accompanying these occurrences, in the form of the derivative dilatometer curve. Such derivative curves should be readily obtainable with modern computerized dilatometers, and can facilitate the interpretation of the data. Firing shrinkages of different type clays are shown in Fig. 12 (40).

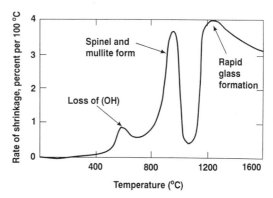

Fig. 11.   Rate of shrinkage of kaolin (40, p. 238).

Many dilatometric curves of different type clays from different sources can be found in the literature (54).

Lawrence (30) quotes an example of dilatometer curves of prefired clay–quartz mixtures that demonstrates clearly the diagnostic ability of dilatometer curves to evaluate the effect of the prefiring temperature on the extent of formation of cristobalite. It was evident from an inspection of the curves that quartz is converted to cristobalite only above 1190°C.

Israeli flint clay is used to produce refractory grains. It consists mainly of kaolinite, and the higher alumina flint clays contain boehmite and diaspore as well. Besides XRD, Fischer (17) attempted to use DTA to distinguish between these minerals and obtain semiquantitative mineralogical analysis, but this approach was not very successful. On the other hand, the dilatometer curves enabled the boehmite content in this raw flint clay to be determined readily, despite the fact that the cause of the expansion of boehmite has not been clarified. Moitron and Kletenik (34) studied the firing of Israeli flint clays containing increasing amounts of alumina. Firing the raw samples

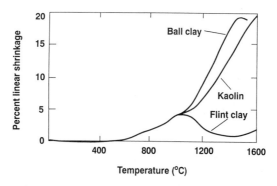

Fig. 12.   Firing/shrinkage curve for three clays (40).

showed the expansion due to boehmite present in the high-alumina sample. The dilatometer curves of the prefired samples revealed that firing to 1070°C was too low, as evidenced by the lack of cristobalite in all samples and the presence of even residual quartz in the one with the highest silica content. Prefiring to 1200°C was sufficient for the formation of cristobalite in all three high-silica samples.

Firing of a typical triaxial body illustrates the complex sequence of events taking place during the firing process. A simplified picture is presented diagrammatically in Fig. 13 and in Table 6 (40, p. 266), based on numerous studies using various techniques, such as thermal analysis (including TDA), XRD, microscopy, and IR spectroscopy. From the diagram it may seem that there is little interaction between the three components below 1000–1100°C, and their contribution is prescribed by their individual properties. Above this range the melting of the feldspar, which acts as the flux, produces a glassy phase, which gradually changes in quantity and composition by the dissolution of the clay constituents and part of the quartz, with recrystallization of mullite. The bulk of the void space (comprising more than 30 vol% in the green body) is being eliminated during this vitrification stage, leading to practically full densification with little closed porosity. Figure 14 shows the linear shrinkage and porosity changes in the vitrification range. Clearly, in practice, many factors will affect the maturing of a triaxial body, for example, processing and properties of the products, such as composition, type, and source of the natural raw materials, type and level of impurities, organic matter, and particle size.

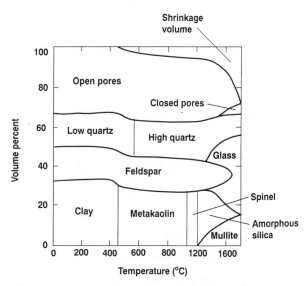

Fig. 13.   Constituents in a triaxial body on firing (40, p. 266).

TABLE 6
Reactions in Firing a Triaxial Body

Temperature (°C)	Reaction
Up to 100	Loss of moisture
100–200	Loss of adsorbed water
450	Breakup of clay structure; increased porosity
500	Organic matter oxidized
573	Inversion of quartz to high form
950	Spinel formed in clay
1100	Mullite formed
1100–1200	Feldspar melts and dissolves clay and cristobalite; shrinkage becomes rapid; porosity decreases
1300	Glass increases; mullite needles larger; only closed pores left

In considering a firing schedule an attempt is made to reduce internal stresses in excess of the strength of the body at a given temperature. Internal stresses may develop at high heating rates as a result of temperature differentials due to heat transfer. These are aggrevated by heat absorption processes which affect the heat transfer or by isothermal volume change. In a triaxial body, in the temperature range of 500–600°C dehydroxylation of kaolinite and the inversion of quartz require slow heating, especially since the dehydroxylation of the clay temporarily weakens the body. Hlavac (25)

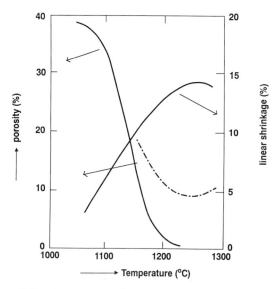

Fig. 14.   The effect of sintering on porcelain-type materials: - - -, actual porosity; —, apparent porosity.

discusses some of these considerations. The effect of temperature and time on shrinkage in the vitrification range of whitewares can be estimated from an empirical relation (38, 39):

$$\log \frac{k_2}{k_1} = A\left(\frac{1}{T_1} - \frac{1}{T_2}\right)$$

where k_1 and k_2 are rate constants at T_1 and T_2, respectively, and A is a constant, close to 10 000. This means that a tenfold increase in time will permit lowering of the maturing temperature by 23°C.

Fast firing has been the subject of many studies. In one example (56), dilatometry and DTA, together with SEM, XRD, and EDS were used. A comparison was made between a commercial triaxial body with ones based on mixtures of kaolin, clay, and wollastonite. From the dilatometric curves it was concluded that the firing range was too short. It appears that the dissolution of quartz in the triaxial body increases the viscosity of the glassy phase and thus provides a longer firing range.

An approach based on dilatometric experiments using the concept of rate-controlled sintering (RCS) to develop optimized firing schedules for whitewares and other systems is discussed later.

The glazing of ceramic bodies and the enameling of metals present the problem of fitting thermal expansions of the glassy coating to the body. The conditions must be arranged such that at the end of the cooling of the glazed body the glaze should be in compression. The stresses in the glaze and body develop upon cooling as a result of the difference in thermal contraction of the glaze and body from the setting temperature of the glaze (approximately the glass transition point) to room temperature. They depend also upon the geometry. For a thin glaze on a cylindrical body, for instance, the stresses are given by (29, p. 609)

$$\sigma_{(gl)} = \frac{E}{1-\mu}(T_0 - T')(TEC_{(gl)} - TEC_{(b)})\frac{A_{(b)}}{A} \tag{9}$$

$$\sigma_{(b)} = \frac{E}{1-\mu}(T_0 - T')(TEC_{(b)} - TEC_{(gl)})\frac{A_{(gl)}}{A} \tag{10}$$

E and μ are the elastic modulus and Poisson ratio, assumed to be the same for body and glaze, T' and T_0 are the setting and room temperatures, respectively, and A, $A_{(b)}$, and $A_{(gl)}$ are the cross-sectional area of the overall cylinders, body, and glaze. Figure 15 (29, p. 611) is an illustration of the dilatometric curves of body and glaze and of the ensuing stresses developed upon cooling. The compressive stresses in the glaze confer higher strength to the ceramic. If tensile stresses result in the glaze, crazing is observed; and if excessive compression occurs then peeling may result, with disruption of the

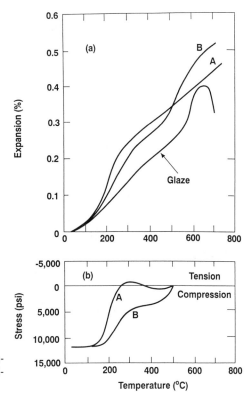

Fig. 15. (*a*) Expansion of glaze and porcelain bodies; (*b*) stresses developed on cooling (29, p. 611).

bond between the body and glaze. It should be noted that in practice the situation is more complex due to the interaction between glaze and body during firing, which should be allowed for in glaze fitting using dilatometer data.

In the pottery industry Steger's method (59) or modifications of it are sometimes used, in which the curvature of a bar of ceramic, glazed on one side, is observed. The degree of curvature is directly related to the stress in the glaze.

VI. MISCELLANEOUS: SILICON NITRIDE

Silicon nitride (SN), Si_3N_4, is of importance in engineering ceramics because of its mechanical properties at room and high temperature and its thermal shock resistance. Examples of its applications or potential applications include high-temperature bearings, tool bits, and parts for heat engines. It is a synthetic material and commercial powders are available from different processes with different levels of impurities (e.g., SiO_2, C) and particle sizes. Silicon nitride bodies are made by sintering of silicon nitride powders (SSN) or by hot-pressing (HPSN), and lower density bodies with useful properties are also made by nitriding silicon powder compacts (RBSN = reaction-bonded SN). Because of its technological importance the system has been extensively studied and reported in the ceramic literature. It is known that in general the sintering of silicon nitride necessitates the presence of a liquid phase, which is supplied by various additives together with impurities present in the powder. It is also presumed that the mechanism of sintering involves a powder rearrangement with the aid of the liquid phases and a solution/precipitation step (in this case solution of the alpha modification and precipitation of the beta modification), and grain growth. The last step, although it may have less of an effect on the densification process, affects the development of the microstructure and thereby the final mechanical properties. The details of the process steps and the resulting properties have to be evaluated for the specific system. In the following, two examples are given where dilatometry was used as the main investigational tool.

Wotting et al. (82) carried out an investigation of the sintering of silicon nitride by dilatometry under high nitrogen pressure, which suppresses the dissociation of silicon nitride and enables the sintering step to be carried out at higher temperatures and thus widens the choice of additives. They employed nitrogen pressure of 50 bars and temperatures up to 2000°C, and used a home-built vertical dilatometer for which they adapted a commercial graphite resistance autoclave. Some parts of the dilatometer were made from boron nitride to withstand the high temperatures. The sensitivity of their linear measurement was about 0.2 μm and they were also able to record the rate of linear change. The samples were in the form of compressed powder pellets of 10 mm diameter and 5 mm height. Wotting et al. studied the densification of some different powders with several additives. The separation of process steps with one additive (Fig. 16a) is shown here, along with the effect of different additives on the same powder (Fig. 16b). In both instances the same powder was used. The additive in the first instance was 2 wt% of a Mg-Al-Si-O-N glass, containing 5.6% nitrogen.

In interpreting Fig. 16a in terms of the sintering mechanism, additional information about the glass should be noted. The glass has been prepared at 1400°C from a mixture containing 15% Si_3N_4, and it may be assumed to be saturated with Si_3N_4 up to this temperature. Its T_g was about 780°C. By DTA it was found that crystallization took place between 1000 and 1100°C, and a strong endothermic effect at 1300°C indicated that above this temperature only molten glass was present. This is consistent with the interpretation of the first, well-separated peak in the rate curve, at 1300°C, as representing only the rearrangement of the powder particles with the aid of the liquid phase. At about 1500°C the solution/precipitation step is initiated, with a peak at 1620°C. It cannot be decided whether dissolution or diffusion is the rate-determining step. Above 1700°C it is claimed that grain growth is the reason for the decrease in densification rate. In any case, under these special conditions clear separation between rearrangement and dissolution/precipitation could be made. In Fig. 16b the effect of 2 wt% MgO and 5 wt% of the

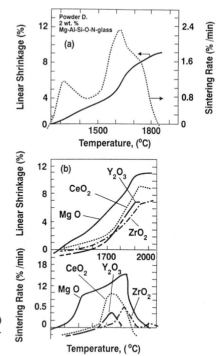

Fig. 16.   Sintering behavior of powder D with (a) 2 wt% Mg–Al–Si–O–N glass; (b) various other additives (82).

other oxides are compared. Only with MgO can there be seen a remnant of the former separation between the first and second peak. The differences in the densification rates and onset densification are related to the temperatures of liquid phase formation in the different systems. The authors show that the trend is consistent with literature data on temperatures of liquid formation.

Rapoport et al. (50) have studied hot-pressing of silicon nitride with mined rare-earth oxide additives, and followed the densification by recording the advancement of the ram and its rate. The emphasis was given to densification taking place during the heating-up stage, before arriving at the isothermal period, and comparison with isothermal hot pressing. They used hot-pressing under vacuum with a ram pressure of 40 MPa. The initial pressure was 4–5 MPa with heating to 900°C to remove volatile impurities, after which full pressure was applied with a heating rate of 15°C/min to the desired temperature, and then cooled or maintained for some time before cooling. Rapoport et al. measured final densities and three-point bending, and examined the samples by XRD and SEM. The densification rate curves for different additives showed in most cases the main peak rate occurs between 1600 and 1700°C, the process starting between 1430 and 1450°C, with a small earlier rate peak near 1330°C. The early peak is attributed to slight rearrangement of particles due to formation of some liquid phase and the second to the formation of rare earth silicon oxynitride liquid with the rearrangement–dissolution–precipitation. With additions of 6–10 wt%, densities of 3.25–3.32 g/cm^3 and flexural strengths of 800–950 MPa were mostly obtained. Alpha–beta conversion of over 80–85% was obtained by reaching 1800°C and cooling, full conversion with grain coarsening by an isothermal dwell at 1900°C, and 85% conversion by heating to the temperature of maximum densification rate with a dwell at this temperature (ca. 1650°C), which yielded a finer grain structure. Rapoport et al. (51) have used this technique in the study of hot-pressing of titanium nitride powders. The technique has been previously described by Brodhag et al. (8).

Bocker and Hausner (5) have made observations on the sintering characteristics of submicrometer silicon carbide powders using a high-temperature graphite dilatometer in an argon atmosphere with additions of carbon and boron. α-SiC was obtained from the Acheson process and was either attrition-milled with steel balls in gasoline using hydrochloric acid to remove the iron contamination followed by a treatment with hot hydrofluoric acid and alcohol, or just treated with hot HF and washed with isopropanol. A β-SiC was prepared by thermal decomposition of CH_3SiH_3 in a graphite reactor at 1600°C, using argon as a carrier gas. The impurity contents and specific surface areas of the powders and the final densities and microstructures after sintering were determined. The effect of the form of the boron addition on the densification and its rate of the β-SiC sample is shown in Fig. 17 and a

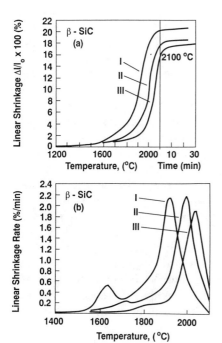

Fig. 17. Densification (a) and densification rate (b) of β-SiC (5).

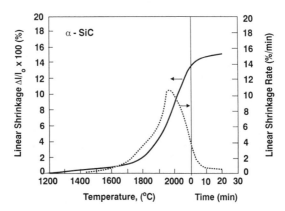

Fig. 18.   Densification and densification rate of α-SiC. Final density is 92.8% T.D. (5).

typical curve of an α-SiC sample in Fig. 18. When β-SiC was sintered above 2050°C, large plate-like crystals appeared in a fine-grained matrix.

VII.   FEEDBACK DILATOMETRY

Conventional thermal analysis uses dynamic temperature programming or ramp and dwell combinations. In all of these programming methods the temperature program is predetermined and controls the process in an open loop; i.e., the temperature moves according to the preset program, irrespective of the effect of that program on the sample. An alternative approach is the closed loop control, in which the response of the sample is being measured and the result fed back to vary the control parameters, e.g., heating rate. The term *feedback dilatometry* (FBD) used in the title of this section, refers to thermaldilatometric experiments in which the temperature/time profile is not predetermined by the experimenter, but results from a preprogrammed time profile of the process rate. This approach requires suitable programming and possibly modification of the equipment. Quasi isothermal dilatometry (QID) and rate-controlled sintering (RCS), described below, employ this approach. In fact, feedback control can be explored in other modes of thermal analysis. An example of parallel feedback dilatometry and gravimetry is also described.

The term feedback is used to avoid some ambiguity revealed in the use of the term quasi-isothermal, which has been used to designate either constant rate-controlled or stepwise-isothermal processing. Wendlandt (80) uses the term QID and quotes papers by Paulik and Paulik (47). Sorensen (57) has a quite different method of implementation. These will be described first.

A. QUASI-ISOTHERMAL METHODS

Paulik and Paulik (47) describe their procedure and equipment with examples of the formation of zinc ferrite, the heating of $BaCl_2 \cdot H_2O$, and thermal decomposition of sodium oxalate monohydrate. The compressed powder sample is situated between a stationary quartz rod fixed to the column of the balance, and another quartz tube fixed to the balance arm. A differential transformer measures the change in length, and a magnet and coil produce a signal proportional to the derivative of the elongation signal. This signal can be fed directly to the recorder for a normal dynamic program or through a sensor controller to the heating programmer for the QID program. The sensor controller reduces the voltage to the heating current as soon as the rate of the process reaches a minute threshold value and immediately increases it when the rate drops below this value. In this way a temperature difference between furnace and sample is established such that the process examined takes place at a minute and constant rate. In effect, this is then a rate-controlled process. The formation of zinc ferrite results from the solid state reaction between ZnO and Fe_2O_3. The compressed pellet expands in length between 570 and 700°C by about 2.7%, followed by about a 5% shrinkage between 700 and 1000°C. Some reaction starts at around 300°C apparently only on the surface of the grains. Only above 570°C does volume diffusion play a dominant role, first in the formation of the spinel structure, accompanied by expansion, and then shrinkage due to recrystallization of the spinel. A comparison of curves indicated that the heating rate strongly affects the rate of spinel formation and only slightly affects the recrystallization process.

A unique feature of the equipment is that identical samples can be examined either by dilatometry or thermogravimetry. In the TG run the sample does not lean against the rod so that the balance is free to record the weight. Figure 19 compares the results of dynamic and QITD and TG experiments on $CaC_2O_4 \cdot H_2O$. The advantage of the QI program in more sharply defining the individual processes is evident. The following observations can be made:

1. The dehydration process in the QID run took place at a constant temperature of 150°C, showing that under these conditions the water vapor pressure was maintained constant in the vicinity of the sample. Moreover, it also proves that the progress of the transformation is influenced solely by the heat transfer, independent of the extent of reaction; i.e., the dehydration is a zero-order process. The expansion during this process is an unexpected result, still unexplained.

2. The second reaction, that of the decomposition of calcium oxalate to produce $CaCO_3$ and CO, appears in the QITG curve as a single process, occurring sharply at 380°C and accompanied by a sharp

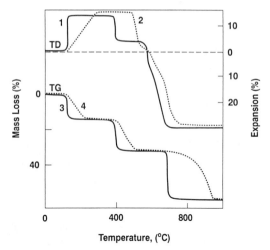

Fig. 19.   TD and TG curves of $CaC_2O_4 \cdot H_2O$. Curves 1 and 3 obtained under quasi-isothermal and curves 2 and 4 under dynamic heating conditions (47).

    shrinkage. But a second shrinkage takes place at a significantly higher temperature of 570–700°C, apparently consisting of two poorly separated steps that are practically complete before the beginning of the thermal decomposition of the $CaCO_3$. The authors suggest recrystallization of calcite as the explanation of this observation.

3.   The decomposition of calcite takes place at constant temperature in the QITG runs at about 700°C without any dimensional change. It is quite possible that due to the slow decomposition of $CaCO_3$, the CaO formed maintained the pseudomorphic form of the calcite without recrystallization to portlandite or sintering below the high temperature of the experiment.

In this mode of operation of QID or QITG the experiment was actually run under rate-controlled processing conditions. The slow rate prolonged the experiment by at least an order of magnitude, having the advantage of maintaining close to equilibrium conditions. The actual rate of length change in this experiment is not stated by the authors but can be estimated to be about 0.02 %/min.

    Sorensen (57, 58) and Ali et al. (1) used a different method QID in their sintering studies. The compacted specimen was heated at constant rate until the dl/dt signal, which was proportional to the shrinkage rate, became larger than a preset limit, at which point the heating was stopped. Shrinkage now continued until the dl/dt signal again became smaller than a second preset limit, at which point the heating was resumed. This procedure was used in a dilatometric study of the influence of CO_2 content on the initial stage of sintering of UO_2 compacts in H_2. From the slope of the straight line in the shrinkage rate against shrinkage a mean value of n, a constant characteristic of the sintering mechanism, of about 0.3 was computed, indicating a grain boundary diffusion-controlled mechanism. From the slope of an Arrhenius type plot of the data, the activation energies were calculated to be 234 kJ/mol for the two different partial pressures of CO_2 used, compared to 390 kJ/mol for pure H_2 atmosphere, obtained in a former study. The authors conclude that the initial sintering stages for slightly nonstoichiometric $UO_{(2+x)}$, which must be expected in the CO_2/H_2 atmosphere used, is controlled by uranium vacancy diffusion along the grain boundaries formed.

B. RATE-CONTROLLED SINTERING

The concept of rate-controlled sintering (RCS) was first proposed by Palmour and Johnson in 1967 (41). The concept and applications have been explored and developed during the last two decades. The first experiments were conducted on a modified TG apparatus (27), but later a purpose-built apparatus with feedback capability using computer control was constructed (4). Recent papers by Palmour and Hare (44) and Palmour (45) summarize the work and concepts involved.

RCS, as generally understood, refers to a sintering process carried out in a preprogrammed densification profile. In a dilatometer this can be accomplished using the measured densification rate signal to control the power input into the furnace so as to follow the preprogrammed rate path (and by integration, the densification path). In general, the program can be arbitrarily chosen to be constant or varying rate. However, RCS as evolved by Palmour et al., involves a more specific, "generic" rate program, empirically optimized as a sintering route.

Figure 20 illustrates the difference between conventional temperature sintering (CTS) and RCS (27). Figure 20a represents a typical CTS temperature/time program calling for a heating rate of 8°C/min followed by a dwell (soak) of 1.5 h at 1550°C, yielding a fractional density of at least 0.99. Figure 20c is the resultant, and dependent, density/time curve determined by dilatometry. It can be seen that the densification proceeds rapidly throughout the initial and intermediate stages and slows down appreciably above $D = 0.9$ even before the soak temperature is reached, so that the additional shrinkage during the long high-temperature soak is very modest. This firing profile has some undesirable consequences, and assists in (1) the entrapment of occluded gases; (2) the entrapment of pores within grains, (3) excessive grain growth.

The RCS procedure is shown in Fig. 20b and d: (b) is the preprogrammed fractional density D vs. time curve, and (d) is the resulting (dependent) temperature/time plot. Visual inspection reveals the significant difference between the corresponding pairs of temperature profiles, ((a) vs. (d)) and densification profiles ((c) vs. (b)). The near optimal D/time path shown in

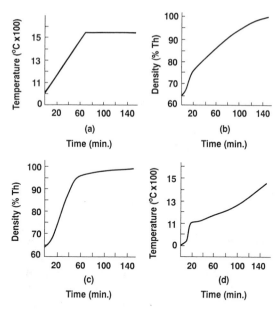

Fig. 20. Comparisons between CTS and RCS firings for chemically prepared 0.1% doped alumina at 0.65 green density: (*a*) CTS temperature–time control profile; (*b*) RCS temperature–time control profile; (*c*) resultant CTS density–time profile; (*d*) resultant RTS density–time profile (27).

(*b*), developed and patented (43), comprises three segments: (1) a relatively fast, constant densification rate, to $D = 0.75$ ($dD/dt = k_1$); (2) a significantly slower constant rate to $D = 0.85$ ($dD/dt = k_2$), approximately $k_1/3$; and (3) a decreasing rate leading to a logarithmic densification with time above $d = 0.85$ ($dD/dt \propto k_3/t$, $D \propto \ln t$). The authors caution that the actual curve is quite specific to a given material and process, but mention that this type of optimized curve has been found to hold for several dissimilar materials. The deliberately slow traverse of the permeable intermediate stage of densification provides a better chance for outgassing. Keeping the temperature as low as possible throughout the RCS firing tends to minimize thermal excitation of undesirable transport processes, which may accompany but not necessarily aid densification (e.g., diminution of surface, pore coarsening, and grain growth). The RCS firing produces finer microstructure. After refining the RCS procedure experimentally, the resulting temperature/time profile can, in principle, be adapted to use in a conventional furnace system.

Further work with the computerized dilatometer (4) led to the development of an integrated software package, which enables experiments to be carried out in an ordinary or feedback mode. From this an extensive data base has been compiled, which can be used to compute a "model" of the sintering behavior of a material and produce a control file for temperature of density path sintering. Refinements of the procedures can then be made by examination of the actual results of these predicted sintering experiments.

REFERENCES

1. Ali El-Sayed, M., O. T. Sorensen, and L. Halldahl, *J. Therm. Anal.* **25**, 175 (1982).

2. ASTM Designation C336-69, *Standard Method of Test for Annealing Point and Strain Point of Glass, Part 13*, 288 (1969).

3. ASTM Designation C338-57, *Standard Method of Test for Softening Point of Glass, Part 13*, 294 (1969).

4. Batchelor, A. D., M. J. Paisley, T. M. Hare, and H. Palmour, *Mater. Sci. Res.* 237 (1984).

5. Bocker, W., and H. Hausner, *Sci. Ceram.* **9**, 168 (1977).

6. Bordia, R. K., and R. Raj, *Adv. Ceram. Mater.* **3**, 122 (1988).

7. Brinker, J. C., D. E. Clark, and D. R. Ulrich, Eds., *Better Ceramics Through Chemistry*, Amsterdam: North Holland, 1984, p. 25.

8. Brodhag, C., M. Bouchacourt, and F. Thevenont, *Silicates Industriele* **4–5**, 91 (1981).

9. Claussen, N., K. L. Weisskopf, and M. Fuhle, *J. Am. Ceram. Soc.* **69**, 288 (1986).

10. Clusner, G. R., *Thermal Expansion*, Proc. 3rd. AIP Conf., New York: American Institute of Physics, 1972, p. 51.

11. Collier, P. W., *J. Am. Ceram. Soc.* **30**, 338 (1947).

12. Daniels, T., *Anal. Proc.* 412 (Oct. 1981).

13. De Jonghe, L. D., and M. N. Rahaman, *Rev. Sci. Instrum.* **55**, 2007 (1984).

14. Dworak, U., H. Olapinski, and W. Burger, *J. Am. Ceram. Soc.* **69**, 576 (1986).

15. Dworak, U., and D. Fincerle, *Br. Ceram. Trans. J.* **86**, 170 (1987).

16. Elmer, T. H., *Ceram. Bull.* **62**, 513 (1983).

17. Fischer, R., *Crystalline Phases Formed During Firing in High Alumina Refractories*, Israel Ceramic and Silicate Institute Proj. 1003, Thesis, Technion IIT, 1977.

18. Ford, R. W., *The Effect of Heat on Ceramics*, London: MacLaren, 1967.

19. Graham, M. G., and H. E. Hagy, Eds., *Thermal Expansion*, Proc. 3rd. AIP Conf., New York: American Institute of Physics, 1972.

20. Hagy, H. E., *J. Am. Ceram. Soc.* **46**, 95 (1963).

21. Hagy, H. E., Rheological behavior of glasses, in L. D. Pye, H. J. Stevens, and W. C. Lacourse, Eds., *Introduction to Glass Science*, New York: Plenum, p. 343.

22. Heckman, R. C., and P. M. Elrick, *Int. Symp. on Therm. Expansion of Solids*, Univ. of Connecticut, Storrs, Connecticut, June 4–6, 1975.

23. Hench L. L., and D. R. Ulrich, Eds., *Ultrastructure Processing of Ceramics, Glasses and Composites*, New York: Wiley, 1984, p. 43.

24. Hench, L. L., and D. R. Ulrich, Eds., *Science of Ceramic Chemical Processing*, New York: Wiley, 1986, p. 37.

25. Hlavac, J., *The Technology of Glass and Ceramics: An Introduction*, Amsterdam: Elsevier, 1983.

26. Hormadaly, J., *J. Non-Crystall. Solids* **79**, 311 (1986).

27. Huckabee, M. L., T. M. Hare, and H. Palmour, in H. Palmour, R. F. Davis, and T. M. Hare, Eds., *Material Science Research: Processing of Crystalline Ceramics*, New York: Plenum, 1978, p. 205.

28. Huttig, G., *Z. Angew. Chem.* **49**, 882 (1936); *Z. Anorg. Allgem. Chem.* **224**, 225 (1935).

29. Kingery, W. D., H. K. Bowen, and D. R. Uhlmann, *Introduction to Ceramics*, 2d ed., New York: Wiley, 1976.

30. Lawrence, W. G., *Ceramic Science for the Potter*, Philadelphia: Chilten, 1972.

31. Lynch, J. F., G. Ruderer, and W. H. Duckworth, Eds., *Engineering Properties of Selected Ceramic Materials*, Westerville, OH: The American Ceramic Society, Inc., 1966.

32. Matveev, M. A., G. M. Matveev, and B. N. Frenkel, *Calculations and Control of Electrical, Optical and Thermal Properties of Glass*, Holon, Israel: Ordentlich, 1975.

33. Mazurin, O. V., A. S. Totesh, M. V. Sterl'tsina, and T. P. Shvadjko-Shvadjkovskaya, *Heat Expansion of Glass*, Leningrad: Academy of Sciences, 1969.

34. Moitron, J., and J. Kletenik, Report No. RCO-7-008-Oct., 1960, Ceramic Research Association (Israel), Technion City, Haifa, Israel.

35. Moitron, J., *Bull. Ceram. Res. Assoc.* (*Israel*) **11**, 16 (1960).

36. Morev, G. W., *The Properties of Glass*, 2d ed., New York: Reinhold, 1954.

37. Morrell, R., *Anal. Proc.* 443 (Oct. 1981).

38. Norton, C. L., *J. Am. Ceram. Soc.* **14**, 192 (1931).

39. Norton, F. H., and Hodgdon, F. B., *J. Am. Ceram. Soc.* **14**, 177 (1931).

40. Norton, F. H., *Fine Ceramics Technology and Applications*, New York: McGraw-Hill, 1970, p. 507.

41. Palmour, H., and P. P. Johnson, in G. C. Kuczynski, N. A. Hooten and C. F. Gibbon, Eds., *Sintering and Related Phenomena*, New York: Gordon & Breach, 1967, p. 779.

42. Palmour, H., and M. L. Huckabee, *Mater. Sci. Res.* **6**, 275 (1973).

43. Palmour, H., and M. L. Huckabee, *Process for Sintering Finely Divided Particulates and Resulting Ceramic Products*, U.S. Patent 3,900,542, 19 Aug 1975.

44. Palmour, H., and T. M. Hare, in G. C. Kuczyski, D. R. Uskovich, H. Palmour, and M. M. Ristic, Eds., *Sintering '85*, New York: Plenum, 1987, p. 17.

45. Palmour, H., *Metal Powder Reports* 573 (Sep. 1988).

46. Paulik, F., and J. Paulik, *The Analyst* **103**, 417 (1978).

47. Paulik, F., and J. Paulik, *J. Therm. Anal.* **16**, 397 (1979).

48. Pearson, W. B., *A Handbook of Lattice Spacings and Structures of Metals and Alloys*, New York: Pergamon, 1958 (Vol. 1), 1967 (Vol. 2).

49. Plummer, W. A., Differential dilatometry, in J. Valentich, Ed., *Tube Type Dilatometers*, Research Triangle Park, NC: Instrument Society of America, 1981.

50. Rapoport, E., C. Brodhag, and F. Thevenot, *Rev. Chim Minerale* **22**, 456 (1985).

51. Rapoport, E., C. Brodhag, and F. Thevenot, *Rev. Chim Minerale* **24**, 697 (1987).

52. Ritland, H. N., *J. Am. Ceram. Soc.* **37**, 370 (1954).

53. Roosen, A., and H. Hausner, *Adv. Ceram.* 714 (1984).

54. Salmang, H., *Ceramics: Physical and Chemical Fundamentals*, New York: Butterworths, 1961 (Engl. trans.).

55. Scherer, G. W., *Relaxation in Glass and Composites*, New York: Wiley, 1986, p. 331.

56. Sletson, L. C., and J. S. Reed, *Am. Ceram. Soc. Bull.* **67**, 1403 (1988).

57. Sorensen, O. T., in H. G. Wiedemann, Ed., *Thermal Analyses, Proc. 6th. ICTA*, Vol. 1, Basel: Birkhauser, 1980, p. 231.

58. Sorensen, O. T., *Thermochim. Acta* **29**, 211 (1979).

59. Steger, W., *Ber. Dtsch. Keram. Ges.*, **9**, 203 (1928): **11**, 124 (1930).

60. Stutzman, R. H., J. R. Selvaggi, and H. P. Kirchner, *Summary Report on an Investigation of the Theoretical and Practical Aspects of the Thermal Expansion of Ceramic Materials, Literature Survey*, Vol. 1, Cornell Aeronautical Laboratory Inc., 31 Aug. 1959.

61. Taylor, D., *Br. Ceram. Trans. J.* **83**, 5 (1984).

62. Taylor, D., *Br. Ceram. Trans. J.* **83**, 32 (1984).

63. Taylor, D., *Br. Ceram. Trans. J.* **83**, 92 (1984).

64. Taylor, D., *Br. Ceramic. Trans. J.* **83** 129 (1984).

65. Taylor, D., *Br. Ceram. Trans. J.* **84**, 9 (1985).

66. Taylor, D., *Br. Ceram. Trans. J.* **84**, 121 (1985).

67. Taylor, D., *Br. Ceram. Trans. J.* **84**, 149 (1985).

68. Taylor, D., *Br. Ceram. Trans. J.* **84**, 181 (1985).

69. Taylor, D., *Br. Ceram. Trans. J.* **85**, 111 (1986).

70. Taylor, D., *Br. Ceram. Trans. J.* **85**, 147 (1986).

71. Taylor, D., *Br. Ceram. Trans. J.* **86**, 1 (1987).

72. Taylor, D., *Br. Ceram. Trans. J.* **87**, 39 (1988).

73. Taylor, D., *Br. Ceram. Trans. J.* **87**, 88 (1988).

74. Taylor, J. R., and A. C. Bull, *Ceramics Glaze Technology*, New York: Pergamon, 1986.

75. Tool, A. D., *J. Am. Ceram. Soc.* **29**, 240 (1946).

76. Touloukian, Y. S., R. K. Kirby, R. E. Taylor, and T. Y. F. Lee, Eds., *Thermophysical Properties of Matter*, Vol. 13, New York: Plenum, 1978.

77. Valentich, J., *Tube Type Dilatometers*, Research Triangle Park, NC: Instrument Society of America, 1981.

78. Venkatachari, K. R., and R. Raj, *J. Am. Ceram. Soc.* **69**, 499 (1986).

79. Wendlendt, W. W., *Thermal Methods of Analysis*, 2d ed., New York: Wiley-Interscience, 1974.

80. Wendlandt, W. W., *Thermal Methods of Analysis*, 3d ed., New York: Wiley, 1986, Chapter 11.

81. Wood, W. D., and H. W. Deem, *Thermal Properties of High-Temperature Materials*, U.S. Army Material Command Report RSIC-202, 1964, p. 399.

82. Wotting, G., R. Peitzsch, and H. Hausner, *Sci. Sintering*, **1–2**, 87 (1985).

Chapter 5

PYROLYSIS TECHNIQUES

By William J. Irwin, *Drug Development Research Group, Pharmaceutical Sciences Institute, Aston University, Aston Triangle, Birmingham B4 7ET, UK*

Contents

Treatise on Analytical Chemistry, Part 1, Volume 13,
Second Edition: Thermal Methods, Edited by James D. Winefordner.
ISBN 0-471-80647-1   © 1993 John Wiley & Sons, Inc.

I. INTRODUCTION

The existence of complex, nonvolatile organic molecules as either major components of the biosphere or as synthetic polymers has necessitated the development of effective, degradative analytical techniques. One of the earliest of these was pyrolysis, in which the sample is subjected to thermal fragmentation to produce a range of smaller molecular species. These products are more amenable to separation, identification, and quantification than the original macromolecule and enable a characteristic profile of the sample to be obtained. Modern systems generally use gas chromatography or mass spectrometry as the analytical device, and the chromatogram (pyrogram) or mass spectrum (mass pyrogram) of the pyrolysis products may generate qualitative information on the identity or composition of the sample, quantitative data on its constitution, or allow mechanistic and kinetic studies of thermal decomposition and fragmentation to be undertaken. When used in this way the technique is referred to as analytical pyrolysis (70) to distinguish it from applied pyrolysis methods which aim to isolate the products of pyrolysis for use as raw materials. The various stages in the technique may be summarized as in Table 1.

Typically, the sample, which may be a synthetic polymer, biological and forensic samples, whole microorganisms, or soil or geochemical fractions, is loaded into the pyrolyzer, usually onto a wire or filament; ideally, in very small (approximately microgram) amounts. The pyrolyzer is rapidly heated to some 600–800°C, which fragments the macromolecules into smaller, volatile fragments. These pyrolysis products are then separated, usually by GC or MS to provide a characteristic profile of the original macromolecular sample (Fig. 1).

The first reported application of analytical pyrolysis was in 1860 when Williams (209) isolated isoprene and dipentene from the pyrolysis of rubber, results that heralded the elucidation of this substance as an isoprene polymer. Pyrolyses were only infrequently reported thereafter (125, 177) due to the inability of contemporary analytical equipment to deal with the complex mixture of pyrolysis products. This situation changed dramatically when the modern era of pyrolysis was heralded in 1948 with the use of mass spectrometry to analyze pyrolysis product mixtures by Madorsky and Straus (108) and Wall (199). In 1952 this technique was proposed as a fingerprint identification method for synthetic polymers by Zemany (216). Pyrolysis–mass spectrometry (Py-MS) followed one year later with the integration of pyrolysis and analysis by Bradt and coworkers (23) when pyrolyses were undertaken within the mass spectrometer to enable characteristic but less stable products to be analyzed. Developments elsewhere, however, eclipsed the further development of Py-MS in the short term.

The main competitor was gas chromatography, which could separate and quantify a complex mixture of pyrolysis products. This technique was adopted

TABLE 1
Summary of the Various Stages in Analytical Pyrolysis

Stage	Objective
PROBLEM?	
↓	
Sample handling	Preparation of sample and loading into pyrolyzer unit in reproducible way
↓	
Pyrolysis	Application of controlled heat to sample to degrade it to smaller fragments in a reproducible way
↓	
Transference	Movement of pyrolysis fragments from vicinity of pyrolyzer to analytical device without loss or further reaction
↓	
Separation	Efficient separation of the various reaction products
↓	
Detection	Detection of all components of the mixture
↓	
Quantification	Measurement of the relative abundances of the various pyrolysis products
↓	
Identification	Reliable identification of the pyrolysis fragments
↓	
Data handling	Mathematical or statistical analysis of the quantitative data to give a basis for reliable comparisons
↓	
SOLUTION!	

by Davison et al. (39) to characterize rubbers and was integrated into the combined pyrolysis–gas chromatography technique (Py-GC) when Radell and Strutz (143), Lehrle and Robb (90), and Martin (109) reported viable systems in 1959. The technique of pyrolysis–gas chromatography–mass spectrometry (Py-GC-MS) to enable reliable identification of pyrolysis products was heralded two years later by Barlow et al. (11) and introduced as an integrated system by Simon et al. in 1966 (195).

In parallel with these advances in analysis, progress in pyrolysis design was evident. The basic requirements of a pyrolyzer, including rapid heating rates and small sample size, were outlined by Jones and Moyles (72) while the Curie-point pyrolyzer was introduced by Giacobbo and Simon in 1964 (52), the boosted heated filament system by Levy in 1967 (97) and the laser system by Folmer and Azarraga in 1969 (46).

Developments in mass spectrometry and data handling led to a resurgence of Py-MS, largely as a result of Meuzelaar and Kistemaker (116), the major

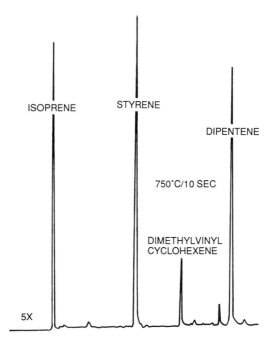

Fig. 1. Pyrogram from the pyrolysis at 750°C for 10 s of an isoprene–styrene copolymer showing monomers and two isoprene dimers. Diagram kindly supplied by Chemical Data Systems, Oxford, PA.

advantage being in analysis time with, typically, 1 min per sample for Py-MS compared with up to perhaps 90 min for Py-GC. Automation soon followed (119) and in 1976 the ultimate, although expensive, automated Py-GC-MS system was in action on the surface of Mars as part of the Viking program (16).

The pyrolysis literature is served by a comprehensive monograph (70), a slim volume on Py-GC specialising on forensic aspects (112), an atlas and compendium of Py-MS pyrograms (121), and a specialized journal (74). Additionally, an international symposium meets every other year and alternates with a biennial Gordon Conference held in New Hampshire (89).

II. PYROLYSIS METHODS

The ease of construction and integration with GC systems led to many varied systems being reported in the earlier phases of analytical pyrolysis. It is now appreciated that considerable control must be exercised at all stages of the technique (Table 1) to ensure adequate reproducibility.

TABLE 2
Effect of Pyrolysis Temperature on Product Distribution
from Poly(Methyl Methacrylate)

Pyrolysis Temperature (°C)	Methanol : Methyl Methacrylate Peak Height Ratio
485	0.78
590	1.48
740	20.5

SOURCE: Fujita and Mizuki (49).

A. CONTROL OF PYROLYSIS

The first consideration of pyrolysis is the temperature at which the pyrolysis is undertaken. When a sample is subjected to a high temperature, bond fission is initiated. Several degradation reactions will generally be possible, each with a different activation energy and temperature profile. To provide reproducible pyrolysis products, a precisely controlled pyrolysis temperature must be used. This temperature is frequently known as the final pyrolysis temperature, but is more properly referred to as the equilibrium temperature T_{eq}.

The effect of temperature may be illustrated by the ratios of methanol to methyl methacrylate following pyrolysis of poly(methyl methacrylate) at different temperatures, shown in Table 2. The major product at the lower temperature is hardly detectable at the higher temperature (49). As a general rule, smaller, less characteristic fragments are favored at higher pyrolysis temperatures. The effect on the pyrogram of poly(ethylene) is displayed in Fig. 2 (201). The α,w-diene is eluted first, followed by the alkene and finally the n-hydrocarbon. The relative yield of diene to alkane depends significantly on the pyrolysis temperature, with diene linearly increasing with temperature.

Because temperature may exert a profound effect on product distribution, the heating profile (temperature–time profile, TTP) of the pyrolysis unit is also of great importance. This arises because pyrolysis reactions occur very rapidly. Teflon, for example, a very stable polymer, has an estimated half-life of 26 ms at 600°C. This means that substantial degradation may occur while the sample is being heated to T_{eq}. To ensure that reproducible pyrolysis is achieved, the TTP should also be highly defined and the temperature rise time (TRT) should be as small as possible. In modern systems the TRT may be reduced to some 8 ms or better. The effect of TRT is illustrated in Table 3 for the degradation of an isoprene–styrene (68:32) copolymer which produces the monomers on pyrolysis (198). The effect of heating rates on the pyrogram of poly(ethylene) is illustrated in Fig. 3. This parallels the earlier illustrated effect of temperature with an increase in alkane component as the heating rate is reduced.

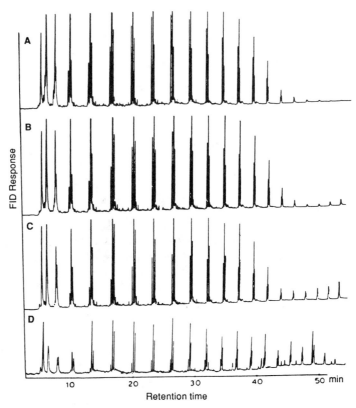

Fig. 2. Effect of temperature on the pyrolysis of poly(ethylene). (*A*) 1000°C; (*B*) 900°C; (*C*) 800°C; (*D*) 700°C. Reproduced from *J. Anal. Appl. Pyrol.* **8**, 154(1985) (201).

TABLE 3
Effect of Temperature Rise Time on Product Distribution
from an Isoprene–Styrene (68 : 32) Copolymer

Temperature Rise Time	Styrene : Isoprene Peak Height Ratio
3 s	2.0
30 ms	1.11

SOURCE: Walker (198).

Depending on the T_{eq} selected, the TRT may allow total pyrolysis to occur nonisothermally during the heating phase or allow significant pyrolysis both during heating and when the equilibrium temperature has been reached. Such competing processes may provide a ready source of pyrogram variability. Additionally, variability in the total heating time (THT) may result in variable volatilization of fragments from the pyrolyzer to the analyzer, and

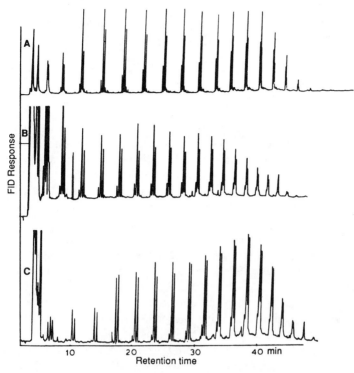

Fig. 3. Effect of heating rate on the pyrolysis of poly(ethylene). (*A*) 60°C/min; (*B*) 30°C/min; (*C*) 5°C/min. Reproduced from *J. Anal. Appl. Pyrol.* **8**, 154 (1985) (201).

high, uncontrolled temperatures may also cause secondary pyrolysis of the primary fragments. Again, such processes endanger reproducibility. For similar reasons, the cooling of the pyrolyzer unit after removal of the energy source should be as rapid as possible. Units with low thermal mass minimize secondary pyrolysis and evaporation during the cooling stage. A half square-wave heating profile is the ideal for pyrolysis performance with heating and cooling as rapid as possible so that the time during which the sample is subjected to T_{eq} is close to the total heating time.

The pyrolyzer must also be interfaced with the analytical device so that loss of volatile samples is reduced to a minimum with no opportunities for adsorptive loss, condensation, or secondary reaction. In Py-GC, pyrolysis should be as close to the column as is possible, whereas with Py-MS, pyrolysis directly in the ion source is ideal.

The final control to ensure reproducible pyrolysis is that of sample size. The temperature dependence of pyrolysis pathways means that the entire sample should experience the same heating profile. The temperature of the sample should follow that of the pyrolyzer unit and is best achieved with intimate contact of sample and wire. Further, the sample should not be

TABLE 4

Effect of Sample Thickness on Extent of Pyrolysis of Poly(Methyl Methacrylate)

Sample Thickness (nm)	160	130	100	96	64	28	24	16
Degradation in 10 s (%)	8	20	27	32	42	47	47	47

SOURCE: Lehrle and Robb (91).

subjected to a temperature gradient through its thickness and therefore very thin samples should be used. Table 4 illustrates this for the degradation of poly(methyl methacrylate) at 375°C (91). It is not until the sample thickness falls to less than 30 nm that reproducible levels of pyrolysis are achieved.

A further disadvantage of thick films is that in a typical configuration the primary pyrolysis products, produced close to the surface of the pyrolysis wire, must diffuse through nondegraded sample before they can escape to the analyzer. This provides scope for uncontrolled, competitive secondary reactions which may threaten pyrogram reproducibility. For example, pyrolysis of a styrene–divinyl benzene (96:4) copolymer at 550°C produces only styrene and toluene from 100 μg of sample, but many other products, including benzene, toluene, and methyl styrene, are formed by secondary reactions when 650 μg is used (167). With sample sizes of the order of 1 μg the temperature–time profile of the sample closely follows that of the pyrolysis unit (45) and loadings in the 5- to 60-μg range are recommended (202), although in practice many reported pyrolyses vastly exceed this loading while providing useful results. One final point of control should be recognized. In even the best systems the ends of the pyrolysis filament are cooler than the center regions, so samples should be placed centrally on the wire and not over its complete length.

Many pyrolyzers, not all of which have used the principles outlined above, have been described in the literature (70, 96, 112) and the commonest systems have used heated filaments, inductive heating, or furnace methods. With each of these it is possible to modify the basic design to ensure adequate performance. Most systems have been developed initially for use in combination with gas chromatography (Py-GC), with pyrolyses being undertaken directly in the carrier gas stream so that volatile products are immediately swept onto the column for separation and identification. More detailed accounts of reproducibility, sample loading and comparisons of pyrolyser units may be found elsewhere (3, 42, 70, 190, 197).

B.   HEATED-FILAMENT PYROLYZERS

In the heated-filament design the temperature increase is achieved by passing an electric current through a resistive wire or ribbon. If operated at a fixed voltage, such systems have very long TRTs (low voltage, 10–30 s); at higher voltages, with TRTs near 1 s, very high temperatures may be generated, which may melt the conductor. To provide control over the heating program, boosted filament systems have been introduced. These provide an initial surge of power, provided by a high current (91) or a capacitor-discharge system (97, 98), to establish a short TRT. Feedback control reduces power to a maintenance current when T_{eq} is reached. A commercial system using the capacitor-discharge principle is the Pyroprobe 120-123 (203) which uses feedback control to define the temperature program very closely. The platinum pyrolysis filament operates as the sample holder, the heating element, and the temperature sensor. This last property is achieved through the increase in resistance of the filament, as the temperature rises, which is detected through a Wheatstone bridge comparator. With the bridge unbalanced, heat is initially supplied as a full surge of power through the filament. This provides a rapid increase in temperature (short TRT), which increases filament resistance and causes the bridge to be nearer the balance point. The reduced bias is used to reduce power to the heater and thus control the heating profile. The system is capable of controlled variable heating rates (5°C/min to 20°C/ms) with a maximum of 75°C/ms to provide claimed TRTs of 8 ms to 600°C and 17 ms to 1000°C. The pyrolysis temperature may be set from ambient up to 1400°C at 1°C intervals, and variation in pyrolysis time from 20 ms to 240 s is also possible. Over these heating periods a two-phase temperature rise may occur and TRTs of 600 ms are more typical (203). To allow a range of samples to be handled, various designs of probe are also available. These range from the standard filament through several coil probes for holding sample in a quartz boat to a direct insertion probe for mass spectrometry.

In this system, temperatures quoted are average values over the whole filament. For mechanistic work it is essential to provide even greater control over the temperature of reaction and a more refined feedback system has been reported (92). In essence, the temperature of a boosted filament, produced by a current of some 30 A through a resistance of 0.5 Ω, is measured near its central point by means of a very fine thermocouple spot-welded to the rear of the filament. The thermocouple output is compared to that of a reference voltage preset to correspond to the required T_{eq}. As the temperature increases and the two voltages converge, the filament current is reduced rapidly to a maintenance level. TRTs of 50 ms to 800°C were achieved in general use, although 15 ms to 500°C was achievable with less robust filaments ($R = 0.12 \ \Omega$). Cooling periods were identified as the limiting factor in achieving the ideal square-wave thermal profile. The pyrolysis unit, which provides elegant control over precise pyrolysis temperature, is illustrated in Fig. 4.

Several other attempts to optimize pyrolyzer performance have been made, with the system introduced by Tyden-Ericsson (192) being among the best. This unit uses two half-sequential square-wave voltage pulses. The first, which controls the TRT, is variable from 8 to 80 ms and induces a rapid heating current of up to 60 A. The second, of much lower power, maintains T_{eq} and may be set for up to 1 min. The precise operating conditions are

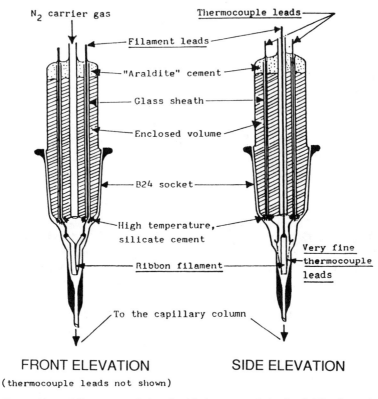

FRONT ELEVATION   SIDE ELEVATION
(thermocouple leads not shown)

Fig. 4.   Boosted heated filament pyrolysis unit with thermocouple feedback. The chromel–alumel thermocouple wires attached to the nichrome ribbon have a small thermal capacity (d, 25 μm) with a fast response time. When mounted on the GC, the chamber is surrounded by a heater jacket. Diagram kindly supplied by R. S. Lehrle, University of Birmingham, UK.

measured by means of a photodiode, which monitors filament temperature; a replaceable pyrolysis cell is also included.

The heated filament pyrolyzers are perhaps the most versatile of the units available, for they also enable stepped pyrolyses to be undertaken. Here, the same sample is repetitively subjected to a series of increasing temperatures. This provides information on the thermal stability of the sample and may provide extra data for identification purposes or yield valuable data for kinetic analysis of polymer degradation (11, 82).

C. CURIE-POINT PYROLYZERS

Curie-point pyrolyzers use the principle of induction heating of a ferro-magnetic wire and were introduced in the anticipation that this system would provide the required heating profile with automatic temperature control (25, 136, 155). A ferromagnetic wire is placed in a radio-frequency oscillator,

TABLE 5
Curie Points of Various Ferromagnetic Alloys

Curie Point (°C)	Alloy Composition (%)			
	Ni	Fe	Co	Cr
358	100			
400	45	55		
400		61.7	38.3	
480	51	48		1
510	49.4	50.6		
590	60	40		
600	41	42	16	
610	70.8	29.2		
700	67		33	
700	33	33	33	
770		100		
800	55		45	
900	40		60	
1128			100	

SOURCE: Levy et al. (98).

typically 500 kHz to 1.2 MHz, which induces an alternating magnetic flux in the conductor. Hysteresis losses and eddy currents cause a rapid increase in temperature. This continues until the Curie point of the material is reached. At this temperature, specific for various alloys (Table 5), a transition from ferro- to paramagnetism occurs, the conductor no longer interacts with the oscillating field, and the temperature is held constant at this value. The system has versatility in that the temperature control may be effected by using alloys of different compositions and sample holders may be of any desired shape including wires, filaments, boats, tubes or a folded foil. A typical setup is illustrated in Fig. 5.

A disadvantage of these units is the inability to choose a precise pyrolysis temperature or to undertake stepped pyrolysis. Also, a poorer description of the thermal profile of the filament is generally available. In particular, the shape of the sample holder and the power of the rf field exert major influences. For example, a 30-W unit provides a TRT of 500 ms to 600°C and 1.35 s to 700°C, whereas a more powerful 1.5-kW system affords TRTs of 70 and 90 ms, respectively (38), and different rf frequencies require different diameter wires for optimal performance. A solvent-free loading method involves compression of wires, which markedly reduces TRTs (62). For example, the TRT of a normal wire at 600°C was 42 ms and that of the flattened wire required $\leq$ 10 ms to equilibrate, while TRTs at 800°C were 90 and 30 ms, respectively. Further, the control of the final temperature to that of the alloy Curie point is overstated. The expected temperature may not be reached in low-power systems and may be significantly exceeded with the higher powers required for short TRTs (98). Additionally, small variations in

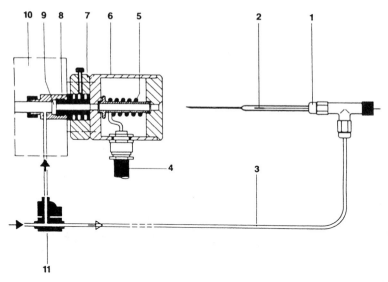

Fig. 5. High-power Curie-point pyrolysis unit: 1, glass pyrolysis injector with stainless steel needle; 2, ferromagnetic wire; 3, Teflon carrier gas tubing; 4, impulse cable; 5, induction coil; 6, aluminium housing; 7, adaptor for fastening; 8, GC inlet; 9, GC septum; 10, GC; 11, carrier gas changeover valve. Diagram kindly supplied by Fischer Labor- und Verfahrenstechnik, Meckenheim bei Bonn, West Germany.

alloy composition may affect the temperature substantially, as do sample loading, the position of the wire within the induction coil, and whether the conductor has been bent to hold an insoluble sample. Of considerable advantage is that there are no electrical connections between the power source and the sample holder. This facilitates automation (33, 119).

D. FURNACE PYROLYZERS

In furnace systems the sample is usually dropped or pushed into a continuously heated zone and in many cases large amounts of sample have been used. There are two problems with such a setup. First, long TRTs, perhaps in the 20- to 50-s range, may be encountered, and second, primary pyrolysis products diffuse through unreacted sample and continue to meet high-temperature conditions on their way to the analytical region. This provides opportunity for secondary reactions to occur, which may increase pyrogram complexity and result in the loss of structural information. In some instances, for example, in the study of burning tobacco or fire retardation, such conditions may be actively sought, but, in general, they lead to confusion and a severe lack of reproducibility. To overcome such problems Simmonds et al. (171) designed a unit that provided a pulse of energy for rapid heating of the sample. Additionally, the pyrolysis temperature could be varied and

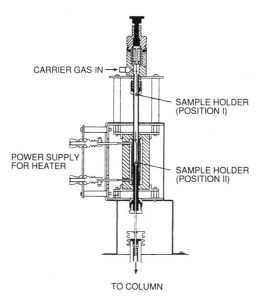

CARRIER GAS IN →

SAMPLE HOLDER
(POSITION I)

POWER SUPPLY
FOR HEATER

SAMPLE HOLDER
(POSITION II)

TO COLUMN

Fig. 6. Vertical microfurnace pyrolyzer. Diagram kindly supplied by S. Tsuge, Nagoya University, Japan.

estimated and samples were held in a removable chamber so that memory effects did not contribute to background contamination. In general, furnace systems have much poorer performances than either the heated filament or Curie-point pyrolyzers, but some designs are outstanding (185, 215). A system introduced by Tsuge and Takeuchi (185) is based on a vertical design of the unit. The sample (50 μg) is held in a platinum bucket that is suspended above the heating zone so that no thermal reactions can occur. When the bucket is released the sample falls into the heated zone, which is progressively narrowed to decrease dead volume and to increase carrier gas flow rates, which rapidly sweeps volatiles away from the heater, thus minimizing secondary reactions. The unit is illustrated in Fig. 6. The small sample size and design of the unit provides a high degree of reproducibility such that standard deviations of < 1% were obtained (186).

E. OTHER PYROLYSIS UNITS

Of more specialized interest is pyrolysis by laser (29, 77). Here, a laser beam is focused onto the surface of the analytical sample to produce a temperature increase. In the case of a ruby (694.3 nm) or neodymium (1040 nm) source, a burst of energy (1 ms) is focused on a spot 1 mm in diameter. Energy dissipation may be of the order of 1 MW/cm^2, which is sufficient to increase the temperature of the sample to in excess of 10 000°C, thus providing pyrolysis fragments. As a result of the high temperatures involved, nonspecific products such as acetylene abound, together with some diagnostic residues and combination products produced by reaction with acetylene. These lasers are not absorbed by transparent samples, which require pretreatment such as mixing with graphite (44, 148) to enable energy absorption to take place.

The use of low-power lasers in the infrared region, for example, CO_2 at 9.1–11 μm, overcomes some of the problems associated with high energy and lack of absorption (32, 107) and may provide product distribution similar to Curie-point pyrolysis. In these systems, power is supplied continuously and the pyrolysis time is controlled by a shutter. Although laser systems offer advantages in that a particular area of a sample may be specifically analyzed and samples requiring high pyrolysis temperatures may be easily handled, the techniques are rather specialized and are not for routine use.

Among other pyrolysis units available are those using infrared (182), photolysis (75), pyrolysis–hydrogenation (40), and systems for liquids (36, 174).

III. PYROLYSIS GAS CHROMATOGRAPHY

Although many of the parameters governing Py-GC are common to gas chromatography itself some special considerations are evident (70).

A. GAS CHROMATOGRAPHY

The decision to use packed or capillary columns is vital. The packed columns offer robustness and the ability to deal with large loadings of pyrolysis products, but suffer from a lack of resolution. This is a disadvantage when a large variety of products are obtained (over 200 products may be encountered in taxonomic work). Overlapping or nonresolved peaks threaten reproducibility and identification and, if possible, should be resolved. Capillary columns are capable of high resolution and may separate very small amounts of products. Thus, they can deal with many more components in the pyrolysis mixture and are to be recommended for complex samples. However, several problems may arise and their use requires careful optimization of pyrolyzer design, flow rate, and sample loading. First, the capacity of these columns is low and hence small loadings (< 10 μg) are demanded. Splitter systems have been described (180) to reduce this problem, but they may not provide a representative sample to the column; small molecules are preferentially vented, larger molecules may not be transferred efficiently. This is important, for example, in mechanistic work where complete product distributions must be known. In these cases it is better to use very small sample sizes, required anyway to provide rapid heating, combined with low dead-volume pyrolyzers (208) to protect resolution. An alternative is cryogenic focusing, in which products are trapped at liquid air temperatures prior to analysis (101, 200). The performance of capillary columns is seriously reduced by the deposition of tarry products (119, 142). Removal of the offending section or back-flushing have been used to overcome this problem (128).

Pyrolysis–hydrogenation–GC using capillary columns has also been reported (40, 193). This technique is used to reduce the complexity of pyrograms from polyalkene degradation. Such compounds produce homologous series comprising alkane, α-alkene, and α,w-dialkene. In-line reduction before chromatography converts each group of homologues to the parent hydrocarbon and greatly simplifies the resulting chromatogram.

B.   COMBINED GAS CHROMATOGRAPHY–MASS SPECTROMETRY

A range of detectors has been used for Py-GC, with the flame ionization detector (FID) systems being most common. Care should be taken to ensure that the detector responds to the pyrolysis products. For example, the FID may be poorly sensitive to chlorinated compounds. More specific or mixed detectors have been used to overcome these problems (79, 81, 83, 132). As the identification of pyrolysis products becomes more important in characterizing pyrograms, the deficiencies of the standard detectors in this respect have led to greater emphasis on mass spectrometric detection and the development of the technique of Py-GC-MS. Electron impact (EI) and chemical ionization (CI) are usually chosen and when combined with a data system the technique offers an extremely powerful analytical methodology for intractable samples (28). In addition to the potential of providing mass pyrograms and mass spectra of individual components, reconstituted data may be selected to remove background contamination or to provide unique recognition of certain fragments. This latter technique may also be used as mass fragmentography to provide specific detection and has been used to provide reliable and sensitive quantification of compounds such as acetylcholine (138, 140). Variations between Py-GC and Py-GC-MS pyrograms may be apparent and may be due to (1) intensity differences as a result of the disparity between response factors for FID and MS detection, (2) resolution differences due to a change in carrier gas, or (3) product distribution discrepancies caused by differential transmission of the pyrolysis products through interfaces. Mass spectra, normally collected at 70 eV, may also vary due to the lower ionization voltages ($\sim$ 20 eV) used to prevent ionization of the helium carrier gas in GC-MS analyses.

C.   THE TUNED APPROACH

When combining a pyrolysis unit with a gas chromatograph it is important to ensure that dead volumes, which threaten resolution and may cause differential loss of pyrolysis products, are minimized and that the temperature of the interfaces is controlled to prevent condensation on cold spots or secondary reactions at hot spots. The difficulties of establishing full control over the system have been considered in a series of correlation trials which have attempted to establish interlaboratory reproducibility (34, 35, 54, 73, 139, 178). Both systems and operating conditions were found to influence pyrograms and care must be taken to remove residual solvent used in loading the sample, to control sample size ($\sim 100\ \mu g$), and to obviate contamination from earlier runs. Moreover, the use of Curie-point and heated-filament systems may cause temperature differences on the pyrolysis and it was noted that as the complexity of the sample increased, interlaboratory reproducibility was more difficult to achieve. To provide the best chance of interlaboratory reproducibility a tuned approach has been proposed. Here, the Py-GC system is adjusted so that standard samples produce pyrograms with prechosen retention times, column performance, and pyrolysis temperature. The pyrolysis of polymer samples for this purpose is advantageous over standard solutions in that they test the entire system, including interface and pyrolyzer dead volume. An ideal method is to use a molecular thermometer (99). A polymer such as Kraton 1107, an isoprene:styrene copolymer, yields isoprene, dipentene, dimethyl vinyl cyclohexane, and styrene. The isoprene:dipentene ratio is severely dependent upon the pyrolysis temperature and ranges from 0.43 at 400°C to 18.75 at 900°C. This variation allows the pyrolysis temperature to be assessed and adjusted, in the case of filament pyrolysers, by examination of the pyrogram. The products may also be used to tune the GC separation to achieve desired retention times and theoretical plates (201). A useful source of chromatogram standardization when using capillary columns may be the pyrolysis of polyethylene which produces homologous triplets of peaks detectable up to at least C_{31} (180).

D. AUTOMATION

The development of Py-GC systems is such that unattended, automatic control of the system is possible. A useful system has been introduced by Meuzelaar et al. (119) in which a Curie-point pyrolyzer is adapted so that pyrolysis wires, precoated with sample and held in quartz tubes on a turntable, are sequentially loaded, pyrolyzed and ejected. The system enables accurate positioning of the sample in the induction coil and was described using high-resolution capillary columns. An alternative Curie-point system used for rubber samples that are enclosed in iron foil has used a commercially available automatic solids injector (33), while the most highly automated system of all was part of the Viking lander expeditions to Mars (15, 151).

IV. PYROLYSIS MASS SPECTROMETRY

Although the bulk of recorded work using analytical pyrolysis techniques is based upon Py-GC and considerable utility in this technique is apparent, a major disadvantage in routine analysis is the throughput of samples. The analysis time is determined by the longest retained component and in microbiological taxonomy, for example, elution may take in excess of 1 h, resulting in few samples being analyzed per instrument. Although such problems may be overcome by automation, an alternative strategy is available. The technique of Py-MS, whereby pyrolysis is immediately followed by mass spectral analysis of the product mixture, has made important advances and is now in direct competition with the chromatographic method (64, 70, 121, 122). Of particular importance are the reproducible mass scale, which characterizes the ions of the mass pyrogram and gives direct chemical information from the pyrogram; the speed at which analyses may be undertaken, perhaps 1 min per sample; and the ability to fully automate the process. This allows the possibility that standards may be incorporated with the test samples to increase reliability. The detection of involatile, polar, or unstable pyrolysis products which are not transmitted through the GC system is also much enhanced when pyrolysis is undertaken directly within the source of the mass spectrometer. It is well to note, however, that a single pyrolysis product may give rise to several ions in the mass pyrogram, that one fragment may be derived from several pyrolysis products, and that isomers will give rise to the same molecular ion although different retention times allow resolution in Py-GC.

A. PYROLYSIS

Curie-point pyrolysis has been the most popular mode of pyrolysis in Py-MS. The same parameters as discussed earlier for Py-GC systems are applicable with the exception that the pyrolysis is undertaken under low-pressure conditions and hence cooling rates may be expected to be even slower than those reported in the presence of a carrier gas. To avoid source contamination and to provide a reservoir from which pyrolysis products may be leaked into the source an expansion or buffer zone may also be included. This must be inert (gold-lined) to reduce losses and the temperature carefully controlled to avoid further thermal reactions.

Available systems for Py-MS include custom-built instruments, which are suitable for Curie-point (116) or laser (88) pyrolysis and dedicated commercial analogues (5, 57), but, additionally, routine analytical mass spectrometers may be converted readily by using pyrolysis probes in place of the direct insertion probe. These may be of dedicated design or adaptations of commercial pyrolyzers normally used for Py-GC and use Curie-point or heated-filament systems. A very low-pressure Curie-point system has also been designed specifically to reduce the potential for secondary reactions that may occur during pyrolysis (156). One further technique is available which enables a dedicated GC-MS system to be used either for Py-GC-MS or for Py-MS (68). Here, the packed column is replaced with a short length of empty column which acts as the buffer zone between pyrolyzer and source. Clearly, in this method, the potential for loss of compounds or secondary reactions is enhanced, but the system provides efficient use of expensive equipment.

To complement these traditional pyrolysis methods, the use of the direct insertion probe to effect pyrolysis has also received much attention, particularly in structural and mechanistic polymer chemistry. Such systems have generally used a temperature ramp on the probe combined with repetitive MS scanning to produce temperature-resolved pyrograms (7, 26, 27, 102, 127, 149, 204). Bacterial taxonomy has been proposed using this technique (4, 146, 147).

B. MASS SPECTROMETRY

Ionization and analysis of the pyrolysis products have been undertaken using a plethora of different systems. This variation may be allowed if only direct chemical information is sought, but if taxonomic fingerprint comparisons or detailed kinetic analyses are required, then standardization of all variables must be ensured. Electron impact (EI) ionization is the standard ionization method, but, in contrast to direct probe work (~ 70 eV) or GC-MS (~ 20 eV), ionization voltages just sufficient to ionize parent molecules (10–15 eV) are generally used for taxonomic comparisons. This is to reduce the complexity and variability of the mass pyrogram by ensuring a high proportion of molecular ions and low numbers of fragmentation ions. In the case of poly(methyl methacrylate) pyrolysis, for example, the molecular ion methyl methacrylate at 70 eV carries less than 5% of the total ion current (TIC). At 15 eV this has risen to about 22% and with 10 eV this ion predominates with some 80% of the ion current (78). High ionization voltages (20) and comparisons between low- and high-voltage mass pyrograms (117) have, however, received study.

Other modes of ionization have also received attention for their ability to provide mild ionization and to decrease the proportion of fragment ions in the mass pyrogram. These include chemical ionization (CI), which has been used for polymers (71, 170), biological samples (146, 147, 175), and forensic work (152). Of more specialized application are systems using field ionization (FI) or field desorption (FD) systems. Field ionization offers a low-energy ionization process enhancing the abundance of molecular ions (134, 141, 157, 161, 164, 165), whereas the FD system offers great potential for simultaneous pyrolysis and ionization under the mildest possible conditions (6, 162, 163). The technique is capable of producing and detecting the largest fragments of all pyrolysis methods and can, for example, detect dimeric units from the pyrolysis of nucleic acids ($M^{+\cdot}$ 730) compared to toluene ($M^{+\cdot}$ 92) in the corresponding Py-GC-MS analysis (14, 159–161). Py-FDMS is of particular application to biological samples where complex, polar, and unstable residues abound. Photoionization has been considered as a reliable means of ensuring reproducible, low-energy ionization (51).

Quadrupole spectrometers have been recommended for taxonomic Py-MS studies (210) due to their linear mass scale and ability for very rapid recycling for data accumulation. Triple quadrupole systems have also been described

TABLE 6
Standardization of Variables for Py-MS Analysis of Biological Samples

Parameter	Comment
Cleaning of wire	Extended heating in a moist hydrogen atmosphere
Solvent	Methanol, pH control of aqueous solutions
Sample size	~ 5 μg
Equilibrium temperature	500°C to maximize both protein and carbohydrate products
Temperature rise time	0.1–1.5 s; little effect on pyrogram
Total heating time	Pyrolysis complete in 0.3 s; little effect on pyrogram
Inlet temperature	150°C

SOURCE: Windig et al. (210).

(183). In combination with ion counting (120, 191) a formidable and sensitive system for routine taxonomic applications is available. Nevertheless, magnetic sector instruments using electron-multiplier or photographic (19) detectors provide extremely useful analyses (69) and are indispensable when high resolution or accurate mass measurements are required. The use of tandem mass spectrometry (37, 38, 196) and positive and negative ion spectra (1) has also been described.

As a further refinement of the Py-MS technique, collisional activation mass spectrometry (CAMS), with a reversed-geometry double-focusing spectrometer, has been used to identify pyrolysis products. This technique allows the structure, rather than merely the molecular formula, of the ionized products to be established by analyzing fragmentation induced by collision with helium alone (41, 95, 130) or in combination with metastable mapping (8).

C. REPRODUCIBILITY

Recommendations to aid reproducibility in Curie-point Py-MS have been detailed (210). These are summarized in Table 6, which reflects results for glycogen and serum albumin. Carbohydrate products are maximized at somewhat lower temperatures than those for proteins, and the recommended pyrolysis temperature is a compromise to characterize both types of macromolecules.

Longer term reproducibility has also been assessed and has been found to be satisfactory, although drift, probably due to contamination of inlet and source regions, was noticed. Tuning of the mass spectrometer to provide a standard spectrum has also been advocated (65). In contrast, results of an interlaboratory correlation trial showed variable results (207) and paralleled the Py-GC tests in that complex samples, particularly those of biological origin, were problematic. However, a large range of sample size (5–10 μg; 20–50 μg), pyrolyzer and temperature (Pyroprobe, 700°C; Curie point, 510°C, 770°C; and direct insertion probe, temperature programmed

60°C/min), and mass spectrometer (quadrupole and magnetic sector) were used with no attempt at standardization. Major sources of variation were identified as the pyrolysis conditions and the problems of reproducing low ionization voltages.

D. AUTOMATION

In contrast to Py-GC, a fully automated Py-MS system must operate entirely within the vacuum of the mass spectrometer. A system for Py-MS that allows this feature has been described by Meuzelaar and coworkers (120) and is based upon a Curie-point pyrolyzer interfaced to a quadrupole spectrometer. Precoated wires are held in quartz reaction tubes, which are transferred from a rotary turntable via an electromagnetic arm into the induction coil. All events are organized by a minicomputer, and a throughput of 30 samples per hour has been achieved (210).

V. DATA HANDLING

The data handling requirements for analytical pyrolysis are largely determined by the application. In kinetic studies of polymer degradation, for example, it is sufficient to be able to identify and quantify the relatively few primary products that are of interest. In cases of taxonomy or classification, where complex pyrograms are compared with standards or with those from similar classes of sample, a detailed statistical analysis, perhaps using multivariate methods is appropriate (58, 70).

A. SCALING DATA

With GC analysis it is frequently impossible to reproduce retention times exactly for long periods of time. When the intensities of corresponding peaks in complex pyrograms are to be compared, this leads to some difficulty in assignment. Retention time normalization, which uses standard compounds, either present in the pyrogram or added as internal or external markers (179), overcomes this problem. A time-rescaled pyrogram may be calculated using the expression

$$t_B^s = t_A^s + \frac{(t_B^r - t_A^r)(t_C^s - t_A^s)}{t_C^r - t_A^r}$$

where t_B^s is the rescaled retention time of component B, occurring between the two standards A and C. Subscripts A, B, and C refer to the species, superscript r (t^r) refers to raw retention times in the original pyrogram, and superscript s (t^s) refers to those in the standardized pyrogram (135).

Due to variations in sample loading, rescaling of intensities is also required. To average out intensity differences it is best to use a fractional total area or total ion current rather than normalization to a, perhaps variable, base peak. This process is described by

$$I_i^n = \frac{I_i}{\sum_{j=1}^{N} I_j}$$

where I_i^n is the fractional intensity of a normalized peak i of initial intensity I_i. Intensity I_j represents the N peaks in the pyrogram.

Peak selection is also a feature of data analysis. Py-GC comparisons usually select peaks if a complex pyrogram is produced, while in Py-MS almost all peaks may be used in the comparison. Typically, three peaks may be sufficient for a polymeric sample, but for taxonomic applications 13–24 or even up to 100 may be used (106, 118, 135, 176). Unless all peaks are included a compromise is sought. For example, peaks that have a high within-sample variability should be eliminated, but this could exclude some diagnostic components (43). Characterization of mass pyrograms, however, has been undertaken using as few as five diagnostic ions.

B. FINGERPRINT COMPARISONS

Various techniques have been used to assess the degree of correspondence between pyrograms. A simple parameter used in Py-GC studies is the similarity coefficient ($S_{i,j}$):

$$S_{i,j} = \frac{\sum_{k=1}^{N} \left(I_i^k / I_j^k \right)}{N}$$

where $S_{i,j}$ is the similarity coefficient between pyrograms i and j, I_i^k is the normalized intensity of the kth peak in pyrogram i, and I_j^k is the corresponding intensity in pyrogram j. The quotient is arranged such that the more intense peak is the denominator ($I_j^k > I_i^k$). N is the number of unique peaks in the pyrograms. A value of 1.0 signifies a perfect match, but values in excess of 0.84 for 13–15 peaks indicate an acceptable degree of similarity for microorganisms (118, 176).

Alternatively, a FIT factor ($F_{i,j}$) is available and has been used for the comparison of Py-MS data (67):

$$F_{i,j} = 1000 \left\{ 1 - \frac{\sum_{k=m_1}^{m_2} \left(I_i^k - I_j^k \right)^2}{\sum_{k=m_1}^{m_2} \left[\left(I_i^k \right)^2 + \left(I_j^k \right)^2 \right]} \right\}$$

where $F_{i,j}$ is the FIT factor between pyrograms i and j, I_i^k is the intensity of the kth peak in pyrogram i, I_j^k is the corresponding intensity in pyrogram j, m_1 is the lower mass limit, and m_2 the higher mass limit. A perfect match is indicated by $F_{i,j} = 1000$, with a good match indicated by values in excess of 975 for a mass range of m/z 25–200.

A related matching coefficient ($M_{i,j}$) has also been proposed for the fingerprint comparison of unknowns with library spectra, with superscript L indicating the intensities of peaks in the library spectra and superscript U indicating those in the unknown spectrum (50):

$$M_{i,j} = 100 \sum_{j=1}^{N} \left(1 - \frac{I_i^L - I_j^U}{I_i^L + I_j^U} \right)$$

These methods are satisfactory if there are gross differences between pyrograms, but when only minor variations are present more sophisticated treatments are necessary. These depend on multivariate methods and chemometrics (111, 214). Methods have included nearest-neighbor (80) and linear-learning machine analysis, cluster analysis (24), factor analysis (105, 184, 211), including target transformation (88), and related methods (19) and nonlinear mapping (43). The dimensionality of the analysis is set by the number of peaks used in the comparison. The nonlinear mapping technique, which uses a two-dimensional representation which most nearly retains the organization of the N-dimensional data set, has shown particular promise in taxonomic applications (63). Mixture analysis has also been considered (212, 213). Illustrative programs that undertake the calculation of similarity coefficient, FIT factor, and linear learning are available (70).

VI. APPLICATIONS

The applications of analytical pyrolysis are extensive and diverse. Wherever problems arise due to the complexity or intransigence of the sample, this technique offers the potential for a rapid, sensitive, and systematic profile to be obtained. Applications range from quantitative measurements of small molecules in biological fluids (76) to the direct classification of whole organisms (59); from the identification of synthetic polymers (50) to modelling the structure of complex biopolymers such as coal (110), petroleum source rock (10, 86), peat (150), and soil fractions (22); and from earthly studies of meteorite components (66) to extraterrestrial samples analyzed at source (151). Derived information may enable the identification of samples, as in forensic analysis (205), or may include detailed kinetic (62) and mechanistic studies (144). Quality control (124), biomass conversion (55, 169), and cigarette smoke have also been dealt with in detail (154). Some specialized applications have also appeared, ranging from the determination of O^{18} enrichment (84) and elemental analysis (113, 181) to the identification of components from thin-layer chromatography (103, 104). As judged by the volume of published material, the major application of analytical pyrolysis is in the analysis of synthetic polymers followed by samples of organic geochemical origin. Significant interest is also found in taxonomic and profiling applications and in forensic studies. This vast range of applications precludes a full review of these topics here; the present discussion is limited to recent illustrative examples. Fuller discussion and references to earlier work may be found elsewhere (70).

A. SYNTHETIC POLYMERS

The properties of synthetic polymers usually preclude analysis by the conventional methods available for small molecules, and degradation techniques are widely employed. Applications of analytical pyrolysis are legion and offer information on many aspects of polymer structure and thermal stability (100).

1. Fingerprint Identification

Fingerprint identification of polymers was one of the earliest applications of the pyrolysis of polymers and relies on a comparison of pyrograms from unknown and reference samples. The identification of peaks is not necessary, although diagnostic products may be used (2, 53), but standardization of retention times and peak intensities is important. Two approaches are possible: (1) a single temperature pyrolysis and the comparison of single pyrograms from a sample, or (2) stepped-temperature pyrolysis, which provides a series of sequential pyrograms and also enables thermal stability to be used in the comparison (62). Both Py-GC and Py-MS techniques are available. For example, Fig. 7 compares Py-CIMS pyrograms (71) to show significant variations, with fragments explainable on the basis of the polymer structure. Poly(methyl methacrylate) shows monomer (M + 1; m/z, 101) as the major fragment, poly(styrene) reveals both monomer (M + 1; m/z, 105) and dimer (M + 1; m/z, 209), while poly(vinyl chloride) has a major residue due to benzene (M + 1; m/z, 79).

To illustrate Py-GC data, Fig. 8, for example, shows pyrograms of two Novolac resins obtained at 700°C, one based on phenol and the other on cresol. An unknown sample is readily identified by visual inspection of retention times, although the structures of the pyrolysis products have not

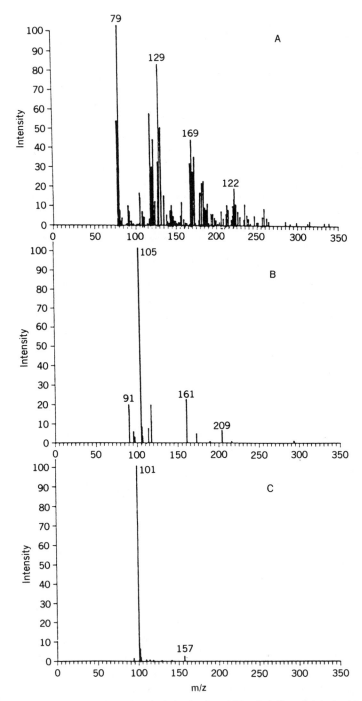

Fig. 7. Mass pyrograms using chemical ionization MS of (*A*) poly(vinyl chloride), (*B*) poly(methyl methacrylate), and (*C*) polystyrene at 1000°C. Reproduced from *J. Macromol. Sci. Chem.* **A22**, 779 (1985) (71).

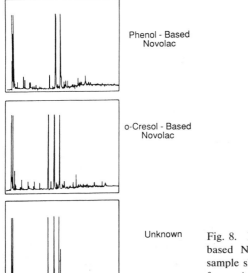

Phenol - Based
Novolac

o-Cresol - Based
Novolac

Unknown

Fig. 8. Pyrograms of a phenol-based and a cresol-based Novolac resin, together with an unknown sample shown to be of the cresol type. Reproduced from *Analytical Pyrolysis: Techniques and Applications*, London: Butterworths, 1984, p. 434 (172).

been determined (172). Reproducibility is now such that comparisons with library data are possible (50, 88, 112).

2. Analysis of Composition

Compositional studies enable the detection, identification, and quantification of components in blends or copolymer systems and as such offer a useful industrial quality control procedure. Figure 9 reveals differences in sulfur-cured rubbers and identifies sulfur-containing heterocyclic residues (132). Benzothiazole is probably derived from pendant terminal groups of the accelerators used (e.g., *N-tert*-butylbenzothiazole-2-sulfenamide), while CS_2 originates from the coaccelerator (tetramethylthiuram disulfide). Thiophene and its methyl analogue originate from the sulfur associated with the cross-linking in the cured rubbers.

Quantification may be effected using calibration curves, analogous to those used in GC analysis, produced by pyrolysis of standards that may either be pure components or samples of known composition. The latter case is preferred because some copolymers have a different pyrolysis profile than the individual monomers. For example, polystyrene–acrylate copolymers yield less monomer from the copolymer than do the homopolymers. The quantitative ability of the Py-MS approach has also been demonstrated with the analysis of triblends composed of natural, butadiene and styrene–butadiene rubbers (88). The mass pyrogram is almost solely the summation of the mass spectra of the individual components of the blend, showing little interaction

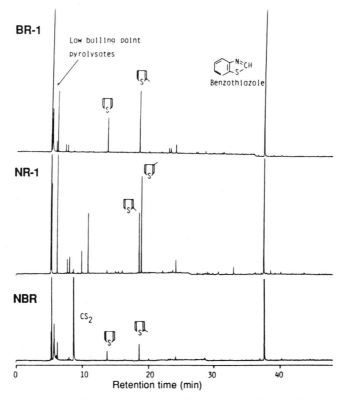

Fig. 9. Pyrograms from sulfur-cured rubbers: BR-1, poly(butadiene); NR-1, natural rubber; NBR, acrylonitrile–butadiene rubber. Reproduced from *J. Anal. Appl. Pyrol.* **10**, 36 (1986) (132).

on pyrolysis, and thus techniques such as factor analysis enable the extraction and quantification of the component spectra from the pyrogram.

3.   Determination of Structure

The structure elucidation of polymers may be limited to identification of the monomeric components, but the technique may also be used to provide detailed information on the microstructure of the polymer (189). Direct fission of the polymer backbone to produce partial sequences of residues (diads and triads) is a useful ploy. Figure 10 shows the presence of styrene monomer, dimer, and trimer, characteristic of styrene sequences, but also of hybrid dimers resulting from the pyrolysis of various chloromethylated styrene–divinylbenzene polymers (133). The intensities of the styrene components decrease as the degree of chlorination increases. Data also show reductions in peak intensities associated with *meta* and *para* divinylben-

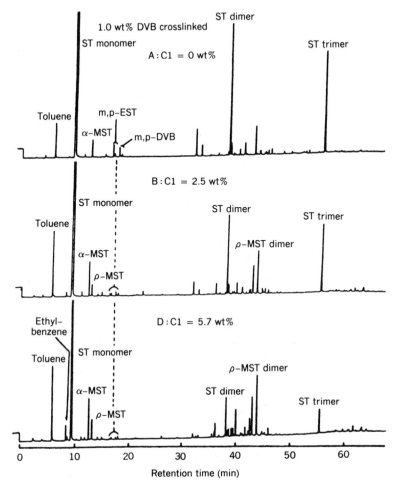

Fig. 10. High-resolution pyrograms of chloromethylated styrene–divinylbenzene copolymers. ST, styrene; MST, chloromethylstyrene. Reproduced from *Macromolecules* **21**, 931 (1988) (133).

zenes, which are accounted for by chloromethylation in these rings, particularly the *meta* isomers.

The branching of poly(alkene) chains may be investigated by pyrolysis, which yields clusters of isomeric peaks, due to various branching modes, between the principal, homologous *n*-hydrocarbon fragments. In-line hydrogenation simplifies the chromatogram by reducing unsaturates and focuses on the essential features of the pyrogram. Figure 11 illustrates the complexity of the C_{11}–C_{12} region from Py-hydrogenation-GC of poly(ethylene) to determine chain branching. This reveals the presence of four methyldecanes and three ethyldecanes, which result from chain branches. Such data show, for example, that with low molecular weight fractions short-chain branching increases and that chain ends have substantial 3-ethyl residues (193).

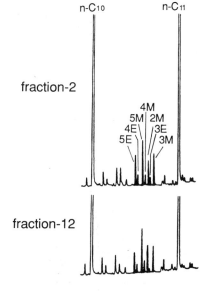

Fig. 11. Expanded partial Py-Hy-GC pyrogram of two samples of low-density poly(ethylene) in the C_{10}–C_{11} region showing branched residues. Fraction 12 has the larger molecular weight and lower contributions of branched residues. 2M, 2-methyldecane; 3M, 3-methyldecane; 4M, 4-methyldecane; 5M, 5-methyldecane; 3E, 3-ethyldecane; 4E, 4-ethyldecane; 5E, 5-ethyldecane. Reproduced from *Macromolecules* **20**, 1559 (1987) (193).

The stereoregularity of the polymer may also be explored and requires high-resolution GC analysis to enable the separation of isomeric fragments. In polypropylene the partial structure and fission on pyrolysis to yield C_{11}–C_{13} fragments may be represented as follows. Olefinic products are reduced (Py-Hy-GC) to simplify analysis. Asterisks (*) indicate geometrical isomeric centers in the pyrolysis products.

This process yields a triplet of peaks (C_{11}, C_{12}, C_{13}) each of which appears as a doublet due to the geometrical isomers, which result by virtue of the stereochemical nature of the polymer. Figure 12 records pyrograms illustrating differences between a largely atactic (random) sample and one that has comparable amounts of isotactic (cisoid) and syndiotactic (alternate) blocks (187).

Evidence may be obtained from intramolecular cyclization reactions (168). For example, poly(chloropropylene) eliminates HCl on thermal degradation to yield mesitylene:

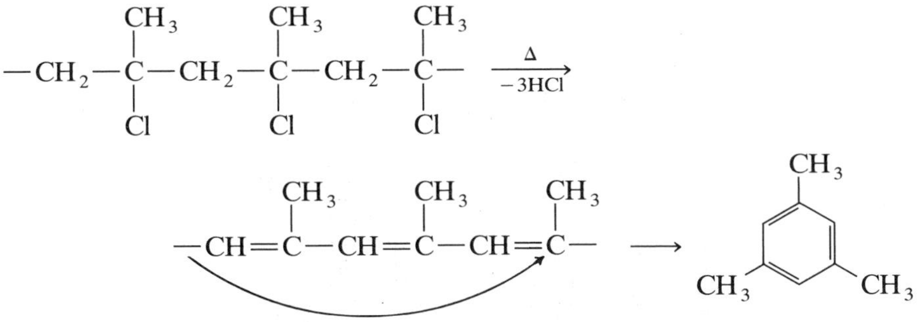

When head-to-head or head-to-tail polymerization has occurred, analogous reaction produces 1,2,4-trimethylbenzene, quantification of which enables an estimation of the amount of inverse monomer to be made:

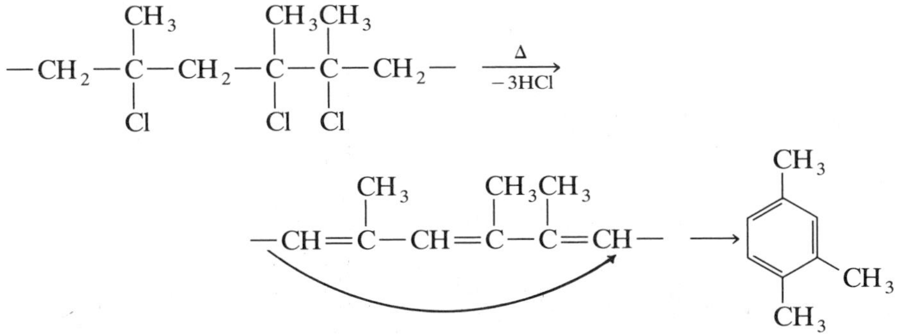

4. Mechanistic Aspects

The thermal degradation of polymeric compounds follows a variety of pathways, and the elucidation of degradation mechanism by pyrolysis techniques is an important application (56, 131, 144). Poly(methyl methacrylate), styrene–methacrylate copolymers, and similar polymers produce essentially monomer by unzipping of the polymer chain. Poly(vinyl chloride) yields benzene via the elimination and the back-biting six-membered transition state illustrated above for the chloropropylene polymer. Cyclization reactions are common and are found, for example, in polysiloxanes (48) and polyesters (102). In contrast, others, such as poly(ethylene), undergo random scission to produce homologous alkane, olefin, and diene fragments. The suppression of volatile degradation products to minimize combustibility (30, 31, 129) and the protecting influence of antioxidants (126) modify degradation mechanisms. Product distribution and a detailed description of the rates of reaction enable these processes to be elucidated. When monomer is the major product initiation, by end or scission reactions, depropagation and termination are

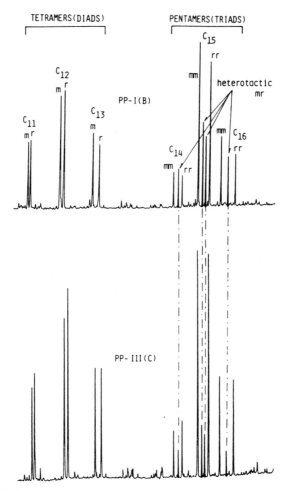

Fig. 12. Detailed Py-Hy-GC pyrograms of two samples of polypropylene showing tetramers (C_{11}, C_{12}, and C_{13} doublets) and pentamers (C_{14}, C_{15}, C_{16}) indicating stereoregularity. PP-1(B), largely atactic; PP-III(C), largely isotactic and syndiotactic blocks. Reproduced from *Analytical Pyrolysis: Techniques and Applications*, London: Butterworths, 1984, p. 416 (187).

the stages involved. Unbuttoning is analogous to unzipping, but involves intramolecular H transfer and leads to oligomeric products (60).

The pyrolysis of poly(methyl acrylate) has been shown to involve significant dimer and trimer production with small amounts of volatiles, monomer, and related esters, including methyl formate, methyl methacrylate, and methyl ethacrylate (56). Some 21 products were identified and the origin of all major components have been derived. Significant products, produced via 1,3-, 1,5-, or 1,7-hydrogen abstraction reactions following homolytic fission of the chain, were of the general formula:

$$CH_2{=}\underset{\underset{COOCH_3}{|}}{C}{-}[{-}CH_2{-}\underset{\underset{COOCH_3}{|}}{CH}{-}]_n{-}CH_2{-}\underset{\underset{COOCH_3}{|}}{CH_2} \qquad n = 0, 1, 2$$

and accounted for some 23% of the yield. The major products, with 29% each of the product abundance, were

$$CH_2{=}\underset{\underset{COOCH_3}{|}}{C}{-}CH{=}\underset{\underset{COOCH_3}{|}}{C}{-}CH{=}\underset{\underset{COOCH_3}{|}}{CH}$$

and

$$\underset{\underset{COOCH_3}{|}}{CH_2}{-}CH_2{-}\underset{\underset{COOCH_3}{|}}{CH}{-}CH_2{-}\underset{\underset{COOCH_3}{|}}{CH_2}$$

The conjugated compound arises from initiation via hydrogen radical loss at the tertiary atom followed by elimination, whereas the saturated homologue results from successive backbone fissions (70, 199) and intermolecular hydrogen abstraction.

5. Kinetic Aspects

Pyrolysis methods are also useful to provide measurements of thermal rate processes. Typically, a polymer may be repetitively pyrolyzed to provide a series of pyrograms of decreasing yield which provide cumulative degradation data (42). To avoid cumulative errors in pyrolysis time and temperature and the possibility of thermal conditioning, an alternative method for polymers that have a limiting yield of product is to pyrolyze identical samples for increasing periods of time. Using this approach rates of degradation of poly(acrylonitrile) have been modelled using the first-order expression (93)

$$-\ln\left(1 - \frac{a_t}{a_\infty}\right) = k_{obs}t$$

where a_t represents the peak area of volatiles after time t and a_∞ is the total peak area after complete pyrolysis.

For polymers that degrade quantitatively to the monomer, such as poly(methyl methacrylate), a further technique is available. Here, the polymer is pyrolyzed twice. First, the sample is partially pyrolyzed at the required temperature (area a_1), which is followed by a further, full pyrolysis (area a_2). The above equation also holds in this instance because $a_\infty = a_1 + a_2$ and

$a_t = a_1$. Various kinetic situations have been evaluated and models presented (61, 94) which may be used to study mechanistic events (93).

B. BIOLOGICAL MOLECULES

The pyrolysis of biological molecules has been examined both to provide structural information per se and to investigate their contribution to pyrograms from complex biological sources (9, 12, 13, 70, 145). Most classes of naturally occurring molecules have received attention both as pure components and as diverse samples. These include both Py-GC and Py-MS studies of amino acids and proteins, polysaccharides, and nucleic acids, all of which yield characteristic fragments, apart from perhaps the Py-GC analysis of nucleic acids, which generally provides carbohydrate residues only. In the case of amino acids, for example, decarboxylation to yield the corresponding amine is a major process. This is accompanied by nitrile, olefin, and, in the case of aliphatic amino acids, aldehyde and imine, to provide characteristic fragments. Peptide pyrolysis appears to proceed largely via side-chain stripping rather than depolymerization, but nevertheless products diagnostic of the parent amino acids are revealed and provide quantitative information on residues. Figure 13, for example, shows the pyrograms from the enzymes lysozyme (3 × Phe, 3 × Tyr, 6 × Trp) and ribonuclease A (3 × Phe, 6 × Tyr, no Trp). Indole and skatole, from tryptophan, are absent in ribonuclease, which reveals more intense phenol and cresol components from tyrosine (188). It should be noted, however, that direct mass spectrometry with fast atom bombardment enables sequencing of oligopeptides, and the structure of large peptides may be deduced from nucleic acid sequences.

Other biological macromolecules also produce characteristic products. Analytical pyrolysis of polysaccharides, for example, yields cyclic oxygenated derivatives with furfural, and the 5-hydroxymethyl derivative the main products from glucose, while furfural, together with levoglucosenone, characterizes cellulose (137). Furnace conditions produce vastly more products. Chitin is a further important biopolymer. It is a polymer of N-acetyl-D-glucosamine and a component of fungal cell walls and crustacean shells. Acetamide has been used as an indicator of this material but, additionally, acetamidofurans are detectable and have been suggested as unique chitin and N-acetylglucosamine products (47). Pyrograms of nucleic acids depend on the analytical conditions, and Py-GC is of limited application. Curie-point Py-MS studies largely detect pyrolysis products of the sugar residues, but the soft ionisation technique of Py-FDMS can detect dinucleotides. Lignin is a biomolecule that has also received much attention. It is the principal nonpolysaccharide polymer of wood and is largely composed of phenolic components. It is important as a source of carbon in soil and its pyrolysis pattern is diagnostic. Series of methoxyphenols characterize the pyrogram, with 4-substituted 2,6-dimethoxyphenols (syringyl), 2-methoxyphenols (coniferyl), and phenols (coumaryl) predominating (115). The major remaining group of natural

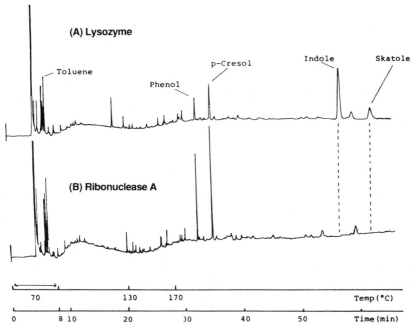

Fig. 13. Pyrograms from lysozyme and ribonuclease A. Indole and skatole arise from trypto-phan, phenol and cresol from tyrosine, and toluene from phenylalanine. Reproduced from *J. Anal. Appl. Pyrol.* **8**, 60 (1985) (188).

macromolecules, the lipids, are not polymeric structures, but are esters of glycerol and various long-chain fatty acids. Waxes are esters of long-chain alcohols and acids. Pyrolysis of such molecules usually liberates the fatty acid component together with acrolein from glycerol. Homologous series are observed comprising saturated and monounsaturated acid, hydrocarbon, and olefin formed by decarboxylation or disproportionation.

C. TAXONOMY

The reliable identification of microorganisms or pathological conditions may require many time-consuming tests of a biochemical, serological, and morphological nature. In some instances, such as Mycobacteria, slow growth necessitates long-term testing. A potential taxonomic tool that is rapid, diagnostic, and requires one sample preparation and analytical system for many diverse samples is attractive (123, 131). When combined with the power of chemotaxonomic data handling to establish relationships and taxonomic groupings the appeal becomes irresistible. Such considerations have led to much interest in the applications of analytical pyrolysis as a taxonomic tool. In particular, the special advantages of Py-MS have led to this becoming the technique of choice in such studies although Py-GC data, particularly with

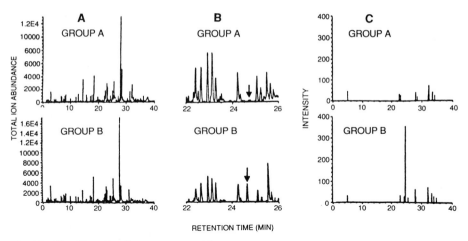

Fig. 14. Total ion abundance pyrogram (1), expanded partial total ion abundance pyrogram (2), and reconstructed ion abundance pyrogram using m/z 86 ions only (3) of group A and group B streptococcus. Pyrograms (2) show the unique fragment found in group B samples at 24.7 min, which is selected in the mass fragmentograms (3). Reproduced from *Anal. Chem.* **59**, 1411 (1987) (173).

component identification, also have useful applications. Essential features of the system are that it be (1) universal, so as to allow standard analytical conditions to be used for a wide variety of samples; (2) specific, so as to group together samples from the same class, while separating different but closely related classes; (3) reproducible, so as to give reliable conclusions without operator or system dependence; (4) rapid, so as to give adequate throughput of samples; (5) sensitive, to minimize culture demands; (6) interpretive, to provide structural information on taxonomic class; and (7) automated, to minimize operator error and time.

Many of the components of bacterial pyrolysis have been identified (114, 157) and can be shown to be largely derived from biological macromolecules. Although different organisms provide characteristic pyrograms, in most cases the same pyrolysis products are obtained and it is only the relative yields that vary. Few unique fragments are found. One example where this is so is in the classification of streptococci by Py-GC-MS (173). Here, as shown in Fig. 14, pyrograms from group B organisms were found to display a unique peak with a retention time of 24.7 min, which was totally absent in those from group A. Total ion current pyrograms showed a high degree of similarity, but the two groups were readily distinguished by mass fragmentography when pyrograms were constructed using ion currents from m/z 86 ions only. Figure 14 also shows the dramatic diagnostic increase with this filter applied. The unique fragment was shown to be produced by pyrolysis of group B antigen and from glucitol, a component of this antigen.

In Py-MS studies an alternative approach to increase discrimination to provide marker components involves tandem mass spectrometry (Py-MS-MS), which has recently been described using a triple quadrupole system with CI (196). This setup allows variable scanning to provide neutral loss scans, parent ion scans, or daughter ion scans to reduce chemical noise in the mass pyrogram and to increase its information content. The first quadruple was used as a mass filter to select the ion of interest, the second as the collision chamber with argon as the collision gas, while the third was scanned to provide a spectrum on each ion in turn. Normal Py-MS pyrograms of *Bacillus cereus*, *B. subtilis*, and *Escherichia coli* were found to provide certain peaks that appeared useful for taxonomic purposes. Simple daughter ion spectra of component parent ions showed no characteristic components suitable for a chemical marker, but when a parent ion scan was undertaken, i.e. to

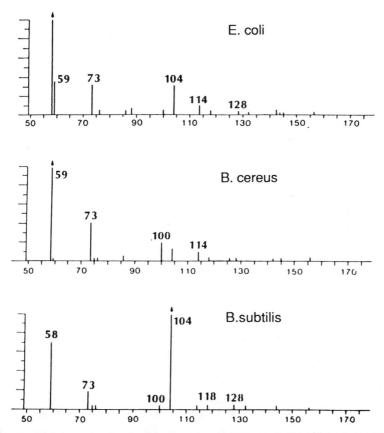

Fig. 15.   Parent ion spectra for daughter ion at m/z 58 from Py-MS-MS analysis of *Bacillus subtilis*, *B. cereus*, and *Escherichia coli*. Reproduced from *J. Anal. Appl. Pyrol.* **14**, 12 (1988) (196).

reconstruct a pyrogram containing all parent ions that provide a single daughter ion, significant variation between the organisms was revealed. A typical set for the daughter ion at m/z 58 is shown in Fig. 15. Using all of such data and optimisation by factor analysis, spectra entirely diagnostic of each organism were obtained.

D. FORENSIC SCIENCE

Forensic science frequently requires the identification of small, complex specimens of unknown composition. Samples may be from various sources (paint, fibers, adhesives) and of various forms (flakes, smears, particles). Frequently, it is necessary to link a suspect to the scene of a crime by proving the identity of a sample found at the crime scene with that associated with the suspect. Pyrolysis methods have been of great use in tackling these problems, because they offer a general sample handling and analytical methodology that can provide high-resolution fingerprinting (206).

Paint samples are an important area of interest and household gloss paints, for example, are generally alkyd-based, consisting of phthalate polyesters with natural drying oils together with modifiers. Py-GC has good discriminatory performance, using acrolein and methacrolein from glycerol, but does not transmit phthalate components through the column. An alternative Py-MS study of such samples has shown that certain fragments are produced with a greater reproducibility than others. Twenty of the most discriminatory ions were chosen and from these were eliminated those that were too variable. This left only five ions (m/z 174, toluene di-isocyanate; m/z 118, vinyltoluene; m/z 105, benzoic acid, $C_6H_5 — C = O^+$; m/z 104, phthalic anhydride, $C_6H_4 — C = O^+$; m/z 77, $C_6H_5^+$) which proved to be sufficient for identification. The glycerol residues, characteristic of Py-GC, are not used at all in this technique and thus the two approaches are complementary (205).

E. ORGANIC GEOPOLYMERS

Complex, condensed organic macromolecules are found in abundance in the biosphere and lithosphere. The insoluble fraction of sedimentary rock may be classified into humic substances, which are insoluble in organic solvents but partially soluble in alkali, and kerogen, which is insoluble in both organic solvents and alkali. The amount of kerogen, the most abundant form of organic carbon, has been estimated to be of the order of 12×10^{15} tonnes compared to a calculation globally of 15×10^{12} tonnes of coal (87). The characterization of these residues has proved difficult due to their intractability, but their study is of great importance. This has both practical implications, as a fingerprint or to evaluate the oil- and fuel-bearing potential of rocks, and theoretical significance, to develop models for the maturation processes, which result in the development of fertile soil from humus or the evolution of oil- and coal-bearing seams. The study of these aspects is much facilitated by analytical pyrolysis, particularly when combined with model systems based on synthetic analogues of the natural polymers.

The varied nature of specimens, together with the frequently low loading of organic matter in rock samples, has necessitated several different pyrolyzer designs, including furnace systems. The applications of kerogen analysis involve oil, gas, and shale evaluation and the assessment of feedstocks. The recognition of source rocks, the characterization of oils and residues, and the analysis of small samples in the laboratory are also important exploitations. Samples vary from those with a high hydrogen content, derived from biolipids (Types 1 and 2), to those that are hydrogen-depleted with a condensed cyclo-organic matrix (Types 3 and 4). Pyrolysis readily reveals this distinction, with Type 1 yielding much low mass hydrocarbon, whereas Type 4 reveals condensed hydrocarbon products. Types 2 and 3 are intermediate, with the Type 3 showing abundant alkylbenzene and alkylphenol components, while the Type 2 has more aliphatic and olefinic hydrocarbon. Although pattern recognition techniques allow elegant classification of such mass pyrograms, a simple triangular diagram, based upon yields of octene, xylene, and phenol from samples, can produce partition of types (87). Selective ion monitoring to enhance variations provides a useful modification (85).

In contrast to kerogen pyrolysis, that of soil provides few hydrocarbon residues and many with oxygen and nitrogen. A peaty podzol, for example, yielded a range of compounds that included furan, acetonitrile, acetic acid, benzonitrile, and phenol (21). Various fractions of soil, including polysaccharide, humic, and fulvic acids may be separated and their structure studied by pyrolysis. A Py-GC-MS study of the A_1 horizon of a Spanish brown soil, for example, was partitioned according to solubility and absorptive behavior into seven fractions. Characteristic pyrograms were obtained for each fraction and, in total, 322 products were identified (153). The origin of these was assigned to various macromolecules (polysaccharide, protein, lignin, lipid) from the known behavior of reference materials, and showed, for example, that the hymatomelanic acid fraction was largely of lipid origin, fulvic acids appeared as lignin and polysaccharide structures, and the humic acid and humin behaved as complex mixtures of all components with major lignin and polysaccharide residues. Some pyrograms were contaminated with phthalates, possibly an artifact due to plasticizers present in the organic solvents. Figure 16 shows the significant differences between the humic acid pyrogram and that of the lipid-based hymatomelanic acid.

The use of Py-FIMS to almost exclusively yield molecular ions in the mass pyrogram has also shown promise in the study of soil fractions. In combination with time- and temperature-resolved pyrolysis modes, efficient characterization of diverse samples with a minimum of preparation is achievable (166). Although isomeric components are not resolved by this technique, typical ion series result from macromolecular components. Thus, the series m/z 60, 82, 96, 110, and 126 shows the presence of polysaccharides, while nitrogen-

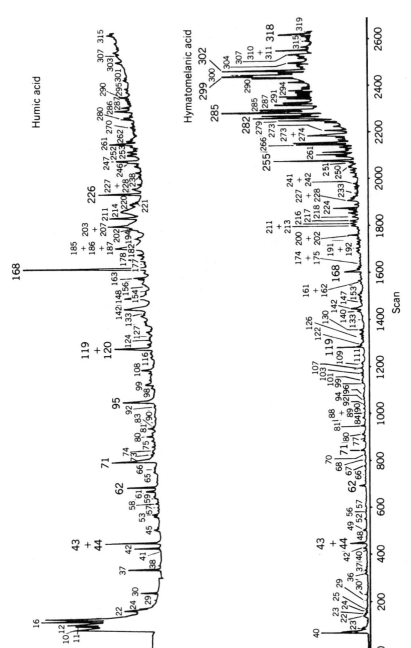

Fig. 16. Pyrograms of humic acid and hymatomelanic acid fractions of the a_1 horizon of a Spanish brown soil. Representative peaks are 43, 2-methylbutanal; 44, benzene; 62, toluene; 71, 2-furaldehyde; 95, 2-methylfuraldehyde; 119, guaiacol; 120, levoglucosenone; 168, vinylguaiacol; 226, 4-vinyl-2,6-dimethoxyphenol; 255, n-C_{14} acid, methyl ester; 282, iso-$C_{16:1}$ acid, methyl ester; 285, n-C_{16} acid, methyl ester; 299, $C_{18:1}$ acid, methyl ester; 302, n-C_{18} acid, methyl ester; 318, n-C_{21} acid, methyl ester. Reproduced from *J. Anal. Appl. Pyrol.* **9**, 103 (1986) (153).

346

containing ions are found at m/z 79, 93, and 117. The technique has also been applied to the differentiation of soil horizons using whole soil samples. Moreover, the technique allowed the identification of the biocide atrazine by the appearance of its molecular ion at m/z 215.

F. EXTRATERRESTRIAL SOURCES

One further, imaginative use of analytical pyrolysis remains. The study of various surface deposits has shown that even samples of desert soil, with some 0.34% organic carbon, were characterized by molecules of biological origin with furan, indicating carbohydrate residues and benzene, toluene, and indole, which is characteristic of proteins (171). In contrast, analysis of meteorite samples showed the presence of significant organic carbon, but the profile of pyrolysis products was quite different. Naphthalene, from the condensed polycyclic–aliphatic complex present in the meteorite, was the major product, together with other aromatic and aliphatic hydrocarbons and a few oxygenated products including butanol and furan. This variation was confirmed with samples from various soils and meteorites and led to the development of a pyrolysis life-detection system based on the search for the biological marker compounds found in earthly soils.

In 1976 this system was twice landed on the surface of Mars as part of the *Viking* expeditions to investigate the Martian surface and atmosphere. The instrumentation consisted of a furnace pyrolysis oven interfaced to a GC and linked to a double-focusing mass spectrometer and data system (15). Although the MS analysis of the Martian atmosphere proceeded smoothly the pyrolysis results were unexpected. No traces of biological marker compounds were found and no indication of other organic matter, including naphthalene and similar products of meteorite origin, were apparent (17, 18).

REFERENCES

1. Adams, R. E., *Anal. Chem.* **55**, 414 (1983).
2. Alekseeva, K. V., *J. Anal. Appl. Pyrol.* **2**, 19 (1980).
3. Andersson, E. M., and I. Ericsson, *J. Anal. Appl. Pyrol.* **2**, 97 (1980).
4. Anhalt, J. P., and C. Fenselau, *Anal. Chem.* **47**, 219 (1975).
5. Aries, R. E., and C. S. Gutteridge, *J. Anal. Appl. Pyrol.* **9**, 81 (1986).
6. Bahr, U., I. Luederwald, R. Mueller, and H.-R. Schulten, *Angew. Makromol. Chem.* **120**, 163 (1984).
7. Ballistreri, A., S. Foti, G. Montaudo, S. Pappalardo, E. Scamporrino, A. Arnesano, and S. Calgari, *Makromol. Chem.* **180**, 2835 (1979).
8. Ballistreri, A., D. Garozzo, M. Giuffrida, and G. Montaudo, *J. Anal. Appl. Pyrol.* **12**, 3 (1987).
9. Baltes, W., *Fresenius' Z. Anal. Chem.* **327**, 220 (1987).
10. Barker, C., and L. Wang, *J. Anal. Appl. Pyrol.* **13**, 9 (1988).

11. Barlow, A., R. S. Lehrle, and J. C. Robb, *Polymer* **2**, 27 (1961).

12. Bayer, F. L., *Adv. Chem. Ser.* **203**, 13 (1983).

13. Bayer, F. L., and S. L. Morgan, *Chromatogr. Sci.* **29**, 277 (1985).

14. Beckey, H. D., and H.-R. Schulten, in C. Merritt, Jr. and C. N. McEwen, Eds., *Mass Spectrometry, Part A*, New York: Dekker, 1979, p. 145.

15. Biemann, K., *Origins of Life* **5**, 417 (1974).

16. Biemann, K., J. Oro, P. Toulmin III, L. E. Orgel, A. O. Nier, D. M. Anderson, P. G. Simmonds, D. Flory, A. V. Diaz, D. R. Rushneck, and J. A. Biller, *Science* **194**, 72 (1976).

17. Biemann, K., J. Oro, P. Toulmin III, L. E. Orgel, A. O. Nier, D. M. Anderson, P. G. Simmonds, D. Flory, A. V. Diaz, D. R. Rushneck, J. E. Biller, and A. L. LaFleur, *J. Geophys. Res.* **82**, 4641 (1977).

18. Biemann, K., A. L. Lafleur, J. E. Biller, and T. Owen, *Rev. Sci. Instrum.* **49**, 817 (1978).

19. Blomquist, G., E. Johansson, B. Soderstrom, and S. Wold, *J. Chromatogr.* **173**, 19 (1979).

20. Boss, B. D., and R. N. Hazlett, *Anal. Chem.* **48**, 417 (1976).

21. Bracewell, J. M., and G. W. Robertson, in C. E. R. Jones, and C. A. Cramers, Eds., *Analytical Pyrolysis*, Amsterdam: Elsevier, 1977, p. 167.

22. Bracewell, J. M., and G. W. Robertson, *Geoderma* **40**, 333 (1987).

23. Bradt, P., V. H. Diebler, and F. L. Mohler, *J. Res. Nat. Bur. Stand.* **50**, 201 (1953).

24. Brosseau, J. D., and J. W. Carmichael, *Mycopathologia* **63**, 67 (1978).

25. Buhler, Ch., and W. Simon, *J. Chromatogr. Sci.* **8**, 323 (1970).

26. Chakravarty, T., H. L. C. Meuzelaar, W. Windig, G. R. Hill, and R. M. Khan, *Prepr. Pap. Am. Chem. Soc. Div. Fuel Chem.* **32**, 211 (1987).

27. Chakravarty, T., W. Windig, G. R. Hill, H. L. C. Meuzelaar, and M. R. Khan, *Energy Fuels* **2**, 400 (1988).

28. Chapman, J. R., *Computers in Mass Spectrometry*, London: Academic, 1978.

29. Charscan, S. S., *Lasers for Industry*, New York: Van Nostrand Reinhold, 1972.

30. Chien, J. C. W., and J. K. Y. Kiang, *Makromol. Chem.* **181**, 45 (1980).

31. Chien, J. C. W., and J. K. Y. Kiang, *Macromolecules* **13**, 280 (1980).

32. Coloff, S. G., and N. E. Vanderborgh, *Anal. Chem.* **45**, 1507 (1973).

33. Coulter, G. L., and W. C. Thompson, in C. E. R. Jones, and C. A. Cramers, Eds., *Analytical Pyrolysis*, Amsterdam: Elsevier, 1977, p. 1.

34. Coupe, N. B., C. E. R. Jones, and S. G. Perry, *J. Chromatogr.* **47**, 291 (1970).

35. Coupe, N. B., C. E. R. Jones, and P. B. Stockwell, *Chromatographia* **6**, 483 (1973).

36. Cramers, C. A. M. and A. I. M. Keulemans, *J. Gas Chromatogr.* **5**, 58 (1967).

37. Dallinga, J. W., N. M. M. Nibbering, and A. J. H. Boerboom, *J. Chem. Soc. Perkin Trans.* 2 281 (1983).

38. Dallinga, J. W., N. M. M. Nibbering, and A. J. H. Boerboom, *J. Chem. Soc. Perkin Trans.* 2 1065 (1984).

39. Davison, W. H. T., S. Slaney, and A. L. Wragg, *Chem. Ind.* 1356 (1954).

40. Dimbat, M., in R. Stock, Ed., *Gas Chromatography 1970*, London: Institute of Petroleum, 1971, p. 237.

41. Egsgaard, H., E. Larsen, and L. R. Carlsen, *J. Anal. Appl. Pyrol.* **4**, 33 (1982).

42. Ericsson, I., *J. Anal. Appl. Pyrol.* **8**, 73 (1985).

43. Eshuis, W., P. G. Kistemaker, and H. L. C. Meuzelaar, in C. E. R. Jones and C. A. Cramers, Eds., *Analytical Pyrolysis*, Amsterdam: Elsevier, 1977, p. 151.

44. Falmer, O. F., Jr., and L. V. Azarraga, *J. Chromatogr. Sci.* **7**, 665 (1969).

45. Farre-Ruis, F., and G. Guiochon, *Anal. Chem.* **40**, 998 (1968).

46. Folmer, O. F., Jr., and L. V. Azarraga, *J. Chromatogr. Sci.* **7**, 665 (1969).

47. Franich, R. A., and S. J. Goodin, *J. Anal. Appl. Pyrol.* **7**, 91 (1984).

48. Fujimoto, S., H. Ohtani, and S. Tsuge, *Fresenius' Z. Anal. Chem.* **331**, 342 (1988).

49. Fujita, K., and K. Mizuki, *Jpn. Anal. Instrum. Bull.* 38 (1975).

50. Garozzo, D., and G. Montaudo, *J. Anal. Appl. Pyrol.* **9**, 1 (1985).

51. Geniut, W., and J. J. Boon, *J. Anal. Appl. Pyrol.* **8**, 25 (1985).

52. Giacobbo, H., and W. Simon, *Pharm. Acta Helv.* **39**, 162 (1964).

53. Goetz, N., P. Lasserre, and G. Kaba, *Cosmet. Sci. Technol. Ser.* **4**, 81 (1985).

54. Gough, T. A., and C. E. R. Jones, *Chromatographia* **8**, 696 (1975).

55. Graham, R. G., and M. A. Bergougnou, *J. Anal. Appl. Pyrol.* **6**, 95 (1984).

56. Gunawan, L., and J. K. Haken, *J. Polym. Sci. Polym. Chem. Ed.* **23**, 2539 (1985).

57. Gutteridge, C. S., A. J. Sweatman, and J. R. Norris, in K. J. Voorhees, Ed., *Analytical Pyrolysis: Techniques and Applications*, London: Butterworths, 1984, p. 324.

58. Gutteridge, C. S., L. Vallis, and H. J. H. Macfie, *Spec. Publ. Soc. Gen. Microbiol.* **15**, 369 (1985).

59. Gutteridge, C. S., *Methods Microbiol.* **19**, 227 (1987).

60. Haken, J. K., D. K. M. Ho, and E. Houghton, *J. Polym. Sci. Chem.* **12**, 1163 (1974).

61. Hammond, T., and R. S. Lehrle, *Br. Polym. J.* **19**, 523 (1987).

62. Hammond, T., and R. S. Lehrle, in C. Booth and C. Price, Eds., *Comprehensive Polymer Science, Vol. 1, Polymer Characterisation and Properties*, Oxford, UK: Pergamon, 1988, p. 1.

63. Harper, A. M., Meuzelaar, H. L. C., Metcalf, G. S. and Pope, D. L., *Analytical Pyrolysis: Techniques and Applications*, K. J. Vorhees, Ed., London: Butterworths, 1984, p. 157.

64. Haverkamp, J., and P. G. Kistemaker, *Int. J. Mass Spectrom. Ion Phys.* **45**, 275 (1982).

65. Hickman, D. A., and I. Jane, *Analyst* **104**, 334 (1979).

66. Holzer, G., and J. Oro, *J. Mol. Evol.* **13**, 265 (1979).

67. Hughes, J. C., B. B. Wheals, and M. J. Whitehouse, *Forensic Sci.* **10**, 217 (1977).

68. Hughes, J. C., B. B. Wheals, and M. J. Whitehouse, *Analyst* **102**, 143 (1977).

69. Hughes, J. C., B. B. Wheals, and M. J. Whitehouse, *Analyst* **103**, 482 (1978).

70. Irwin, W. J., *Analytical Pyrolysis: A Comprehensive Guide*, New York: Dekker, 1982.

71. Israel, S. C., W. C. Yang, and M. Bechard, *J. Macromol. Sci. Chem.* **A22**, 779 (1985).

72. Jones, C. E. R., and A. F. Moyles, *Nature* **189**, 222 (1961).

73. Jones, C. E. R., S. G. Perry, and N. B. Coupe, in B. Stock, Ed., *Gas Chromatography 1970*, London: Institute of Petroleum, 1971, p. 399.

74. *Journal of Analytical and Applied Pyrolysis*, Amsterdam: Elsevier, 1979 et. seq.

75. Juvet, R. S., Jr., J. L. S. Smith, and K.-P. Li, *Anal. Chem.* **44**, 49 (1972).

76. Khandelwal, J. K., P. I. Szilagyi, L. A. Barker, and J. P. Green, *Eur. J. Pharmacol.* **76**, 145 (1981).

77. Kimelt, S., and S. Speiser, *Chem. Rev.* **77**, 437 (1977).

78. Kistemaker, P. G., A. J. H. Boerboom, and H. L. C. Meuzelaar, *Dynamic Mass Spectrom.* **4**, 139 (1975).

79. Kubat, J., and J. Zachoval, *Chem. Vlastnosti Zprac.* **S14**, 227 (1986).

80. Kullik, E., M. Kaljurand, and M. Koel, *J. Chromatogr.* **126**, 249 (1976).

81. LaFleur, A. L., and K. M. Mills, *Anal. Chem.* **53**, 1202 (1981).

82. Lai, S. T., and D. C. Locke, *J. Chromatogr.* **255**, 511 (1983).

83. Lai, S. T., and D. C. Locke, *J. Chromatogr.* **314**, 283 (1984).

84. Larsen, E., H. Egsgaard, and N. Bjerre, *J. Trace Microprobe Technol.* **1**, 387 (1983).

85. Larter, S. R., H. Soli, and A. G. Douglas, *J. Chromatogr.*, **167**, 421 (1978).

86. Larter, S. R. and A. G. Douglas, *J. Anal. Appl. Pyrol.* **4**, 1 (1982).

87. Larter, S. R., in K. J. Voorhees, Ed., *Analytical Pyrolysis: Techniques and Applications*, London: Butterworths, 1984, p. 212.

88. Lattimer, R. P., K. M. Schur, W. Windig, and H. L. C. Meuzelaar, *J. Anal. Appl. Pyrol.* **8**, 95 (1985).

89. Lattimer, R. P., *J. Anal. Appl. Pyrol.* **12**, 189 (1987).

90. Lehrle, R. L., and J. C. Robb, *Nature* **183**, 1671 (1959).

91. Lehrle, R. S., and J. C. Robb, *J. Gas Chromatogr.* **5**, 89 (1967).

92. Lehrle, R. S., J. C. Robb, and R. Suggate, *Eur. Polym. J.* **18**, 443 (1982).

93. Lehrle, R. S., R. E. Peakman, and J. C. Robb, *Eur. Polym. J.* **18**, 517 (1982).

94. Lehrle, R. S., *J. Anal. Appl. Pyrol.* **11**, 55 (1987).

95. Levsen, K., and H.-R. Schulten, *Biomed. Mass Spectrom.* **3**, 137 (1976).

96. Levy, R. L., *Chromatogr. Rev.* **8**, 48 (1966).

97. Levy, R. L., *J. Gas Chromatogr.* **5**, 107 (1967).

98. Levy, R. L., D. L. Fanter, and C. J. Wolf, *Anal. Chem.* **44**, 38 (1972).

99. Levy, E. J., and J. Q. Walker, *J. Chromatogr. Sci.*, **22**, 49 (1984).

100. Liebman, S. A., and E. J. Levy, *Pyrolysis and GC in Polymer Analysis*, New York: Dekker, 1985.

101. Lloyd, R. J., *J. Chromatogr.* **284**, 357 (1984).

102. Luderwald, I., M. Przybylski, and H. Ringsdorf, *Adv. Mass Spectorm.* **7B**, 1437 (1978).

103. Lyle, S. J., and M. S. Tehrani, *J. Chromatogr.* **236**, 25, 31 (1982).

104. Lyle, S. J., and M. S. Tehrani, *J. Chromatogr.* **240**, 209 (1982).

105. MacCarthy, P., S. J. DeLuca, K. J. Voorhees, R. L. Malcolm, and E. M. Thurman, *Geochim. Cosmochim. Acta* **49**, 2091 (1985).

106. MacFie, H. J. H., C. S. Gutteridge, and J. R. Norris, *J. Gen. Microbiol.* **104**, 67 (1978).

107. McClennen, W. H., J. M. Richards, H. L. C. Meuzelaar, J. B. Pausch, and R. P. Lattimer, *Polym. Mater. Sci. Eng.* **53**, 203 (1985).

108. Madorsky, S. L. and S. Straus, *Ind. Eng. Chem.* **5**, 848 (1948).

109. Martin, S. B., *J. Chromatogr.* **2**, 272 (1959).

110. Marzec, A., *J. Anal. Appl. Pyrol.* **8**, 241 (1985).

111. Massart, D. L., B. G. M. Vandeginste, S. N. Deming, Y. Michotte, and L. Kaufman, *Chemometrics: A Textbook*, Amsterdam: Elsevier, 1987.

112. May, R. W., E. F. Pearson, and D. Scothern, *Pyrolysis–Gas Chromatography*, London: Chemical Society, 1977.

113. Mazzeo, P., and A. Mazzeo-Farina, *Microchem. J.* **28**, 137 (1983).

114. Medley, E. E., P. G. Simmonds, and S. L. Manatt, *Biomed. Mass Spectrom.* **2**, 261 (1975).

115. Metzger, J., *Z. Anal. Chem.* **295**, 45 (1979).

116. Meuzelaar, H. L. C., and P. G. Kistemaker, *Anal. Chem.* **45**, 587 (1973).

117. Meuzelaar, H. L. C., M. A. Posthumus, P. G. Kistemaker, and J. Kistemaker, *Anal. Chem.* **45**, 1546 (1973).

118. Meuzelaar, H. L. C., P. G. Kistemaker, and A. Tom, in C. A. Heden and T. Illeni, Eds., *New Approaches to the Identification of Micro-organisms*, New York: Wiley, 1975, p. 165.

119. Meuzelaar, H. L. C., H. G. Ficke, and H. C. den Harink, *J. Chromatogr. Sci.* **13**, 12 (1975).

120. Meuzelaar, H. L. C., P. G. Kistemaker, W. Eshuis, and A. J. H. Boerboom, *Adv. Mass Spectrom.* **7B**, 1452 (1978).

121. Meuzelaar, H. L. C., J. Haverkamp, and F. D. Hileman, *Curie-Point Pyrolysis Mass Spectrometry of Recent and Fossil Biomaterials; Compendium and Atlas*, Amsterdam: Elsevier, 1982.

122. Meuzelaar, H. L. C., W. Windig, A. M. Harper, S. M. Huff, W. H. McClennen, and J. M. Richards, *Science* **226**, 268 (1984).

123. Meuzelaar, H. L. C., S. M. Huff, W. Windig, and P. Kenemans, *Tijdschr. Ned. Ver. Klin. Chem.* **9**, 47 (1984).

124. Meuzelaar, H. L. C., W. Windig, S. M. Huff, and J. M. Richards, *Anal. Chim. Acta* **190**, 119 (1986).

125. Midgley, T., Jr., and A. L. Henne, *J. Am. Chem. Soc.* **51**, 1215 (1929).

126. Milina, R., *J. Anal. Appl. Pyrol.* **3**, 179 (1981).

127. Mischer, G., *Adv. Mass Spectrom.* **7B**, 1444 (1978).

128. Mitchell, A., and M. Needleman, *Anal. Chem.* **50**, 668 (1978).

129. Mol, G. J., R. J. Gritter, and G. E. Adams, *Appl. Polym. Spectrosc.* 257 (1978).

130. Montaudo, G., E. Scamporrino, C. Puglisi, and D. Vitalini, *J. Anal. Appl. Pyrol.* **10**, 283 (1987).

131. Montaudo, G., C. Puglisi, D. Vitalini, and Y. Morishima, *J. Anal. Appl. Pyrol.*, **13**, 161 (1988).

132. Nakagawa, H., S. Tsuge, and K. Murakami, *J. Anal. Appl. Pyrol.* **10**, 31 (1986).

133. Nakagawa, H., S. Tsuge, S. Mohanraj, and W. T. Ford, *Macromolecules* **21**, 930 (1988).

134. Narang, S. C., R. Malhotra, and P. E. Harisiades, *Polym. Prepr.* (*Am. Chem. Soc. Div. Polym. Chem.*) **28**, 274 (1987).

135. Needleman, M., and P. Stuchbery, in C. E. R. Jones, and C. A. Cramers, Eds., *Analytical Pyrolysis*, Amsterdam: Elsevier, 1977, p. 77.

136. Oertli, Ch., Ch. Buhler, and W. Simon, *Chromatographia* **6**, 499 (1973).

137. Ohnishi, A., K. Kato, and E. Takagi, *Polym. J.* **7**, 431 (1975).

138. Ohya, K., and M. Sano, *Biomed. Mass Spectrom.* **4**, 241 (1977).

139. Perry, S. G., *J. Chromatogr. Sci.* **7**, 193 (1969).

140. Polak, R. L., and P. C. Molenaar, *J. Neurochem.* **32**, 407 (1979).

141. Posthumus, M. A., N. M. M. Nibbering, A. J. H. Boerboom, and H.-R. Schulten, *Biomed. Mass Spectrom.* **1**, 352 (1974).

142. Quinn, P. A., *J. Chromatogr. Sci.* **12**, 796 (1974).

143. Radell, E. A., and H. C. Strutz, *Anal. Chem.* **31**, 1890 (1959).

144. Radhakrishnan, T. S., and M. Rao, *J. Anal. Appl. Pyrol.* **9**, 309 (1986).

145. Reiner, E., and T. F. Moran, *Adv. Chem. Ser.* **203**, 705 (1983).

146. Risby, T. H., and A. L. Yergey, *J. Phys. Chem.* **80**, 2839 (1976).

147. Risby, T. H., and A. L. Yergey, *Anal. Chem.* **50**, 327A (1978).

148. Ristau, W. T., and N. E. Vanderborgh, *Anal. Chem.* **44**, 359 (1972).

149. Roche, R. S., D. R. Salomon, and A. A. Levinson, *Appl. Geochem.* **1**, 619 (1986).

150. Roy, C., E. Chornet, and C. H. Fuchsman, *J. Anal. Appl. Pyrol.* **5**, 261 (1983).

151. Rushneck, D. R., A. V. Diaz, D. W. Howarth, J. Rampacek, K. W. Olsen, W. D. Dencker, P. Smith, L. McDavid, A. Tomassian, M. Harris, K. Bulota, *Rev. Sci. Instrum.* **49**, 817 (1978).

152. Saferstein, R., and J. J. Manura, *J. Forensic Sci.* **22**, 748 (1977).

153. Saiz-Jimenez, C., and J. W. de Leeuw, *J. Anal. Appl. Pyrol.* **9**, 99 (1986).

154. Schlotzhauer, W. S., and O. T. Chortyk, *J. Anal. Appl. Pyrol.* **12**, 189 (1987).

155. Schmid, J. P., P. P. Schmid, and W. Simon, *Chromatographia* **9**, 597 (1976).

156. Schmid, P. P., and W. Simon, *Anal. Chim. Acta* **89**, 1 (1977).

157. Schulten, H.-R., H. D. Beckey, H. L. C. Meuzelaar, and A. J. H. Boerboom, *Anal. Chem.* **45**, 191 (1973).

158. Schulten, H.-R., H. D. Beckey, A. J. H. Boerboom, and H. L. C. Meuzelaar, *Anal. Chem.* **45**, 2358 (1973).

159. Schulten, H.-R., in C. A. Heden and T. Illeni, Eds., *New Approaches to the Identification of Microorganisms*, New York: Wiley, 1975, p. 155.

160. Schulten, H.-R., in C. E. R. Jones and C. A. Cramers, Eds., *Analytical Pyrolysis*, Amsterdam: Elsevier, 1977, p. 17.

161. Schulten, H.-R., and W. Gortz, *Anal. Chem.* **50**, 428 (1978).

162. Schulten, H.-R., and H. J. Duessel, *J. Anal. Appl. Pyrol.* **2**, 293 (1981).

163. Schulten, H.-R., *Fuel* **61**, 670 (1982).

164. Schulten, H.-R., N. Simmleit, and R. Mueller, *Anal. Chem.* **59**, 2903 (1987).

165. Schulten, H.-R., B. Plage, H. Ohtani, and S. Tsuge, *Angew. Makromol. Chem.* **155**, 1 (1987).

166. Schulten, H.-R., *J. Anal. Appl. Pyrol.* **12**, 149 (1987).

167. Sellier, N., C. E. R. Jones, and G. Guiochon, in C. E. R. Jones and C. A. Cramers, Eds., *Analytical Pyrolysis*, Amsterdam: Elsevier, 1977, p. 309.

168. Senoo, H., S. Tsuge, and T. Takeuchi, *Makromol. Chem.* **161**, 185 (1972).

169. Shafizadeh, F., *J. Anal. Appl. Pyrol.* **3**, 282 (1982).

170. Shimizu, Y., and B. Munson, *J. Polym. Sci. (Chem. Ed.)* **17**, 1991 (1979).

171. Simmonds, P. G., G. P. Shulman, and C. H. Stembridge, *J. Chromatogr. Sci.* **7**, 36 (1969).

172. Smith, C. G., in K. J. Voorhees, Ed., *Analytical Pyrolysis. Techniques and Applications*, London: Butterworths, 1984, p. 428.

173. Smith, C. S., S. L. Morgan, C. D. Parks, A. Fox, and D. G. Pritchard, *Anal. Chem.* **59**, 1410 (1987).

174. Smith, D. F., and J. R. Durig, *J. Anal. Appl. Pyrol.* **13**, 63 (1988).

175. Snyder, A. P., J. H. Kremer, H. L. C. Meuzelaar, and W. Windig, *Anal. Chem.* **59**, 1945 (1987).

176. Stack, M. V., H. D. Donoghue, J. E. Tyler, and M. Marshall, in C. E. R. Jones and C. A. Cramers, Eds., *Analytical Pyrolysis*, Amsterdam: Elsevier, 1977, p. 57.

177. Staudinger, H., and A. Steinhofer, *Liebig's Ann. Chem.* **517**, 35 (1935).

178. Stewart, W. D., Jr., *J. Ass. Offic. Anal. Chem.* **59**, 35 (1976).

179. Stoev, G., and I. Mladenov, *Dokl. Bolg. Akad. Nauk.* **32**, 1385 (1979).

180. Sugimura, Y., and S. Tsuge, *Anal. Chem.* **50**, 1968 (1978).

181. Sullivan, J. F., and R. L. Grob, *J. Chromatogr.* **268**, 219 (1983).

182. Truett, W. L., *Am. Lab.* **9**, 33 (1977).

183. Tsao, R., and K. Voorhees, *Anal. Chem.* **56**, 368 (1984).

184. Tsao, R., and K. Voorhees, *Anal. Chem.* **56**, 1339 (1984).

185. Tsuge, S., and T. Takeuchi, *Anal. Chem.* **49**, 348 (1977).

186. Tsuge, S., Y. Sugimura, and T. Nagaya, *J. Anal. Appl. Pyrol.* **1**, 221 (1980).

187. Tsuge, S., in K. J. Voorhees, Ed., *Analytical Pyrolysis: Techniques and Applications*, London: Butterworths, 1984, p. 407.

188. Tsuge, S., and H. Matsubara, *J. Anal. Appl. Pyrol.* **8**, 49 (1985).

189. Tsuge, S., *Bunseki Kagaku* **35**, 417 (1986).

190. Tsuge, S., H. Ohtani, H. Matsubara, and M. Ohsawa, *J. Anal. Appl. Pyrol.* **11**, 181 (1987).

191. Tuithof, H. H., A. J. H. Boerboom, P. G. Kistemaker, and H. L. C. Meuzelaar, *Adv. Mass Spectrom.* **7B**, 838 (1978).

192. Tyden-Ericsson, I., *Chromatographia* **6**, 353 (1973).

193. Usami, T., Y. Gotoh, S. Takayama, H. Ohtani, and S. Tsuge, *Macromolecules* **20**, 1557 (1987).

194. Venema, A., and J. Veurink, *J. Anal. Appl. Pyrol.* **7**, 207 (1985).

195. Vollmin, J., P. Kriemler, I. Omura, J. Seibl, and W. Simon, *Michrochem. J.* **11**, 73 (1966).

196. Vorhees, K. J., S. L. Durfee, J. R. Holtzclaw, C. G. Enke, and M. R. Bauer, *J. Anal. Appl. Pyrol.* **14**, 7 (1988).

197. Walker, J. Q., *Chromatographia* **5**, 547 (1972).

198. Walker, J. Q., *J. Chromatogr. Sci.* **15**, 267 (1977).

199. Wall, L. A., *J. Res. Nat. Bur. Stand.* **41**, 315 (1948).

200. Wampler, T. P., and E. J. Levy, *J. Anal. Appl. Pyrol.* **8**, 65 (1985).

201. Wampler, T. P., and E. J. Levy, *J. Anal. Appl. Pyrol.* **8**, 153 (1985).

202. Wampler, T. P., and E. J. Levy, *J. Anal. Appl. Pyrol.* **12**, 75 (1987).

203. Wells, G., K. J. Voorhees, and J. H. Futrell, *Anal. Chem.* **52**, 1782 (1980).

204. Westall, W. A., and A. J. Pidduck, *J. Anal. Appl. Pyrol.* **11**, 3 (1987).

205. Wheals, B. B., *J. Anal. Appl. Pyrol.* **2**, 277 (1981).

206. Wheals, B. B., *J. Anal. Appl. Pyrol.* **8**, 503 (1985).

207. Whitehouse, M. J., J. J. Boon, J. M. Bracewell, C. S. Gutteridge, A. J. Pidduck, and D. J. Puckey, *J. Anal. Appl. Pyrol.* **8**, 515 (1985).

208. Whiton, R. S., and S. L. Morgan, *Anal. Chem.* **57**, 778 (1985).

209. Williams, C. G., *J. Chem. Soc.* **15**, 110 (1862).

210. Windig, W., P. G. Kistemaker, J. Haverkamp, and H. L. C. Meuzelaar, *J. Anal. Appl. Pyrol.* **1**, 39 (1979).

211. Windig, W., P. G. Kistemaker, J. Haverkamp, and H. L. C. Meuzelaar, *J. Anal. Appl. Pyrol.* **2**, 7 (1980).

212. Windig, W., J. Haverkamp, and P. G. Kistemaker, *Anal. Chem.* **55**, 81 (1983).

213. Windig, W., and H. L. C. Meuzelaar, *Anal. Chem.* **56**, 2297 (1984).

214. Wolff, D. D., and M. L. Parsons, *Pattern Recognition Approach to Data Interpretation*, New York: Plenum, 1983.

215. Wright, D. W., K. O. Mahler, L. B. Ballard, and E. Dawes, *J. Chromatogr. Sci.* **24**, 13 (1986).

216. Zemany, P. D., *Anal. Chem.* **24**, 1709 (1952).

Chapter 6

APPLICATION OF THERMAL ANALYSIS TO PROBLEMS IN CEMENT CHEMISTRY

By Javed I. Bhatty, *Construction Technology Laboratories Inc., Skokie, Illinois*

Contents

Treatise on Analytical Chemistry, Part 1, Volume 13,
Second Edition: Thermal Methods, Edited by James D. Winefordner.
ISBN 0-471-80647-1   © 1993 John Wiley & Sons, Inc.

I. INTRODUCTION

The discovery of the technique of differential thermal analysis (DTA) is usually credited to Le Chatelier (58), as a result of his report in 1887 on heat on the constitution and decomposition of clays and limestones. Since then the application of DTA has become more versatile in both industrial and scientific fields, such as ceramics, glass, polymers, pharmaceuticals, foods, oils, and cementitious materials. During the last two decades the use of DTA in the cement industry has found diverse application. The ability to detect minor chemical changes and provide relevant analytical information for portland cement and its hydration products makes it one of the foremost analytical methods, along with x-ray diffraction and scanning electron micro-copy. The technique is convenient, fast, and accurate and provides additional information that is not readily available using other techniques. In addition, it has special applications in monitoring the hydration behavior and the formation of hydration products during the setting of individual cement phases, such as tricalcium silicate, dicalcium silicate, tricalcium aluminates, and tetracalcium alumino ferrite, with or without the presence of inorganic additives like fly ash, slags, and silica fumes and organic admixtures such as accelerators, retarders, and superplasticizers. DTA has also found applica-tions in studying other hydraulic cements, such as gypsum plasters, high alumina cements, and pozzolanas.

Other thermoanalytical techniques frequently used in cement analysis are thermogravimetric analysis (TG), differential thermogravimetery (DTG), derivative differential thermal analysis (DDTA), differential scanning calorimetry (DSC), and isothermal calorimetry. Reviews by Ramachandran (93), Barta (2), Mackenzie (60), and more recently Ben-Dor (8) have covered the applications, techniques, and interpretation of thermal analysis in cement sciences.

To have detailed understanding of the application of thermal analysis in cement chemistry, it would be appropriate to briefly describe the manufactur-ing of portland cement and various raw materials that are used in the process.

A. RAW MATERIALS IN CEMENT MAKING

Portland cement is manufactured by burning an intimate mixture of limestone and clay at 1500°C which chemically react to form clinker. The clinker is then interground with up to 5% gypsum, added as set regulator, to a Blaine fineness of 3500 cm^2/g to form portland cement. The silicate phases

TABLE 1
Raw Materials for Portland Cement Industry

Carbonates	Limestone, chalk, marble, sea shell, and marl
Alumino silicates	Clay, shale, and slate
Corrective materials	Sand and sandstones, bauxite, iron ore, laterite
Mineralizers	CaF_2, Na_2SiF_6, $Ca_5(PO_4)_3$, $CaSO_4 \cdot 2H_2O$, etc.
Set retarder	Gypsum
Hydraulic blending materials	Natural pozzolanic rocks, blast-furnace slags, fly ash

SOURCE: Chatterjee (23).

formed in the clinker are the principal strength developers in cement and are present up to 90% by weight.

The worldwide production of portland cement is over one billion tons (77), which requires more than 1.5 billion tons of raw materials and some 200 million tons of fuel (23). There are 143 cement plants that operate in the United States alone and produce about 80 million tons of cement annually. This figure also includes about 4 million tons of masonry cement and minor quantities of high-alumina cement (77).

Limestone and clay are the most common sources of lime (calcium oxide) and silica, respectively, required in cement making. Secondary sources of lime such as chalk and shell deposits are also used if necessary. Other sources of silica are silts, shales, and slates. The presence of aluminates and iron-bearing minerals in the raw materials result in the formation of other phases in clinkers such as tricalcium aluminate and tetracalcium aluminoferrite. Additives and corrective materials are therefore added to the raw mix to compensate for any compositional deficiency or to correct marginal deviations from the desired compositions (23). Additives are generally of argillaceous type such as mudstones and siltstones, whereas the corrective materials are the ferruginous types such as ironstone and bauxites. Natural gypsum is later added as set regulator during the grinding of clinker. A list of various raw materials used in cement making is given in Table 1.

The reaction between calcium and silica results in the formation of two major phases, tricalcium silicate and dicalcium silicate, whereas alumino silicate and iron minerals present in the raw material will form tricalcium aluminate and tetracalcium aluminoferrite. The reactions that lead to the formation of these compounds take place as follows:

$$3CaO + SiO_2 \longrightarrow 3CaO \cdot SiO_2 \tag{1}$$

$$2CaO + SiO_2 \longrightarrow 2CaO \cdot SiO_2 \tag{2}$$

$$3CaO + Al_2O_3 \longrightarrow 3CaO \cdot Al_2O_3 \tag{3}$$

$$4CaO + Al_2O_3 + Fe_2O_3 \longrightarrow 4CaO \cdot Al_2O_3 \cdot Fe_2O_3 \tag{4}$$

TABLE 2
Approximate Composition Limits of Portland Cement

Oxides	Content (%)
CaO	60–67
SiO_2	17–25
Al_2O_3	3–8
Fe_2O_3	0.5–6.0
MgO	0.1–4.0
Alkalis	0.2–1.3
SO_3	1–3

TABLE 3
Typical Compound Composition of Ordinary Portland Cement

Chemical Name	Chemical Formula	Cement Notation[a]	Weight (%)
Tricalcium silicate	$3CaO \cdot SiO_2$	C_3S	50
Dicalcium silicate	$2CaO \cdot SiO_2$	C_2S	25
Tricalcium aluminate	$3CaO \cdot Al_2O_3$	C_3A	12
Tetracalcium aluminoferrite	$4CaO \cdot Al_2O_3 \cdot Fe_2O_3$	C_4AF	8
Calcium sulfate dihydrate (gypsum)	$CaSO_4 \cdot 2H_2O$	$C\bar{S}H_2$	3.5

[a]$C = CaO, S = SiO_2, A = Al_2O_3, F = Fe_2O_3, \bar{S} = SO_3, H = H_2O$.

The formation of these compounds is largely dependent on the heat treatment of raw material and cooling history of the clinker. An approximate chemical composition of a portland cement is given in Table 2, and a typical compound composition along with the notations frequently used in cement chemistry are given in Table 3.

II. THERMAL ANALYSIS OF MAJOR RAW MATERIALS

Owing to the nature of the technique, DTA can readily be used for simulating closely the conditions existing in the rotary kiln during clinker formation (87). The level and rate of heating of the samples as well as the atmospheric conditions in DTA can be adjusted and matched to that of the kiln, thereby suggesting DTA and also TG to be powerful tools for characterizing not only the raw material but also the finished product.

Raw materials of high purity and uniform composition are required to ensure the quality production of cement. Deleterious materials in the raw materials are always minimized to avoid production of cement of variable compositions and unpredictable properties. The use of thermal analysis (DTA and TG in particular) in the qualitative analysis of these materials is of useful technical advantage because it can detect the presence of any deleterious material prior to the preparation of raw feed.

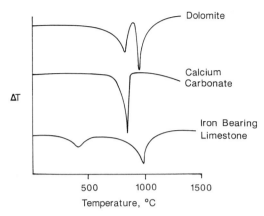

Fig. 1.   DTA curves for some carbonates important to the cement industry.

A.   LIMESTONE

The presence of alkalies such as Na, K, and Mg in limestone can easily add to the complexity of reaction during clinker formation and give inferior clinker that forms less durable cements. Typical DTA and TGA plots for limestone ($CaCO_3$) are given in Fig. 1. An endothermic peak between 850 and 950°C results from the decomposition of $CaCO_3$ with the evolution of CO_2 and is accompanied by a weight loss. The presence of large amounts of MgO as a contaminant in limestone, however, produces an additional endothermic peak. Dolomite exhibits two endothermic peaks. The first endothermic peak is caused by the decomposition of $MgCO_3$, and the second is caused by the decomposition of $CaCO_3$. If dolomite is present as contaminant, the peaks will be slightly shifted (93). In carbonate rocks dolomite can be estimated by using DTA even if present as low as 0.3% (10). Pure magnesite, $MgCO_3$, is identified by an endothermic peak around 700°C due to the decomposition and release of CO_2, whereas brucite (magnesium hydroxide, $Mg(OH)_2$) bearing limestones can show an endothermic effect around 475°C as a result of dehydroxylation. Iron-bearing limestones show an additional endothermic peak between 350 and 400°C. Selected DTA peaks of dolomite and iron-bearing limestones are also shown in Fig. 1.

B.   CLAY

Clays, shales, slates, and schists provide ingredients such as SiO_2, Al_2O_3, Fe_2O_3, and alkalies that take part in the formation of the essential phases, such as silicates, aluminates, and aluminoferrites in cement clinker. The most desirable clay minerals in cement making are kaolinite, which is a hydrous aluminum silicate $Al_4Si_4O_{10}(OH)_8$, and montmorillonite, essentially $Al_4Si_8O_{20}(OH)_4 \cdot nH_2O$ with the substitution of Mg for part of Al (108). These clays contain mainly Al_2O_3 and SiO_2 with much less alkali content and

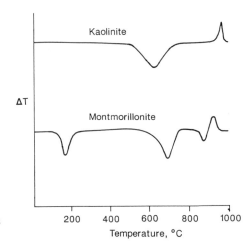

ΔT

Fig. 2. DTA curves for some clay minerals important to the cement industry.

also satisfy other regulatory requirements. Kaolins are preferred over pure quartz because they are already naturally finely ground, whereas the use of quartz involves expensive grinding (93) and may require extra heat during burning. Natural clays may require less heat due to the presence of some natural fluxes.

DTA has been effectively used in identifying clay minerals in the past as reported by Ramachandran (93) and Mehta (68). Typical DTA plots of two major clay minerals, kaolinite and montmorillonite, are given in Fig. 2. A highly crystalline kaolinite mineral shows a characteristic endothermic peak due to dehydroxylation at 550–600°C and an exothermic peak around 1000°C due to the formation of γ-Al$_2$O$_3$ or mullite nucleation. A poorly crystalline kaolinite can show an additional small endothermic peak at about 100°C due to dehydration.

Montmorillonite clays definitely show an endothermic peak around 100°C as a result of dehydration, followed by another peak between 600 and 700°C due to loss of lattice water. Another endothermic peak noted at about 900°C is due to lattice breakdown, and a final exothermic peak at 955°C is due to the formation of spinel.

Ampian and Flint (1) have used the DTA and TG techniques in monitoring the effects of mineralizers such as silicofluorides on the formation of major clinker components such as C$_3$S, C$_2$S, C$_3$A, and C$_4$AF. They estimated the extent of their effects by measuring the free lime content in the corresponding reactions and substantiating the results using x-ray diffraction data.

Chen (25) has used DTA and TG to troubleshoot the clinker-burning process problems of some of the raw materials containing minor constituents such as K, Na, S, Cl, and F. These are most troublesome constituents, for they tend to form low-melting volatile salts such as CaF$_2$, KCl, K$_2$SO$_4$, Na$_2$SO$_4$, and CaSO$_4$. These salts condense on cooler refractory surfaces and

also cause plugging problems in the suspension preheaters. Chen (25) empha-
sized the application of DTA in the identification of spurrite ($2CaO \cdot SiO_2 \cdot CaCO_3$), which forms during the clinker-burning process, by referring to the
determination of melting/freezing points and the operational problems
caused by the volatability of this problem compound.

III.   THERMAL ANALYSIS IN THE CHARACTERIZATION
OF CEMENT PHASES

Major phases in the portland clinker are tricalcium silicate (C_3S) and
β-dicalcium silicate (β-C_2S). C_3S is by far the most abundant compound,
present up to 60% by weight, and is regarded as the primary strength
developer when hydrated with water. C_2S is present up to 25% by weight and
is vital in contributing to the ultimate strength of cement. Another phase that
can be present up to 15% by weight in cement is tricalcium aluminate (C_3A).
C_3A reacts violently with water and undergoes flash setting. To avoid this
and to regulate the setting properties of the cement, a small quantity of
gypsum is added during clinker grinding. C_3A is responsible for the early
stiffening of cement paste. Tetracalcium aluminoferrite (C_4AF) is another
phase present in a reasonable quantity (up to 8% by weight) and affects the
early hydration properties of cement. The C_4AF phase is also helpful in
reducing chemical attack on cement-based concretes.

A.   TRICALCIUM SILICATE (C_3S)

At room temperature C_3S occurs in three crystallographic forms: monoclinic, triclinic, and trigonal. The monoclinic form is present in the cement
clinker and is close in structure to the mineral alite. DTA plots shown in Fig.
3 of C_3S and alite differ significantly, however. An alite plot shows endothermic peaks at 825 and 1427°C due to the monoclinic–trigonal transition and to
α transitions, respectively. The DTA plot of C_3S shows a number of endothermic peaks at 464, 923, 980, and 1465°C. The peak at 465°C is due to
dehydroxylation of $Ca(OH)_2$ formed from the hydration of free lime present

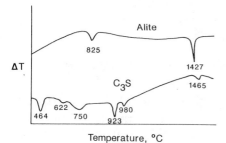

Temperature, °C   Fig. 3.   DTA curves for alite and C_3S.

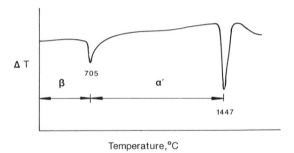

Fig. 4.   DTA curve of β-C$_2$S.

in the preparation of C$_3$S. Reactions at 920 and 980°C are typical of C$_3$S and could be the result of triclinic–monoclinic transition and trigonal modifications. A peak at 1465°C can be due to an α transition. C$_2$S is present as a contamination, which shows peaks at 622, 750, and 1465°C due to the transitions of triclinic to monoclinic and other similar solid state transitions often accompanied by anion rotation (93).

Maki and Kato (61) have also applied the DTA and TG techniques in identifying the different forms of alite phase (i.e., the C$_3$S) in a variety of cement clinkers. They have also suggested that the endothermic peaks in the temperature region 790–850°C are due to the transitions of alite.

B.   DICALCIUM SILICATE (C$_2$S)

The most desirable form of dicalcium silicate in the cement clinker is the β-C$_2$S, because its hydraulic properties adequately influence the mechanical strength. Also occasionally present in the ore clinker is γ-C$_2$S, which most probably forms due to an inadequate presence of such additives as P$_2$O$_5$, BaO, Mn$_2$O$_3$, etc. These additives generally prevent the $\beta \rightarrow \gamma$ inversions. γ-C$_2$S does not develop significant mechanical strength when hydrated.

A typical DTA plot of β-C$_2$S stabilized by CaO is given in Fig. 4, which shows two sharp endothermic peaks at 705 and 1447°C, which are, respectively, the result of $\beta \rightarrow \alpha'$ and $\alpha' \rightarrow \alpha$ transitions. A DTA plot of γ-C$_2$S also shows two endothermic peaks at 780 and 1447°C. These peaks are due to transitions of $\gamma \rightarrow \alpha'C_2$S and $\alpha' \rightarrow \alpha$C$_2$S, respectively.

C.   TRICALCIUM ALUMINATE (C$_3$A) AND TETRACALCIUM ALUMINOFERRITE (C$_4$AF)

The other important phases in the cement clinker, C$_3$A and C$_4$AF, require the presence of gypsum to prevent flash set during hydration. Their thermal and hydraulic characteristics are described in forthcoming sections.

TABLE 4
Heats of Hydration of Different Cement Compounds

Compounds	J/g
C_3S	500
C_2S	250
C_3A	1340
C_4AF	420
Free CaO	1150
Free MgO	840

IV. THERMAL CHARACTERIZATION OF HYDRATING CEMENT AND INDIVIDUAL PHASES

A. ISOTHERMAL CONDUCTION CALORIMETRY OF THE CEMENT–WATER SYSTEM

Although this chapter is primarily devoted to the use of DTA and TG techniques in cement chemistry, a brief description of isothermal conduction calorimetry and its applications in estimating the heats of hydration of major phases is also presented.

Upon contact with water, cement undergoes hydration reactions with a detectible evolution of heat, which in fact, is the resultant of the heats of hydration released by individual cement phases. Heats of hydration of individual cement phases are given in Table 4. The degrees of hydration at given curing times are proportional to their heats of hydrations. The C_3A phase, being most reactive, and the C_2S phase, being least reactive, show high and low degrees of hydration, respectively. The hydration reactions of portland cement are all exothermic. The contribution of an individual compound to the overall rate of heat evolution is a function of how much of it there is in the cement, its heat capacity, and the rates of hydration.

The heat evolution plots for C_3S and C_3A are given in Figs. 5 and 6. These compounds, being the most abundant and reactive components, respectively, in cement, are the compounds of major interest. When mixed with water, a rapid evolution of heat occurs (stage 1), followed by a dormant period of relative inactivity (stage 2). For C_3A the heat evolution is much higher and the dormant periods are shorter that those of C_3S. At the end of dormant periods both C_3S and C_3A react with renewed vigor and continue to hydrate, reaching maximum hydration toward the end of the acceleration period (stage 3). The rates of reaction slow down during a deceleration period (stage 4), eventually reaching a steady state toward the end of the hydration reaction, which finally becomes diffusion controlled. It may be noted that the maximum hydration rate for C_3A is again higher than for C_3S. Thus, although the calorimetric plot looks qualitatively similar to that of C_3A, the underlying reactions are different and the amount of heat evolved

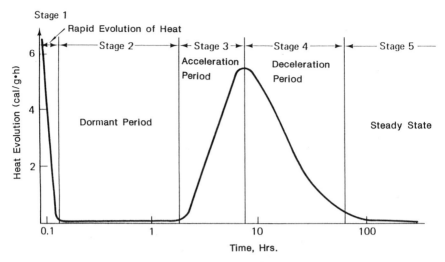

Fig. 5.   Heat evolution as a function of time for hydrating C_3S.

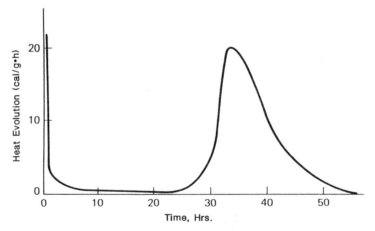

Fig. 6.   Heat evolution as a function of time for hydrating C_3A.

is much less. It is possible to estimate the heat of hydration of cement at any given time provided the composition, the degrees of hydration, and respective ΔH of the reaction are known. A compound curve of cement containing both C_3S and C_3A is shown in Fig. 7, which clearly distinguishes between the contributions of both these components.

Further elaboration on isothermal conduction calorimetry in cement hydration is beyond the scope of this chapter.

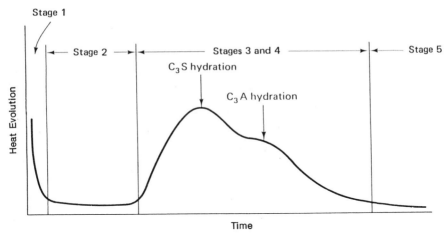

Fig. 7.   Heat evolution as a function of time for a mixture of C_3S and C_3A.

B.   DIFFERENTIAL THERMAL ANALYSIS

The use of thermal analysis techniques in monitoring the hydration behavior of portland cement phases has given rise to a better understanding of cement hydration and appreciation of numerous practical situations. The work done on cement hydration is well documented in various reviews by Mackenzie (60), Ramachandran (93), Barta (2), and Ben-Dor (8).

Numerous workers have contributed to the use of thermal analysis in studying the hydration charcteristics of cement with or without the presence of admixtures. Ramachandran and coworkers (47, 65, 94, 95, 97–99, 101, 102, 104, 105) studied the hydration of cement pastes (and of individual phases) in the presence of admixtures; Ben-Dor and coworkers (3, 4, 6, 7) and Bensted (9, 10, 12–14) studied the effect of accelerators; El-Jazairi and Illston (42–44) devised a semiisothermal approach to study cement hydration; Midgley and coworkers (69, 74) proposed a method for deriving free $Ca(OH0_2$ of hydrated cement; Maycock, Jawed, Skalny, and coworkers (51, 53, 67) studied individual hydrating phases; and Singh (117, 118), Dollimore, Bhatty, and coworkers (16, 18–21, 32–34), and Mitsuda, Taylor, Dent-Glasser, and coworkers (30, 48, 52, 79, 80) studied the formation of $Ca(OH)_2$ (and also the calcium silicate hydrate gel) during the hydration of cement pastes with or without the presence of admixtures.

1.   Tricalcium Silicate (C_3S) and Dicalcium Silicate (C_2S)

C_3S is the major component and undergoes a steady hydration reaction with water, giving moderate heat of reaction. Both C_3S and C_2S give similar hydration products upon hydration. The reaction takes place according to the following approximate chemical equations:

$$2C_3S + 6H \longrightarrow C_3S_2H_3 + 3CH \tag{5}$$

$$2C_2S + 4H \longrightarrow C_3S_2H_3 + CH \tag{6}$$

The products formed are a calcium hydrate $C_3S_2H_3$, commonly referred to as C–S–H gel, and calcium hydroxide (CH). The formula for C–S–H $(Ca_3Si_2O_7 \cdot 3H_2O)$ is only a rough approximation, because the product is nonstoichiometric and more than one variety of C–S–H is formed. Less CH is formed on a per molecular basis in the case of C_2S hydration, which has certain advantages for strength development. CH may be strength limiting because of its tendency to cleave under shear. It can also limit the durability of cement because of its greater solubility relative to C–S–H in aggressive chemicals.

Both the C–S–H and CH phases can effectively be detected by a combination of available DTA and TG methods. Thermal studies on C_3S hydration and the estimation of C–S–H formed during hydration have been conducted by numerous workers, such as Ben-Dor (3, 4, 6, 7), Ramachandran, Mascolo, and coworkers (65, 94, 95, 102, 105), Bensted (9, 10, 12–14), Midgley (74), and Mitsuda, Taylor and coworkers (48, 52, 79, 80).

A neat cement paste (i.e., cement–water paste without admixture) gives typical DTA and TG plots, as shown in Fig. 8. Three dominant hydrothermal peaks appear. For simplification these plots can be divided into three broad temperature regions giving major decomposition reactions as follows:

1. An endotherm occurs between 105 and 440°C showing a dehydration reaction ascribed mostly to the loss of water by calcium silicate hydrates. The peak in this region is less well defined and may also include dehydroxylation reactions and some desorption.

2. Between 440 and 580°C an endotherm occurs exclusively due to the dehydroxylation of calcium hydroxide formed by the hydrolysis of the

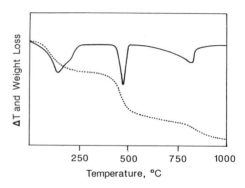

Fig. 8. TG and DTA curves of a cement–water paste.

calcium silicates originally present. The peak in this region is well defined and any other reactions are unlikely to contribute here.

3. Between 580 and 1000°C, decarbonation of calcium carbonate, formed during exposure to air, occurs together with possible solid–solid phase transitions.

Typical weight losses associated with these decomposition regions are derived from TG plots and are also shown in Fig. 8. The rates of these reactions and the formation of relevant amounts of hydration products depends on various factors, such as hydration temperatures, surface areas of starting materials, and water–cement ratios (87). At a given temperature and water–cement ratio, tricalcium silicate hydrates faster than dicalcium silicate and forms three times as much calcium hydroxide per molecule of original silicate (see Eqs. 1 and 2). Hydration products contain bound water incorporated within the calcium silicate hydrate gel and free calcium hydroxide that remains unreacted to the atmospheric carbon dioxide, both of which are widely regarded as a measure of the degree of hydration for a given cement paste (40, 43, 74,). At normal temperature, hydration products have a very low degree of crystallinity and are best detected by DTA and TG, which, because of their ability to detect the chemical changes occurring during hydration, are extensively used in cement hydration analysis.

The rate of the hydration reaction and the formation of the resultant products are also reflected by the peak areas and/or the peak heights. The degree of hydration increases with time and is best estimated from the amounts of chemically bound water and free calcium hydroxide formed.

A typical example of weight losses associated with the DTA peaks for the two cements designated as "tailing" and ordinary type 1 portland cements is given in Table 5. The tailing cement was developed using local iron and copper–nickel mineral tailings (16, 17).

a. ESTIMATION OF CHEMICALLY BOUND WATER

From the TG plots the amount of bound water in hydrating cement has been calculated by El-Jazairi and Illston (43, 44) by taking into account the weight losses during dehydration and dehydroxylation periods. The actual water loss between 105 and 1000°C is given by the sum of dehydration (L_{dh}) and dehydroxylation (L_{dx}) losses. The water loss, however, is less than the total ignition loss, since it also accounts for the decarbonation loss in the temperature region 580–1000°C. From these values, estimation of bound water can be carried out by the following expression (43, 44):

$$\text{Chemically bound water} = L_{dh} + L_{dx} + 0.41 L_{de} \qquad (7)$$

where L_{de} is the decarbonation loss and the factor 0.41 corrects for water loss equivalent to that of decarbonation occurring in this region, assuming that the carbonate is formed by carbon dioxide reaction with calcium hydrox-

TABLE 5

Weight Losses During Various Decomposition Regions and Values of Total Water Loss and
Ignition Loss Derived from TG Plots of Both the Tailing Cement and
Type 1 Cement Cured for Varying Times

| Curing Time (days) | % Weight Loss During | | | Actual Water Loss $(L_{dh} + L_{dx})$ | Ignition Loss $(L_{dh} + L_{dx} + L_{de})$ |
	Dehydration 105–440°C (L_{dh})	Dehydroxylation 440–550°C (L_{dx})	Decarbonation 580–1000°C (L_{de})		
			Tailing Cement		
1	5.0	4.2	1.8	9.2	11.0
3	6.7	4.5	2.0	11.2	13.2
7	9.1	5.2	2.0	15.0	17.0
28	10.8	6.0	2.5	16.8	19.3
120	12.0	6.2	3.4	18.2	21.6
200	12.4	6.5	3.8	18.9	22.6
			Type 1 Cement		
1	3.6	1.2	3.6	4.8	8.4
3	5.6	2.4	3.9	8.0	11.9
7	7.0	3.2	4.0	10.2	14.2
28	9.4	4.2	4.4	13.6	18.0
120	10.0	4.6	4.7	14.6	19.3
200	10.5	5.0	5.0	15.5	20.5

SOURCE: Bhatty and Reid (16).

ide present during hydration. Examples of calculated values of chemically bound water for both the tailing and type I cement are given in Table 6.

b. ESTIMATION OF FREE CALCIUM HYDROXIDE

Two major products of the reaction between portland cement and water are colloidal C–S–H gel and crystalline CH. In a fully hydrated cement CH comprises approximately 20% of the volume. Methods of CH determination are many, including references to thermal analysis.

TABLE 6

Calculated Values of Chemically Bound Water, Free Calcium Hydroxide,
and Degree of Hydration of Tailing and Type 1 Cements

| Curing Time (days) | Tailing Cement | | | Type 1 Cement | | |
	Chemically Bound Water (wt%)	Free Calcium Hydroxide (wt%)	Degree of Hydration (%)	Chemically Bound Water (wt%)	Free Calcium Hydroxide (wt%)	Degree of Hydration (%)
1	9.94	17.26	47.33	6.28	4.93	29.90
3	12.02	18.50	57.24	9.64	9.86	45.95
7	15.12	21.37	72.00	11.64	13.15	55.43
28	17.83	24.66	84.90	15.36	17.26	73.14
120	19.59	25.48	93.33	16.53	18.91	78.21
200	20.46	26.72	97.62	17.55	20.55	83.57

SOURCE: Bhatty and Reid (16).

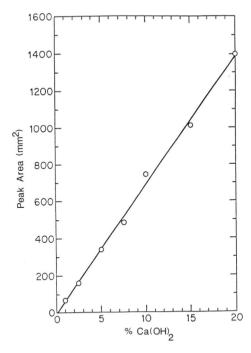

Fig. 9. Calibration curve for calcium hydroxide taken from hydrated cement pastes (18).

Midgley and Rosaman (69, 74), Ramachandran (104), Dollimore and coworkers (33), and Bhatty and coworkers (16, 18, 21) have used the DTA technique for estimating the CH content by constructing a calibration curve from the endothermic peak areas obtained in the temperature range 410–560°C for mixtures of alumina and CH or C_3S. The endothermic peak is caused by the decomposition of CH to lime (CaO) and H_2O. A good linear relationship for samples containing up to 20% CH verifies the applicability of the DTA technique (34). Ramachandran (105) showed that a linear relationship can exist even up to 33% of CH. A calibration plot of CH against peak area is given in Fig. 9.

The decomposition temperature of CH can vary between 460 and 560°C, depending on the grain size of CH and the experimental variables such as sample mass and the heating rate during the DTA run. Rate of growth and morphology of CH is also influenced by temperature, water/cement ratio, and the presence of admixtures.

In other methods El-Jazairi and Illston (43, 44) estimated the amounts of CH by taking into account both the dehydroxylation and decarbonation losses during the DTA run of a neat cement paste. The amount of CH produced during the hydration of cements is derived from both the dehydroxylation and decarbonation losses by using the following expression:

$$\text{Calcium hydroxide (CH)} = 4.11 L_{dx} + 1.68 L_{dc} \qquad (8)$$

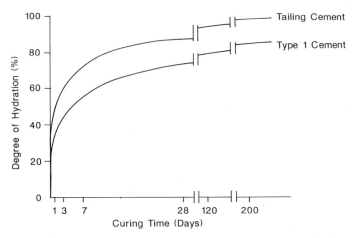

Fig. 10.   Degree of hydration of cements as a function of time (16).

Factors 4.11 and 1.68 are corrections for CH formed during hydroxylation and decarbonation, respectively. The first part of Eq. 8 gives the amount of free CH formed during hydration.

Calculated values of free CH for the tailing and type I cements are also given in Table 6.

c.   DEGREE OF HYDRATION

The degree of hydration is calculated as the ratio of chemically bound water to the corresponding values at complete hydration (40). Bound water at complete hydration of tailing cement is estimated to be 21.00% and is taken as the reference value in calculating degrees of hydration. This value is in general agreement with the value of 21.74% calculated by Copeland et al. for a year-old cement cured at a water–cement ratio of 0.40 (29). The degree of hydration as a function of time is shown in Fig. 10. Values of the degree of hydration for both type I cement and the tailing cement are also given in Table 6. For hydraulic cements the variation in the degree of hydration and strength development show parallel behavior. For instance, the calculated data on hydration of tailing cement shows that after 1 day the hydration was at least 47%, after 7 days it was 72%, and after 28 days hydration exceeded 84% (16). The corresponding compressive strengths on the cement paste were 1600, 5100, and 8150 psi, respectively. Another thermogravimetric method for calculating the degree of hydration in the cement pastes containing admixtures as reported by Rahman and Double (92) is based on the method of Berger et al. (15).

d.   SAMPLE PREPARATION

Hydrated cement samples need to be prepared by a careful procedure if they are to be used for quantitative DTA and TG experiments. Neat paste portland cement samples are preferably prepared by mixing with deionized water at a given water–cement ratio and kept for varying curing times (1, 3, 7, and 28 days) in a 100% relative humidity chamber, controlled at $21 \pm 1°C$. After given curing times the samples required for thermal analysis are collected, partially dried, and ground to -200 mesh in a CO_2-free nitrogen-purged cabinet to avoid carbonation. The samples are dried again at 105°C to a constant weight in a standard oven to drive off the unbound moisture and subjected to thermal analysis by heating at a constant rate from ambient to 1000°C. The experiments are generally conducted in an atmosphere of CO_2-free nitrogen flowing at 50 cm^3/min. Alumina is used as the reference material.

Mascolo's (64) sample preparation involves grinding the cement with acetone followed by desiccation for 15 h. The sample is then ground to -150 mesh and subjected to thermal analysis in static air at 30°C/min using alumina as reference material. The sample is first heated to 1000°C to eliminate thermal effects that were found to interfere with the endotherm at 925°C caused by any remaining unhydrated C_3S phase. This was followed by cooling to 700°C and then determining the DTA curve up to 1000°C. The peak occurring at 925°C is then taken as the proper measure of undehydrated C_3S. In another method on the sample preparation, Dollimore (35) has suggested the use of methanol instead of acetone to avoid interference with hydration characteristics of cement paste due to the presence of any residual water in acetone.

2. Tricalcium Aluminate (C_3A)

C_3A normally occurs in portland cement at 5–15% and, being the most reactive phase, contributes significantly toward the early stiffening properties. It is also the C_3A phase that gives rise to the phenomenon of flash set. Addition of gypsum is made during the clinker grinding to control the flash set.

C_3A hydrates quickly with water to form the hexagonal plate hydrates C_2AH_8 and C_4AH_{13} at ambient temperature, which convert to stable cubic hydrate C_3AH_6 with time or if the temperature is sufficiently high ($> 40°C$). The reaction takes place as follows:

$$2C_3A + 21H \xrightarrow{\text{Room temp.}} \underset{\text{Hexagonal plates}}{C_2AH_8 + C_4AH_{13}} \tag{9}$$

$$\xrightarrow{>40°C} \underset{\text{Cubic hydrate}}{2C_3AH_6 + 9H} \tag{10}$$

The application of thermal analysis to study the hydration products of C_3A pastes and the rate of conversion of the metastable hexagonal hydrates has been carried out by numerous workers, for example, Midgley and Rosaman

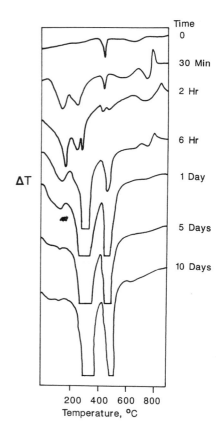

Fig. 11. DTA curves of C_3A hydrated at room
temperature (45).

(69), Ramachandran and Feldman (97–99), Satava and Veprek (112, 114, 115), Singh (117), and others (24, 59).

The most common DTA thermograms of hydrated C_3A at room temperature are shown in Fig. 11. In the early hydration, say up to 2 h, endothermic peaks at 170 and 245°C indicate the formation of hexagonal hydrates. Beyond that the intense endothermic peaks at 310 and 490°C are due to the presence of the cubic C_3AH_6 phase. These peaks become significantly larger with hydration time, clearly showing the conversion of hexagonal hydroaluminates to the cubic hydrate C_3AH_6.

The development of cubic C_3AH_6 in C_3A hydration is preceded by the formation of hexagonal C_2AH_8 and C_4AH_{13}. A distinct endothermic bulge normally occurs at 315–330°C due to C_3AH_6; but the DTA plots may differ slightly from worker to worker. This difference is primarily caused by such variables as sample preparation, sample porosity, period of hydration, temperature and consistency of the paste, and sensitivity of the DTA apparatus.

Thermal studies on the conversion of the C_3A hydration products, carried out at different hydration temperatures, have made it clear that the conversion is temperature dependent. Although both the hexagonal hydrates

(C_2AH_6 and C_4AH_{13}) and cubic hydrates (C_3AH_6) increase in quantity, generally the hexagonal hydrates convert to C_3AH_6 faster (93).

a. HYDRATION OF TRICALCIUM ALUMINATE IN THE PRESENCE OF GYPSUM

In the presence of gypsum, which is generally added to clinkers during grinding to avoid flash setting, the compound C_3A begins its rapid reaction to form the calcium aluminate hydrate C_4AH_{13}, which immediately reacts with calcium sulfate to form the compound ettringite $C_3A \cdot 3C\overline{A}S \cdot H_{32}$. The reaction takes place as follows:

$$C_3A + CH + 12H \longrightarrow C_4AH_{13} \tag{11}$$

Calcium aluminate hydrate

$$C_4AH_{13} + 3C\overline{S}H_2 + 14H \longrightarrow C_3A \cdot 3C\overline{S}H_{32} + CH \tag{12}$$

Ettringite

Ettringite is very insoluble in alkaline medium and is deposited on the surface of hydrating C_3A where it provides an effective barrier to further rapid setting. After all the gypsum is used up, the ettringite reacts back with calcium aluminate hydrate to form monosulfate according to the following equation:

$$C_3A \cdot 3C\overline{S}H_{32} + 2C_4AH_{13} \longrightarrow 3\left(C_3A \cdot C\overline{S}H_{12}\right) + 2CH + 20H \tag{13}$$

Monosulfate

Study of these two forms of calcium aluminate sulfate hydrates is of interest in cement chemistry and has been a subject of thermal study for many years. Midgley and Rosaman (69) discovered the formation of a solid compound partway $C_3A \cdot 3C\overline{S} \cdot$ aq. and $C_3A \cdot 3CH \cdot$ aq. Feldman and Ramachandran (45, 46) have studied C_3A–gypsum mixtures containing 0.25–25% gypsum cured at 2, 12, 23, and 50°C for up to 34 days. Typical DTA plots at room temperature are shown in Fig. 12 in which the endothermic peaks in the 170–200°C range are attributed to high sulfoaluminate. The endothermic effects in the range 140–150°C and 240–285°C are attributed to hexagonal C_2AH_8 and C_4AH_{13} hydrates, respectively, and the two endothermic effects at 290–300°C and 460–500°C are attributed to the cubic C_3AH_6 hydrates. The compound C_4AH_{13} may also be associated with low sulfoaluminates.

The formation of hexagonal hydroaluminates and their conversion to cubic C_3AH_6 increases with increasing temperatures. A comparison of these with corresponding studies of pure C_3A reveals that addition of gypsum retards the formation of hexagonal hydroaluminates and their conversion to C_3AH_6.

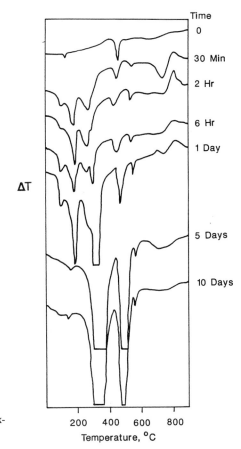

Fig. 12. DTA curves of C_3A–gypsum mixtures hydrated at room temperature (45).

The rate of gypsum consumption is best studied by its characteristic endothermic effects at 120 and 135°C, which persist only up to 30 min in a C_3A + 2.5% gypsum hydrated at room temperature (Fig. 13). The hydration proceeds faster at higher temperatures. Gypsum disappears after 24 h at 23°C, whereas it takes 5 days to disappear at 2 or 12°C.

It has been suggested (93) that the degree of retardation of C_3A with gypsum at a given temperature depends upon (1) the concentration of sulfate ions on and around the C_3A surface, (2) the rate of reaction of sulfate ions with the hexagonal hydroaluminates, and (3) the thickness of this layer around the C_3A grains as the rate of reaction gradually becomes diffusion controlled.

3. Tetracalcium Aluminoferrite (C_4AF)

The ferrite phase in cement is present up to 8%. This phase reacts with water at a slower rate than the C_3A in a manner that has not always been clearly understood. In the absence of gypsum, C_4AF reacts with lime and

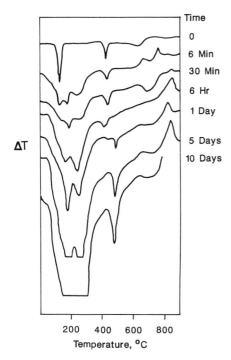

Fig. 13. DTA curves of C_3A + 2.5% gypsum hydrated at room temperature (45).

water to form iron-substituted C_4AH_{13}, which ultimately converts to a solid solution of C_3AH_6 and C_3FH_6. At higher temperatures of about 50°C, this reaction is much quicker. In the presence of gypsum, iron-substituted ettringite is formed, which alters and converts to iron-substituted monosulfate when there is no more gypsum available for reaction. The reactions take place as follows:

$$C_4AF + 2CH + 11H \longrightarrow C_4AH_{13} \qquad (14)$$

$$\xrightarrow{\sim 50°C} C_3AH_6 + C_3FH_6 \qquad (15)$$

In the presence of gypsum,

$$C_4AF + CH + 3C\bar{S}H_2 + 25H \longrightarrow C_3(A, F) \cdot 3C\bar{S}H_{32} \qquad (16)$$

Iron-substituted ettringite

When the supply of gypsum is exhausted this ettringite transforms to monosulfates of the type $C_3(A, F) \cdot C\bar{S}H_{12}$.

Ramachandran and Beaudoin (100), and Collepardi and coworkers (26–28) have conducted a series of DTA investigations on the hydration behavior of C_4AH with and without the presence of lime and gypsum. Figure 14 shows

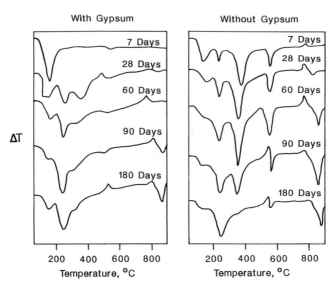

Fig. 14.   DTA curves of C_4AF with and without gypsum hydrated at room temperature (52).

that the C_4AF sample hydrated for 7 days leads to the formation of C_3AH_6 and C_4AH_3, corresponding to endothermic peaks at 360 and 220°C, respectively. The effects are at a maximum at 60 days and disappear after 180 days (52).

In the presence of gypsum, after 7 days curing the only endothermic effect noticeable is around 150°C, which is probably due to the presence of high iron-substituted calcium sulfoaluminate. After 28 days two more endothermic peaks appear around 250 and 350°C, which are due to the formation of Al_2O_3 and Fe_2O_3 solid solutions (93). After 90 days the dominant endothermic peak present is around 250°C, which could be due to the formation of iron-substituted monosulfoaluminates. No evidence of the presence of C_3AH_6 is observed in the C_4AF hydrated in the presence of gypsum. The hydration of both aluminate and ferrite phases is retarded by gypsum and lime.

V.   EFFECTS OF ADMIXTURES

Most of the cement-based concretes produced these days contain at least one admixture. According to the ASTM definition, an admixture is any material other than water, aggregate, and cement used as an ingredient of concrete or mortar and added to the batch immediately before or during mixing. Admixtures are generally classified as accelerators, retarders, plasticizers, water reducers, and air-entraining agents, etc., and they impart

TABLE 7
Production and Consumption of Admixtures and Portland Cement
in the World During 1971–1981

Geographical Area	Estimated Cement Consumption (million tons)		Estimated Admixture Consumption[a] ('000 tons)	
	1971	1981	1971	1981
USA	70	70	200	250
South America	40	60	40	40
Western Europe	155	160	100	220
Africa	25	40	10	15
Middle East	N/A	35	10	50
Far East	105	140	80	165
Australia	5	6	5	15

[a] Excluding calcium chloride accelerators.

SOURCE: Rixom and Mailvaganam (109).

numerous beneficial effects, such as improved workability without increasing water content, early strength development, accelerating or delaying setting times, and better chemical resistance. An account of global consumption of cement and admixtures during the years 1971–1981 is presented in Table 7 (109).

Although a variety of admixtures is available commercially, it is possible to categorize them according to their basic chemical structures (107, 109). Since retarders also have water-reducing ability, they can come under both categories. Most of the retarders are derived from wood resins, petroleum by-products, or hydroxylated polymers and sugars. The list is as follows:

Accelerators

- Calcium chlorides
- Calcium formates
- Triethanolamine

Retarders

- Superplasticizers
- Hydroxylated polymers
- Formaldehydes
- Sugars

DTA techniques have been used extensively in studying the reactions occurring in different cement–admixture systems. Reviews by Ramachandran (93) and Ben-Dor (8) refer to these frequently. Studies by conduction calorimetry and x-ray diffraction supplement the DTA findings. The studies are generally applied in determining the role of accelerators, retarders, water reducers, and superplasticizers in hydrating portland cement and its minerals. The effects of admixture on cement hydration is usually measured in terms of CH formation.

A. ACCELERATORS

Accelerators such as calcium chlorides and formates effectively increase the rate of hydration, thereby increasing the formation of CH in the cement paste. The rate generally increases with increasing concentration of the accelerators (28). DTA investigations conducted by Rahman (91), Ramachandran (96, 106), and Singh and Ojha (119) indicate that the C_3S phase hydrating in the presence of chloride shows peaks that differ from those containing no admixtures. The curves shown in Fig. 15 are typical thermo-

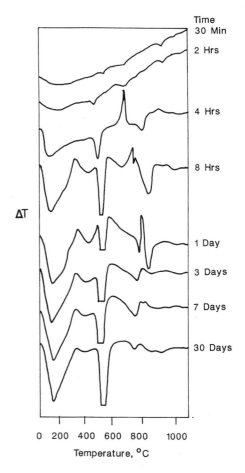

Fig. 15.   Effect of chloride addition on the DTA curves of hydrating C_3S (106).

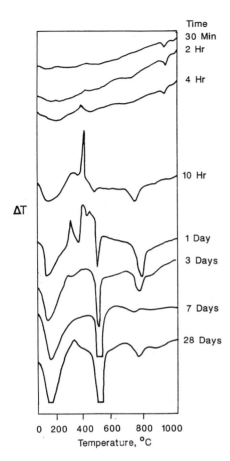

Time
30 Min
2 Hr
4 Hr
10 Hr
1 Day
3 Days
7 Days
28 Days

ΔT

0   200   400   600   800   1000
Temperature, °C

Fig. 16.   DTA curves of hydrated C_3S in the presence of TEA (97).

grams. Similarly, the effect of another type of accelerator, triethanolamine (TEA), on a hydrating C_3S phase is evident from the endothermic peaks in the range of 480 to 500°C caused by the formation of CH. Decomposition appears after about 10 h when 0.5% TEA is used. Figure 16 shows thermograms of C_3S containing TEA and hydrated for 28 days. It is evident that the peak at 480–500°C grows larger with hydrating time. TEA may also act as a retarder to the C_2S phase, which gives two endothermic peaks at 480–500°C. The higher temperature peak may be ascribed to crystalline CH and the other to the formation of noncrystalline CH. Combined TG, DTG, and DTA studies have also been performed on cement pastes containing $CaCl_2$ and also CaF_2, CdI_2, and $CrCl_3$ as accelerators (5, 11, 75, 119, 120). Commonly, the DTA plots show endothermic peaks between 450 and 550°C due to CH dissociation.

Ramachandran (106) obtained a good correlation from DTA and TG results for the rate of formation of CH and the rate of C_2S disappearance for increasing additions of $CaCl_2$. Midgley and Illston (75) measured the chloride

TABLE 8
DTA Peak Temperatures of Hydrated C_3S in the Presence of CaF_2

Time of Hydration (h)	Concentration of Calcium Formate (%)	DTA Peak Temperature (°C)
8	0	560
16		570
24		580
8	2	570
16		580
24		590
16	4	580
24		580
16	6	580
24		580

SOURCE: Singh and Abha (120).

ion penetration into hardened cement pastes and worked out a correlation between water/cement ratios, porosity, and chloride penetration.

Shift in peak temperature due to the presence of accelerators may also be observed due to the change in the amount and morphology of CH. According to Singh and Abha (120) the absence of calcium formate as an accelerator increases the peak temperature for CH, suggesting that it becomes more crystalline. DTA peak temperatures of hydrated C_3S in the presence of calcium formate are given in Table 8.

B.   RETARDERS

One of the generally accepted explanations of the effects of retardation on the cement paste is that the retarders either chemisorb on the most reactive compounds like C_3S and/or C_3A, or on the nuclei of their hydration products, particularly the CH and the hexagonal hydrates, C_2AH_6 and C_4AH_{13}, respectively, therefore poisoning their growth. Thermal analysis is a very useful method to monitor the effectiveness of a retarder by estimation of the rate of formation of CH during hydration.

Cement paste containing lignosulfonates (both sugar-free and commercial) show the endothermic peaks at 450–500°C resulting from dehydroxylation of CH. The intensity of these peaks indicates the extent of hydration of C_3S component, as shown in Fig. 17 for the DTA plots for a paste containing 76% C_3S, 19% C_3A, and 5% gypsum hydrated with sugar-free lignosulfonate at different hydration periods.

Available adsorption data have shown that the phase C_3A, compared to other phases, such as C_3S and C_2S present in cement, adsorbs a maximum amount of lignosulfonate (124). Therefore, the adsorption of lignosulfates on aluminates is primarily responsible for the retarding action on cement pastes.

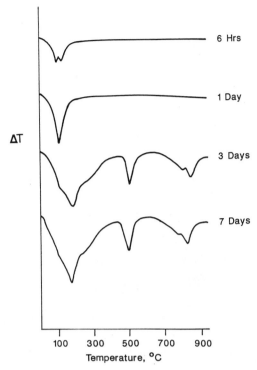

ΔT

6 Hrs

1 Day

3 Days

7 Days

Temperature, °C

Fig. 17.   DTA plots showing the influence of sugar-free sodium lignosulfate (0.3%) on hydrated pastes of C_3S (76%) + C_2S (19%) + gypsum (5%) (107).

Young (124) had earlier demonstrated the changes in the hydration products of C_3A in the presence of lignosulfonates by using thermal analysis along with x-ray diffraction and microscopy. Ramachandran (103) and Tenoutasse and Singh (122) used thermoanalytical and microcalorimetery combined with DTA and x-ray diffraction, respectively, in studying the effects of lignosulfonates and other sugar-based retarders on the hydration of cement pastes.

Some of the DTA plots obtained by Ramachandran (96) for C_3A compound hydrated for different periods are shown in Fig. 18. The plots show two distinct endothermic peaks at 150–200°C and 200–280°C. These peaks are indicative of the presence of a hexagonal C_3AH_8 phase, which increases with curing time. The cubic phase shows two endotherms at about 300–350°C and 500–550°C. Cubic C_3AH_6 is more stable and requires 14 days before the endotherm starts to appear; it appears as a main phase at 6 months, as is evident from the endotherms in Fig. 18. The peaks become intensive with hydration, which infers that the lignosulfates stabilize C_4AH_{13} and C_2AH_8 with respect to C_3AH_6.

Milestone (76) used the DTA technique in identifying the retarders before use to avoid any possibility of adverse reactions during hydration. He demon-

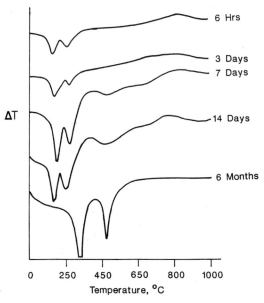

Fig. 18.   DTA curves of C_3A containing lignosulfate (96).

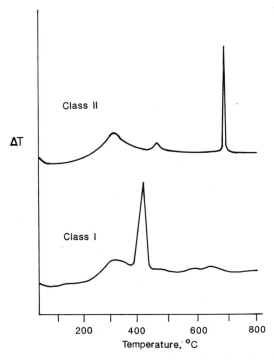

Fig. 19.   DTA curves of admixtures based on lignosulfates (class I) and hydroxycarboxylic acid groups (class 11) (76).

strated that the retarders containing a hydroxycarboxylic acid group show an exothermic peak between 500 and 700°C, whereas those containing lignosulfonates and sugars show the exotherm between 200 and 400°C. Typical DTA plots for these classes of retarders are shown in Fig. 19.

Gupta and Smith (49) also conducted an investigation on the thermoanalytical aspects of certain concrete admixtures containing salts of lignosulfonates, gluconates, and melamine formaldehyde condensates. The effects of hydration rates and purging gas atmosphere on their thermal properties were observed. Supplementary information was obtained from infrared spectroscopy.

The mechanism of action of a water-reducing agent based on naphthalene sulfonate formalin condensate was studied by Dezhen and coworkers (31) using DTA, x-ray diffraction, and infrared techniques on concrete samples. Their studies were based on water estimation formed during different stages of hydration.

C. SUPERPLASTICIZERS

The effects of superplasticizers depend largely on the hydration characteristics of individual cement components, such as C_3A, C_3S, C_2S, and C_4AF. According to the thermal analysis conducted by Ramachandran (106) the superplasticizers such as sodium salts of sulfonated melamine formaldehyde condensate (SMF) retards the hydration of C_3A and C_3S. However, it accelerates the formation of ettringite in a gypsum-added C_3A–water system. DSC curves for the C_3S phase hydrated in the presence of SMF for 1, 8, and 24 h are shown in Fig. 20. It seems that the formation of CH in the C_3S–SMF system does not start until after 8 h, therefore effectively retarding its setting for that length of time. However, there is evidence of CH formation at 24 h hydration.

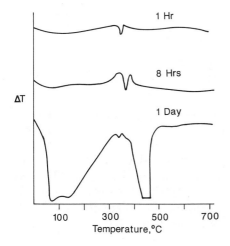

Fig. 20. DSC curves for the C_3S phase hydrated in the presence of sulfonated melamine formaldehyde (106).

VI. OTHER CEMENTITIOUS MATERIALS

A. HIGH-ALUMINA CEMENTS

High-alumina cements, also called aluminous cements, are known for their advantages of developing high early strength, resistance to sulfate attack, and better performance at higher temperatures. But their rapid loss of strength, which occurs due to adverse chemical reaction, has precluded its uses in certain environments.

High-alumina cements harden rapidly and can gain very high strength within 24 h, comparable to the strength of portland cement at 28 days. Calcium aluminate (CA), which is the major component of high-alumina cement, hydrates to form a mixture of CAH_{10} and C_2AH_8 along with AH_3 gel according to the following equation:

$$CA + H \begin{cases} \xrightarrow{<10°C} CAH_{10} & (17) \\ \xrightarrow{10-25°C} C_2AH_8 + AH_3 & (18) \end{cases}$$

At higher temperatures ($> 25°C$) both CAH_{10} and C_2AH_8 are metastable and tend to convert to C_3AH_6:

$$CA + H \xrightarrow{>25°C} C_3AH_6 + 2AH_3 \qquad (19)$$

Exposure to hot and humid conditions is conducive to such conversion.

The decahydrate (CAH_{10}) gives a typical broad endothermic peak around 110–120°C. DTA thermograms of the conversion products show that the stable C_3AH_6 component gives an endothermic peak between 320 and 350°C, whereas AH_3 gel gives a sharper endothermic peak around 295–310°C. Figure 21 shows DTA plots for partially converted high-alumina cements. From the peaks shown in Fig. 21 the degree of conversion can be calculated by the following expression, assuming that the mass of substance present is proportional to the DTA peak height (87):

$$\text{Conversion}(\%) = \frac{\text{Peak height of } AH_3 \times 100}{\text{Peak height of } AH_3 + \text{Peak height of } CAH_{10}} \qquad (20)$$

However, since the C_3AH_6 conversion may also be shown on the DTA plot, Midgley (70) proposed that peak height of C_3AH_6 be taken instead of AH_3 to measure the mass of conversion product.

Midgley (71–73) has also used thermal techniques to identify different phases during the hydration of high-alumina cement and concretes subjected to chemical attack. Bradbury and coworkers (22), El-Jazairi (41), Mitra and coworkers (78), and Quon and Malhotra (88) have consistently used DTA along with other techniques to investigate the effects of water/cement ratios

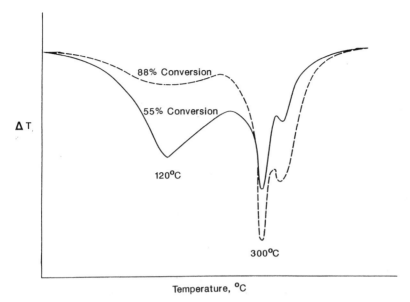

Fig. 21.   DTA curves for high-alumina cement at different stages of conversion.

and temperature on conversion. They have also studied the effects on morphologies during conversion and have reinforced their findings with microscopical observations. Recently Edmonds and Majumdar (39) have also quoted the use of thermal analysis in the chemistry of hydration of monocalcium aluminate, which is the major component of the high-alumina cement. They have conducted these investigations at varying curing temperatures and have substantiated their findings with the aid of conduction calorimetry and x-ray diffraction data. According to them the hydration of CAH_{10} to C_2AH_8 increases with increasing temperatures, followed by a conversion period of weeks before the conversion to the C_3AH_6 phase takes place.

B. GYPSUM PLASTERS

Gypsum plaster products have a widespread use in the building industry as surface finish and interior linings. They have the advantage of being quick-setting materials that develop rapid strength. Plaster is β-calcium sulfate hemihydrate ($C\bar{S} \cdot H_{1/2}$) formed by controlled heating of gypsum. The process of gypsum calcination can readily be simulated by DTA techniques (87).

Various stages of gypsum dehydration may be explained by the following expression:

$$CaSO_4 \cdot 2H_2O \overset{150°C}{\rightleftharpoons} CaSO_4 \cdot 1/2H_2O \overset{170°C}{\longrightarrow} CaSO_4 \cdot \varepsilon\text{-}H_2O \overset{300°C}{\longrightarrow} CaSO_4$$

β-Hemihydrate
or plaster   (21)

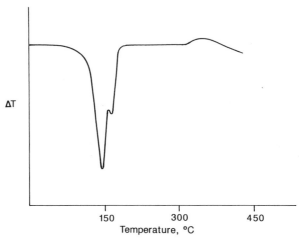

ΔT

150   300   450
Temperature, °C

Fig. 22. DTA plot of gypsum.

Dehydration of gypsum gives rise to an endothermic peak around 112–115°C and another at 140–150°C. According to Murat (81), two forms of hemihydrates exist, which have clearly defined peaks at 144 and 167°C, respectively. But the dehydration product, especially the β-CaSO$_4 \cdot \varepsilon$-H$_2$O, still contains some chemically bound water, which is driven off between 300 and 400°C. A complete DTA plot of the gypsum is shown in Fig. 22. Khalil and coworkers (54–56) found that the hemihydrates were present below 100°C, and the γ-anhydrate above 100°C. On mixing with water, β-hemihydrate, also known as gypsum plaster, is reconstituted to gypsum by rehydration, which develops strength quite rapidly.

Recently, the application of DTA and TG techniques has also established that the ancient Greek and Egyptian civilizations made use of gypsum in their structures (89, 90, 111). Specimen DTA and TG plots of ancient Egyptian mortars samples (see Fig. 23) show that an additional broad endotherm observed in the range 650–840°C is due to the decomposition of calcium carbonate, which was regularly used as a vital mortar ingredient probably in the form of limestone.

C. SLAG CEMENTS

Blast furnace slags are the by-products of the iron and steel industry. They are mostly composed of lime, silica, and alumina and have been regularly used as potential cementitious materials, particularly when used as a partial cement replacement. They also act as an activator.

Dorner and Setzer (36) used quantitative DTA in studying the slag cements between −60 and 20°C at ages between 1 and 28 days. The hydration of blast furnace slag in the presence of gypsum was investigated by Massida and Sanna (66). Typical TG and DTA plots of slag, both unhydrated and hydrated, at different curing temperatures are shown in Fig. 24. An

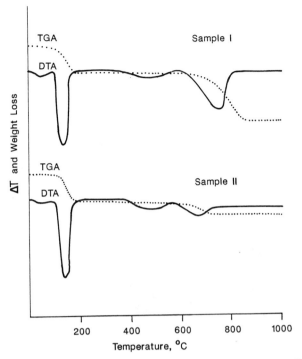

Fig. 23.   TG–DTA curves of samples of ancient Egyptian mortars (90).

exothermic peak at 920°C is due to the recrystallization of a glassy phase of the anhydrous slag. Endothermic peaks at 120–160°C accompanied by weight loss are most probably due to dehydration.

According to Ramachandran (93), the DTA of slag cements activated with lime and gypsum or portland cement would show a low-temperature endothermic effect due to ettringite formation at 150°C, a peak around 250°C for C_4AH_{13}, and another at 520°C for CH. Nurse and Midgley (82) used the DTA technique to estimate the amount of lime adsorbed in slag cement at varying lime concentrations. They noted a change in adsorption at 15–20% lime concentration, which corresponded to the maximum strength development. Figure 25 shows the relationship between lime added and lime adsorbed in slag cement pastes.

VII.   POZZOLANIC MATERIALS

A.   FLY ASH

Recycling of fly ash reduces pollution, conserves energy, and decreases the disposal problem. Use of fly ash as a pozzolanic material and a partial substitute for cement in the construction industry has been practiced for a

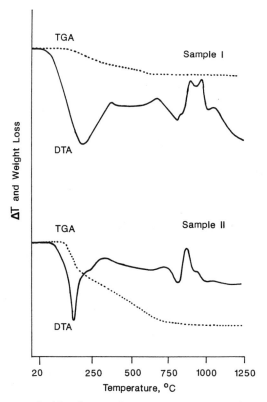

Fig. 24.   TG–DTA curves for blast furnace slag cements steam cured for 15 h. Sample I cured at 70°C; sample II cured at 180°C (66).

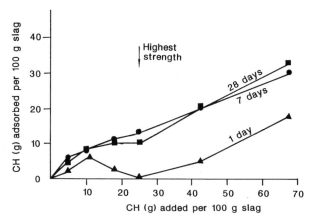

Fig. 25.   Graphical presentation of the relationship between lime added and lime adsorbed in slag–cement pastes (82).

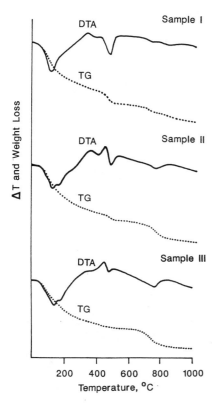

Fig. 26. TG–DTA curves of cement–blast furnace slag mixtures after 28 days normal curing. Sample I, neat cement; sample II, 25% BFS; sample III, 40% BFS (57).

long time and a considerable amount of pertinent literature is now available. To assess the pozzolanic activity of the fly ashes several methods are available, including DTA (57, 63, 84, 85, 116). The degree of reactivity of fly ashes has been studied by incorporating fly ashes into cement pastes, hydrating for different periods, and estimating the hydration products, such as C–S–H and CH. DTA, TG, and DTG analysis on neat and fly ash-incorporated cement pastes indicates a similarity between hydration products, although the relative proportions are different.

A typical example of thermal analysis techniques applied to fly ash–portland cement mixtures obtained after 2, 7, and 28 days curing was reported by Kovacs (57) and is shown in Fig. 26. It was stated by Kovacs that with increasing fly ash contents the amount of C-S-H gel increases while CH decreases.

Pozzolanic and cementitious reactions of fly ash in blended cement pastes have recently been studied by Marsh and Day (63) using thermogravimetry. High-lime fly ash derived from lignite and low-lime fly ash from subbituminous coals have been used at replacement levels of 30 and 50% by weight and the pastes cured up to one year. Marsh and Day suggested a possible correlation between the major hydration products CH and other hydrates

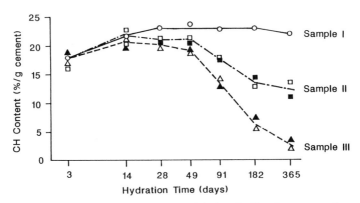

Fig. 27.   Variation of calcium hydroxide content with time for fly ash–cement mixtures. Sample I, 0% fly ash; sample II, 30% fly ash; sample III, 50% fly ash. The fly ash was obtained from two different sources (63).

measured by the weight loss during dehydration between 105 and 1000°C. They also concluded that in the presence of CH the fly ashes containing high lime contents possess both cementitious and pozzolanic reactivities, whereas the low-lime ashes have only pozzolanic reactivity. Their results indicate that the fly ashes undergo significant pozzolanic reaction after 14 days hydration, as shown by the depletion of CH, which increases with time and is larger at high fly ash contents (see Fig. 27). The CH depletion is apparently independent of the fly ash chemistry.

B.   CONDENSED SILICA FUME

Condensed silica fume (CSF) is a by-product of the ferrosilicon and silicon metal industry. It is formed by heating quartz, coal, and coke to 2000°C in an electric submerged-arc furnace. CSF mainly consists of amorphous SiO_2 (95% by weight) of very high surface area and an average particle size of 0.1 μm.

During the last decade a lot of attention has been given to the use of CSF in the concrete industry as a partial replacement for cement, as discussed by Malhotra (62). The CSF first acts as an efficient filler, occupying the voids between the cement grains, and later becomes nuclei for C–S–H. CSF exhibits pozzolanic properties by reacting with calcium hydroxide formed during cement hydration. It therefore influences both the physical and mechanical properties of fresh and hardened cement pastes or concretes.

Durekovic and Popovic (37, 38) have used DTA and TG techniques along with gas–liquid chromatography in studying the influence of silica fume on silicate polymerization during the hydration of cement-silica fume blended pastes. They determined the free lime contents of the pastes as a function of CSF content and hydration times. Typical data on free lime content (ex-

TABLE 9
Variations in Free Lime Content with Hydration Time of Cement
Pastes Containing Different Amounts of Silica Fumes

Hydration Time (days)	Free lime as CaO (% by wt) in Cement Pastes Containing Levels of Silica Fumes		
	0%	5%	10%
1	11.91	11.00	10.00
3	14.50	13.80	12.25
7	15.80	14.05	12.69
28	16.55	15.27	13.42

SOURCE: Durekovic and Popvic (37).

pressed as CaO%), as determined by the DTA and TG methods for different
CSF–cement pastes hydrated for different durations, are given in Table 9.

C.   RICE HUSK ASH

Rice husk ash (RHA) is largely composed of silica with minor amounts of
alkalies and other elements, and is finding use as a constituent of low-cost
cements. By controlled thermal decomposition, the rice husks produce an ash
composed of reactive silica, which exhibits pozzolanic properties in lime
conditions. Recently, James and Subba Rao (50) have investigated the
reaction product of lime and silica from RHA using thermogravimetry and
other analytical techniques, such as x-ray diffraction and electron microscopy.

Typical thermograms of RHA : lime mixtures (1 : 3, lime excess; 2 : 1, lime
deficient) before and after setting and their comparison with those of
synthetic C–S–H are shown in Fig. 28. In a lime-deficient mixture before
setting, the endotherm accompanied by a weight loss in the region of
490–550°C represents the decomposition of CH to calcium oxide. The
lime-deficient mixture, after setting, exhibits an endothermic peak around
840°C. In the lime-excess mixture the mass of free lime and calcium carbonate in the final product is much less than the initial mass of lime.

VIII.   NEW METHODS

Over the past few years some new applications of thermal techniques have
been developed which can be applied to the cement chemistry. Smallwood
(121) made use of differential scanning calorimetry (DSC) to resolve the
complex hydrothermic peaks found in the temperature range 50–250°C
during the hydration of cement and ettringite. The degree of resolution of
these peaks is shown to depend largely on the sample weight, packing, and
the sample holder. Wiedemann and Roessler (123) used thermomechanical
analysis (TMA) to characterize some binding and construction materials.

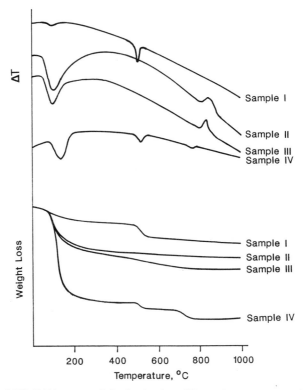

Fig. 28.   Typical TG–DTA curves of rice husk ash and lime mixtures compared with a synthetic C–S–H mixture. Sample I, 2 : 1 RHA–lime mix before setting; sample II, 2 : 1 RHA–lime mix after setting: sample III, synthetic C–S–H mixture; sample IV, 1 : 3 RHA–lime mixture after setting for 28 days (50).

Piasta (86) used complex thermal analysis techniques including thermo-dilatometry (TD) and differential thermodilatometry (DTD) in observing the thermal deformation of different phases and their effects on hardened cement pastes. He studied the thermal deformation in the temperature range of 20–800°C.

Satava and Veprek (113) used a differential hydrothermal analysis technique to observe hydrothermal processes in cement clinkers. Their studies included the determination of the heat of reaction, kinetics, and reversibility. They also demonstrated a method of "rapid quenching" to study the hydrothermal process in clinkers.

Another development on the heating procedure of the sample by El-Jazairi and Illston (43, 44) involves semi-isothermal heating. In their studies on high-alumina cement concrete, the method avoids the overlapping of adjacent peaks by minimizing the heating rate at a given temperature range for a given time until the full DTA peak is resolved before the appearance of the next peak.

IX. CONCLUSIONS

The application of thermal analysis is now an accepted technique in the study of the chemistry of cement and related materials. The technique is fast and reliable and can detect minor physicochemical changes that occur not only during the manufacturing of cement but also during its hydration process, even in the presence of admixtures. Although this technique provides information that is not always readily available using other methods, the results and interpretation should still be substantiated by other techniques.

REFERENCES

1. Ampian, S. G., and E. P. Flint, *Bull. Am. Ceram. Soc.* **52**, 604 (1973).

2. Barta, R., Cements, in R. C. Mackenzie, Ed., *Differential Thermal Analysis*, Vol. 2, London: Academic, 1972, p. 207.

3. Ben-Dor, L., and D. Perez, *Thermochim. Acta* **12**, 81 (1975).

4. Ben-Dor, L., D. Perez, and S. Sarig, *J. Am. Ceram. Soc.* **58**, 87 (1975).

5. Ben-Dor, L., D. Perez, and S. Sarig, *J. Am. Ceram. Soc.* **58**, 87 (1975).

6. Ben-Dor, L., and D. Perez, *Spectrochim. Acta* **18**, 81 (1979).

7. Ben-Dor, L., and Y. Rubinsztain, *Thermochim. Acta* **30**, 9 (1979).

8. Ben-Dor, L. Thermal methods, in S. N. Ghosh, Ed., *Advances in Cement Technology*, Oxford, UK: Pergamon, 1983, pp. 673–710.

9. Bensted, J., and S. P. Varma, *Cem. Technol.* **51**, 440 (1974).

10. Bensted, J., *Tonind-Ztg.* **102**, 544 (1978).

11. Bensted, J., *Cemento* **75**, 13 (1978).

12. Bensted, J., *Cem. Concr. Res.* **9**, 97 (1979).

13. Bensted, J., *Cemento* **76**, 117 (1979).

14. Bensted, J., *Cemento* **88**, 169 (1980).

15. Berger, R. L., J. H. Kung, and J. F. Young, *J. Test. Eval.* **4**, 435, (1978).

16. Bhatty, J. I., and K. J. Reid, *Thermochim. Acta* **91**, 95 (1985).

17. Bhatty, J. I., J. Marrignissen, and K. J. Reid, *Cem. Concr. Res.* **15**, 501 (1985).

18. Bhatty, J. I., D. Dollimore, G. A. Gamlen, R. J. Mangabhai, and H. Olmez, *Thermochim. Acta* **106**, 115 (1986).

19. Bhatty, J. I., *Thermochim. Acta* **106**, 93 (1986).

20. Bhatty, J. I., *Thermochim. Acta* **119**, 235 (1987).

21. Bhatty, J. I., K. J. Reid, D. Dollimore, G. A. Gamlen, R. J. Mangabhai, P. F. Rogers, and T. H. Shah, *Compositional Analysis by Thermogravimetery*, ASTM STP 997, C. M. Earnest, Ed., ASTM, Philadelphia, 1988, p. 204.

22. Bradbury, C., P. M. Callaway, and D. Double, *Mater. Sci. Eng.* **23**, 43 (1976).

23. Chaterjee, A. K., Chemico-physico-mineralogical characteristics of raw materials of portland cement, in S. N. Ghosh, Ed., *Advances in Cement Technology*, Oxford, UK: Pergamon, 1983, pp. 39–68.

24. Chebotnikov, V. M., and A. B. Shalinets, *Chem. Abstr.* **86**, 59552k (1977).

25. Chen, H., in B. Miller, Ed., *Thermal Analysis, Proc. 7th Int. Conf. Therm. Anal.*, Vol. 2, Chichester, UK: Wiley 1303 (1982).

26. Collepardi, M., G. Baldini, and M. Pauri, *J. Am. Ceram. Soc.* **62**, 33 (1979).

27. Collepardi, M., S. Monosi, G. Moriconi, and M. Corradi, *Cem. Concr. Res.* **9**, 431 (1979).

28. Collepardi, M., S. Monosi, G. Moriconi, and M. Corradi, *Cem. Concr. Res.* **10**, 455 (1980).

29. Copeland, L. E., D. L. Kantro, and G. Verbcck, *Proc. 4th Int. Symp. Chem. Cem.*, Vol. 1, Washington, DC, 1962, p. 440.

30. Dent-Glasser, L. S., E. E. Lachowski, K. Mohan, and H. F. W. Taylor, *Cem. Concr. Res.* **8**, 733 (1978).

31. Dezhen, G., D. Xiong, and L. Zhang, *J. Am. Concr. Inst.* **79**, 378 (1982).

32. Dollimore, D., G. A. Gamlen, G. R. Heal, and P. F. Rodgers, *Proc. 2nd Eur. Symp. Therm. Anal.*, D. Dollimore, Ed., Heyden, London, 1981, p. 489.

33. Dollimore, D., G. A. Gamlen, and R. J. Mangabhai, *Proc. 2nd Eur. Symp. Therm. Anal.*, D. Dollimore, Ed., Heyden, London, 1981, p. 485.

34. Dollimore, D., G. A. Gamlen, R. J. Mangabhai, and H. Olmez, *Proc. 14th NATAS Conf.*, B. B. Chowdhury, Ed., San Francisco: North American Thermal Analysis Society, 1985, p. 360.

35. Dollimore, D., Unpublished work, Department of Chemistry, University of Toledo, OH, 1988.

36. Dorner, H., and M. J. Setzer, *Cem. Concr. Res.* **10**, 403 (1980).

37. Durekovic, A. and K. Popvic, *Cem. Concr. Res.* **17**, 108 (1987).

38. Durekovic, A., *Cem. Concr. Res.* **18**, 185 (1988).

39. Edmonds, R. N., and A. J. Majumdar, *Cem. Concr. Res.* **18**, 311 (1988).

40. El-Hemaly, S. A. S., R. EI-Sheikh, F. H. Mosalamy, and H. El-Didamony, *Thermochim. Acta* **78**, 147 (1984).

41. El-Jazairi, B., *Proc. 1st. Eur. Symp. Therm. Anal.*, D. Dollimore, Ed., Heyden, London, 1976, p. 378.

42. El-Jazairi, B., *Thermochim. Acta* **21**, 381 (1977).

43. El-Jazairi, B., and J. M. Illston, *Cem. Concr. Res.* **7**, 247 (1977).

44. El-Jazairi, B., and J. M. Illston, *Cem. Concr. Res.* **10**, 361 (1980).

45. Feldman, R. F., and V. S. Ramachandran, *Mag. Concr. Res.* **18**, 185 (1966).

46. Feldman, R. F., and V. S. Ramachandran, *J. Am. Ceram. Soc.* **49**, 268 (1966).

47. Feldman, R. F., and V. S. Ramachandran, *Cem. Concr. Res.* **4**, 155 (1974).

48. Gard, J. A., and H. F. W. Taylor, *Cem. Concr. Res.* **6**, 667 (1976).

49. Gupta, J. P., and J. I. H. Smith, *Proc. 2nd Eur. Symp. Therm. Anal.*, D. Dollimore, Ed., Heyden, London, 1981, p. 483.

50. James, J., and M. Subha Rao, *Cem. Concr. Res.* **16**, 67 (1986).

51. Jawed, I., W. A. Klemm, and J. Skalny, *J. Am. Ceram. Soc.* **62**, 461 (1979).

52. Kalousek, G. L., T. Mitsuda, and H. F. W. Taylor, *Cem. Concr. Res.* **1**, 305 (1977).

53. Kalyoncu, J. N., M. E. Tadros, A. M. Baratta, and J. Skalny, *J. Therm. Anal.* **9**, 233 (1976).

54. Khalil, A. A., A. T. Hussain, and G. M. Gad, *J. Appl. Chem. Biotechnol.* **21**, 314 (1971).

55. Khalil, A. A., and G. M. Gad, *J. Appl. Chem. Biotechnol.* **22**, 697 (1972).

56. Khalil, A. A., and A. T. Hussain, *Trans. J. Br. Ceram. Soc.* **71**, 67 (1972).

57. Kovacs, R., *Cem. Concr. Res.* **5**, 73 (1975).

58. Le Chatelier, H., *Bull. Soc. Franc. Miner.* **10**, 204 (1887).

59. Lorprayoon, V., and D. R. Rossington, *Cem. Concr. Res.* **11**, 267 (1981).

60. Mackenzie, R. C., Differential Thermal Analysis, in H. F. W. Taylor, Ed., *The Chemistry of Cements*, Vol. 2, London: Academic, 1964, p. 271.

61. Maki, I., and K. Kato, *Cem. Concr. Res.* **12**, 93 (1982).

62. Malhotra, V. M., Ed., *Fly Ash, Silica Fume, Slag and Natural Pozzolans in Concrete, Proc. 2nd Int. Conf.*, Madrid, Spain, American Concr. Inst. Special Publication 91, 1986.

63. Marsh, B. K., and R. L. Day, *Cem. Concr. Res.* **18**, 301 (1988).

64. Mascolo, G., *J. Therm. Anal.* **8**, 69 (1975).

65. Mascolo, G., and V. S. Ramachandran, *Mater. Constr.* (*Paris*) **8**, 373 (1975).

66. Massida, L., and U. Sanna, *Cem. Concr. Res.* **9**, 127 (1979).

67. Maycock, J. N., and J. Skalny, *Thermochim. Acta* **8**, 167 (1974).

68. Mehta, P. K., *Pit and Quarry* 141, (1964).

69. Midgley, H. G., and B. Rosaman, *4th Int. Symp. Chem. Cem.*, Washington, DC, Vol. 1, 1962, p. 259.

70. Midgley, H. G., *Trans. Br. Ceram. Soc.* **66**, 161 (1967).

71. Midgley, H. G., *Proc. 1st Eur. Symp. Therm. Anal.*, D. Dollimore, Ed., Heyden, London, 1976, p. 378.

72. Midgley, H. G., *J. Therm. Anal.* **13**, 515 (1978).

73. Midgley, H. G., *Thermochim. Acta* **27**, 281 (1978).

74. Midgley, H. G., *Cem. Concr. Res.* **9**, 77 (1979).

75. Midgley, H. G., and J. M. Illston, *Cem. Concr. Res.* **14**, 546 (1984).

76. Milestone, N. B., *Cem. Concr. Res.* **14**, 207 (1984).

77. *Mineral Commodity Summaries*, Bureau of Mines, U.S. Department of the Interior, 1988.

78. Mitra, A. K., P. K. Podder, and A. K. Chatterjee, *Indian J. Technol.* **16**, 20 (1978).

79. Mitsuda, T., *Cem. Concr. Res.* **3**, 71 (1973).

80. Mitsuda, T., and H. F. W. Taylor, *Cem. Concr. Res.* **5**, 203 (1975).

81. Murat, M., *J. Therm. Anal.* **3**, 259 (1971).

82. Nurse, R. W., and H. G. Midgley, Slag cements, in H. F. W. Taylor, Ed., *The Chemistry of Cement*, Vol. 2, London: Academic, 1964, p. 52.

83. Olmez, H., D. Dollimore, G. A. Gamlen, and R. T. Mangabhai, *The Use of Fly Ash, Slag and Other Mineral By-products in Concrete, Proc. 1st. Int. Conf.*, Montebello, Canada, American Concr. Inst. Special Publication 79, Detroit, **1**, 607 (1983).

84. Pako, H. J., and R. Kovacs, *4th Int. Conf. Therm. Anal.*, Vol. 3, I. Buzas, Ed., Budapest: Akademiai Kiado, 1975, p. 459.

85. Papayianni, J., *Mag. Concr. Res.* **39**, 19 (1987).

86. Piasta, J., *Mater. Struct.* **17**, 415 (1984).

87. Pope, M. I., and M. D. Judd, *Differential Thermal Analysis*, London: Heyden, 1977.

88. Quon, D. H. H., and V. V. Malhotra, *J. Can. Ceram. Soc.* **48**, 7 (1979).

89. Ragai, J., *Cem. Concr. Res.* **18**, 9 (1988).

90. Ragai, J., *Cem. Concr. Res.* **18**, 179 (1988).

91. Rahman, A. A., *Thermal Analysis, Proc. 7th Int. Conf. Therm. Anal.*, Vol. 2, B. Miller, Ed., Chichester, UK: Wiley 1982, p. 1310.

92. Rahman, A. A., and D. Double, *Cem. Concr. Res.* **12**, 33 (1982).

93. Ramachandran, V. S., *Application of Differential Thermal Analysis in Cement Chemistry*, New York: Chemical 1969.

94. Ramachandran, V. S., *Mater. Struct.* **4**, 3 (1971).

95. Ramachandran, V. S., *Cem. Concr. Res.* **2**, 179 (1972).

96. Ramachandran, V. S., *Thermochim. Acta* **3**, 343 (1972).

97. Ramachandran, V. S., *Cem. Concr. Res.* **3**, 41 (1973).

98. Ramachandran, V. S., and R. F. Feldman, *Cem. Concr. Res.* **3**, 729 (1973).

99. Ramachandran, V. S., and R. F. Feldman, *J. Appl. Chem. Biotechnol.* **23**, 625 (1973).

100. Ramachandran, V. S., and J. J. Beaudoin, *J. Mater. Sci.* **11**, 1893 (1976).

101. Ramachandran, V. S., *Calcium Chloride: Science & Technology*, Applied Scientific Publication, UK, 1976.

102. Ramachandran, V. S., and J. J. Beaudoin, *J. Mater. Sci.* **11**, 893 (1976).

103. Ramachandran, V. S., *Zem. Kalk. Gips.* **31**, 206 (1978).

104. Ramachandran, V. S., *Cem. Concr. Res.* **9**, 677 (1979).

105. Ramachandran, V. S., *Cem. Concr. Res.* **9**, 699 (1979).

106. Ramachandran, V. S., *Thermal Analysis*, *Proc. 7th Int. Conf. Therm. Anal.*, Vol. 2, B. Miller, Ed., Chichester, UK: Wiley 1982, p. 1296.

107. Ramachandran, V. S., Ed., *Concrete Admixture Handbook: Properties, Science and Technology*, Park Ridge, NJ: Noyes, 1984.

108. Read, H. H., *Rutley's Elements of Mineralogy*, 26th ed., London: Thomas Murby, 1970.

109. Rixom, M. R., and N. P. Mailvaganam, *Chemical Admixtures for Concrete*, 2d ed., London: E. & F. N. Spon, 1986, p. 50.

110. Rowland, R. A., and C. W. Beck, *Am. Mineral.* **37**, 76 (1952).

111. Roy, D. M., and Langton, C. A., Material Research Laboratory, Pennsylvania State University, prepared for the Office of Nuclear Waste Isolation, Battella Memorial Institute, Columbia, OH, 1983, p. 501.

112. Satava, V., and O. Veprek, *Zem. Kalk. Gips.* **28**, 170 (1975).

113. Satava, V., and O. Veprek, *Zem. Kalk. Gips.* **28**, 170 (1975).

114. Satava, V., and O. Veprek, *Zem. Kalk. Gips.* **28**, 424 (1975).

115. Satava, V., and O. Veprek, *J. Am. Ceram. Soc.* **58**, 537 (1975).

116. Sauman, Z., *Silkaty* **19**, 193 (1975).

117. Singh, N. B., *Cem. Concr. Res.* **5**, 545 (1975).

118. Singh, N. B., *Indian J. Technol.* **15**, 256 (1977).

119. Singh, N. B., and P. N. Ojha, *J. Mater. Sci.* **16**, 2675 (1981).

120. Singh, N. B., and K. M. Abha, *Cem. Concr. Res.* **13**, 619 (1983).

121. Smallwood, T. B., *Proc. 2nd Eur. Symp. Therm. Anal.*, D. Dollimore, Ed., Heyden, London, 1981, p. 494.

122. Tenoutasse, N., and N. B. Singh, *Indian J. Technol.* **16**, 184 (1978).

123. Wiedemann, H. G., and M. Roessler, in B. Miller, Ed., *Thermal Analysis-Proc. 7th Int. Conf. Therm. Anal.*, Vol. 2, Chichester, UK: Wiley, 1982, p. 1318.

124. Young, J. F., *Mag. Concr. Res.* **14**, 137 (1962).

Subject Index